Sharing the Fish

TOWARD A NATIONAL
POLICY ON INDIVIDUAL
FISHING QUOTAS

Committee to Review Individual Fishing Quotas

Ocean Studies Board

Commission on Geosciences, Environment, and Resources

National Research Council

NATIONAL ACADEMY PRESS
Washington, D.C. 1999

NATIONAL ACADEMY PRESS • 2101 Constitution Avenue, N.W. • Washington, DC 20418

NOTICE: The project that is the subject of this report was approved by the Governing Board of the National Research Council, whose members are drawn from the councils of the National Academy of Sciences, the National Academy of Engineering, and the Institute of Medicine. The members of the committee responsible for the report were chosen for their special competences and with regard for appropriate balance.

This report was supported by contracts from the National Oceanic and Atmospheric Administration (NOAA). The views expressed herein are those of the authors and do not necessarily reflect the views of the NOAA or any of its sub-agencies.

Library of Congress Cataloging-in-Publication Data

Sharing the fish: toward a national policy on individual fishing quotas / Committee to Review Individual Fishing Quotas, Ocean Studies Board, Commission on Geosciences, Environment, and Resources, National Research Council.
    p. cm.
  Includes bibliographical references and index.
  ISBN 0-309-06330-2 (casebound)
  1. Fishery management—United States. 2. Fishery policy—United States. 3. Fishery management.
4. Fishery policy. I. National Research Council (U.S.). Committee to Review Individual Fishing Quotas.
  SH221.S47 S87 1999   98-58059
  333.95´617—dc21

Sharing the Fish: Toward a National Policy on Individual Fishing Quotas is available from the National Academy Press, 2101 Constitution Ave., N.W., Box 285, Washington, DC 20055; 1-800-624-6242 or 202-334-3313 (in the Washington metropolitan area); http://www.nap.edu.

The National Academy of Sciences is a private, nonprofit, self-perpetuating society of distinguished scholars engaged in scientific and engineering research, dedicated to the furtherance of science and technology and to their use for the general welfare. Upon the authority of the charter granted to it by the Congress in 1863, the Academy has a mandate that requires it to advise the federal government on scientific and technical matters. Dr. Bruce Alberts is president of the National Academy of Sciences.

The National Academy of Engineering was established in 1964, under the charter of the National Academy of Sciences, as a parallel organization of outstanding engineers. It is autonomous in its administration and in the selection of its members, sharing with the National Academy of Sciences the responsibility for advising the federal government. The National Academy of Engineering also sponsors engineering programs aimed at meeting national needs, encourages education and research, and recognizes the superior achievements of engineers. Dr. William A. Wulf is president of the National Academy of Engineering.

The Institute of Medicine was established in 1970 by the National Academy of Sciences to secure the services of eminent members of appropriate professions in the examination of policy matters pertaining to the health of the public. The Institute acts under the responsibility given to the National Academy of Sciences by its congressional charter to be an adviser to the federal government and, upon its own initiative, to identify issues of medical care, research, and education. Dr. Kenneth I. Shine is president of the Institute of Medicine.

The National Research Council was organized by the National Academy of Sciences in 1916 to associate the broad community of science and technology with the Academy's purposes of furthering knowledge and advising the federal government. Functioning in accordance with general policies determined by the Academy, the Council has become the principal operating agency of both the National Academy of Sciences and the National Academy of Engineering in providing services to the government, the public, and the scientific and engineering communities. The Council is administered jointly by both Academies and the Institute of Medicine. Dr. Bruce Alberts and Dr. William A. Wulf are chairman and vice-chairman, respectively, of the National Research Council.

Acknowledgement of Reviewers

This report has been reviewed in draft form by individuals chosen for their diverse perspectives and technical expertise, in accordance with procedures approved by the NRC's Report Review Committee. The purpose of this independent review is to provide candid and critical comments that will assist the NRC in making the published report as sound as possible and to ensure that the report meets institutional standards for objectivity, evidence, and responsiveness to the study charge. The review comments and draft manuscript remain confidential to protect the integrity of the deliberative process. We wish to thank the following individuals for their participation in the review of this report:

Robin Allen, Inter-American Tropical Tuna Commission, San Diego, California

Paul Durrenberger, Pennsylvania State University, State College

Lewis Haldorsen, University of Alaska, Fairbanks

Kai N. Lee, Williams College, Williamstown, Massachusetts

Elinor Ostrom, Indiana University, Bloomington

R. Bruce Rettig, Oregon State University, Corvallis

Joseph Sax, University of California, Berkeley

James Wilen, University of California, Davis

While the individuals listed above have provided many constructive comments and suggestions, it must be emphasized that responsibility for the final content of this report rests entirely with the authoring committee and the NRC.

Preface

The Ocean Studies Board (OSB) has provided advice to Congress and the National Marine Fisheries Service (NMFS) on a variety of fishery issues in the past 6 years. In the area of stock assessments, committees of the OSB reviewed Atlantic bluefin tuna data and models (NRC, 1994) and more recently reviewed the NMFS assessments of groundfish stocks off New England (NRC, 1998b) and performed a broad review of fish stock assessment methods (NRC, 1998a). The OSB recently issued a report about how ecosystem principles can be used to sustain marine fisheries (NRC, 1999b) and is about to embark on a study of fisheries data (particularly for summer flounder stocks) at the request of Congress. OSB committees have reported not only on fisheries science issues, but also on matters of fishery management. Many of the features of the 1996 amendments to the Magnuson-Stevens Fishery Conservation and Management Act were recommended in an OSB report (NRC, 1994). Finally, in the 1996 amendments to the Magnuson-Stevens Act, Congress asked the National Academy of Sciences to examine two types of quota programs used in U.S. fisheries management: community development quotas (NRC, 1999a) and the subject of this report, individual fishing quotas.

The committee wishes to thank the following individuals who provided comments to the committee at its meetings (many traveling at their own expense) or by mail and electronic mail:

Tom Able, Robert Allen, Tom Alspach, Bob Alverson, Phil Anderson, Wilma Anderson, Alan Austerman, Greg Baker, Chris Berns, David Berry, Will Bland, Bernie Bohn, Jay Brevik, Donald Brown, John Bruce, Jean Bumpus, Chuck Bundrant, John Bundy, Barry Callaghan, Dana Carros, Tom Casey, Freddie Christiansen, Gordon Colvin, George Conway, Scott Coughlin, Felix Cox, Will

Daspit, Laura Deach, Christopher Dewees, Alex Diaz, Al Didier, Niaz Dorey, Joe Easley, Michelle Eder, Sherry Egle, Penn Esterbrook, Dan Falvey, Lance Farr, John Finley, Jean Flemma, Barney Frank, James Franzel, Kaitilin Gaffney, Steve Ganney, Graciela Garcia-Moliner, John Gauvin, James Gilmore, Mike Gonzales, Carmine Gorga, Shari Gross, Richard Gutting, Peter Halmay, Dyan Hartill, Marcus Hartley, Lee Hilde, David Hillstrand, Ken Hinman, Paul Howard, Jake Jacobson, Ted Jenks, David Keifer, Andrew Kemmerer, Charlie King, Martha King, John Kingeter, George Kirk, Gunnar Knapp, Dave Krusa, Gerry Leape, Justin LeBlanc, Larry LeDoux, Fred Lentz, Jennifer Lincoln, Mike Lopez, Joe Macinko, Jerry Mackie, Roy Madsen, Gary Matlock, Thomas McCloy, Gerry Merrigan, Frank Miles, Joe Mills, Karl Moore, Rod Moore, Jere Murray, Chris Nelson, Jeanette Ness, Mike Nussman, Tom Oakey, Virginia Olney, James O'Malley, Brent Paine, Donna Parker, Clarence Pautzke, Patricia Phillips, Joseph Plesha, Sam Pooley, Gretchen Pullar, Mark Raphael, Jerry Ray, Steve Rebuck, Robin Reichers, Bill Robinson, Eddie Rose, Andrew Rosenberg, John Sanchez, Angela Sanfilippo, Jerry Sansom, Cynthia Sarthou, Jerry Schill, Peter Schonberg, Peter Shelley, Susan Shipman, Robert Shipp, Tamara Shrader, Kitty Simonds, Larry Six, Glen Spain, Kristin Stahl-Johnson, Barbara Stevenson, Patrick Sullivan, Toby Sullivan, Tom Suryan, Wayne Swingle, Robin Taylor, Mike Theirry, Arni Thompson, Peter Thompson, Daniel Tucker, Bruce Turris, Hector Vega, George Veneroso, David Walker, Robert Ward, John Warner, Donald Waters, Greg Waugh, John Webb, Rick Weber, Wayne Werner, Suzanne West, David Whaley, David Whitmire, and Kay Williams.

The committee could not have conducted this study without the diligence and interest of these individuals. The staff of the NMFS Restricted Access Management Division (Phil Smith, Jessica Gharrett), NMFS Alaska Region (Jay Ginter), the Alaska Commercial Fisheries Entry Commission (Elaine Dinneford, Ben Muse, Kurt Schelle), and the U.S. Coast Guard (LCDR Busch) were invaluable in providing data for the committee's analyses. The staffs of each regional council provided important documents for the committee's review, and most of the council executive directors made presentations at the committee's meetings.

Special thanks are deserved by the National Oceanic and Atmospheric Administration (NOAA) Advisory Panels set up by Congress to "assist in the preparation of the report." Lee Anderson and Beth Stewart were able chairs of the East and West Coast Panels, respectively. Amy Buss-Gautam was extremely helpful and creative as the NOAA coordinator for the panels. Members of the panels included Dick Allen, Ted Ames, Linda Behnken, Francis Christy, Harriett Didricksen, Dave Fraser, Rod Fujita, Walter Gordon, Tom Hill, Ralph Hoard, Doug Hopkins, John Iani, Jan Jacobs, Pete Jensen, Jim Kendall, Linda Kozak, Mark Lundsten, Miles Mackeness, Scott Matulich, Tom Morrison, Ben Muse, Howard Nickerson, Jim Ponts, Ken Roberts, Paul Seaton, David Wallace, Roy Williams, and Bob Zales II. The committee sincerely appreciated the panels' willingness to meet with us twice. The panels' inputs were very helpful to the committee as it probed the difficult issues related to individual fishing quotas and alternative management methods.

Committee members deserve special thanks for the many weeks of time they volunteered to this task over the past year. Their product is remarkable for its breadth and perception. Special thanks are due to our able project assistant, Jennifer Wright, and to Glenn Merrill, whose work as research associate was insightful and diligent. Finally, the study director, Ed Urban, deserves plaudits for his diligence, patience, and diplomacy. Without his efforts and expertise, this report could never have been completed.

Jan S. Stevens
Chair

Contents

Executive Summary

For centuries, fish in the sea were assumed to be a limitless resource, available to all for the taking. More recently, however, depleted stocks and increasing competition for fish have led to a reexamination of this assumption and a search for new ways to manage marine fisheries. The challenge has been to maintain fisheries at sustainable levels, with due regard to productivity, employment, and the cherished way of life in many coastal communities.

With passage of the Fishery Conservation and Management Act of 1976 (now the Magnuson-Stevens Fishery Conservation and Management Act, MSFCMA), Congress for the first time mandated a national program for the conservation and management of fishery resources, to be developed by eight regional fishery management councils and implemented by the Department of Commerce through the National Marine Fisheries Service (NMFS). Councils have implemented measures to limit inputs *to* the fisheries and outputs *from* fisheries. *Input* controls limit such things as the number of participants in fisheries, the type and amount of gear, and methods of fishing. They may close certain areas to fishing and restrict the length of fishing seasons. *Output* controls use various means to limit catch to some level determined to be sustainable over the long term. Limits on overall catch, including total allowable catch (TAC), are set by the regional fishery management councils based on recommendations of stock assessment scientists. A range of input and output controls can be used separately or together, and many of these measures are discussed in Chapter 4.

Output controls typically include some mechanism for closing the fishery after the target harvest level has been achieved. One form of output control is the individual fishing quota (IFQ), a system under which harvesting privileges are allocated to individual fishermen. The Magnuson-Stevens Act defines an IFQ as "a Federal permit under a limited access system to harvest a quantity of fish, expressed by a unit or units representing a percentage of the total allowable catch of a fishery that may be received or held for exclusive use by a person" (MSFCMA, Sec. 3[21]). Individual fishing quotas have been used worldwide since the late 1970s. A few countries, particularly Canada, New Zealand, and Iceland, have significant experience in the benefits and problems of developing, implementing, and managing IFQs. This tool has been adopted in four U.S. fisheries (Alaskan halibut and sablefish, wreckfish, and surf clams/ocean quahogs), and programs were about to be implemented in two other fisheries when Congress intervened through enactment of the Sustainable Fisheries Act of 1996, establishing a moratorium on new programs. Congress asked the National Academy of Sciences to study a wide range of questions concerning the social, economic, and biologic effects of IFQs and other limited entry systems and to make recommendations about existing and future IFQ programs.

A committee with expertise in fisheries biology and management, anthropology, economics, law, political science, and business was established to study all aspects of IFQs in response to the request from Congress. Over a seven-month period, the committee held hearings in Anchorage, Seattle, New Orleans, Washington, D.C., and Boston. It heard testimony from fishermen, processors, state and federal regulators, academicians, environmental groups, and members of the public, and received a large amount of written material. This report is the result of the committee's deliberations.

The many witnesses who addressed the committee at its five hearings provided a broad view of the real and perceived effects of existing and proposed IFQ programs. Just as there is tremendous variation among U.S. fisheries, their regulations vary according to perceived necessities in each region and the dynamics of the regional fishery management councils. Again and again, the committee was warned against a "one-size-fits-all" approach. The committee was entreated to respect the individual needs of fisheries, fishing communities, and fishing regions, and to refrain from endorsing rigid blueprints at the expense of hard-won measures, carefully crafted to address unique local biologic and social conditions.

Critics as well as supporters of IFQs recognized that this tool arose in response to real and pressing fishery problems—situations in which other types of regulation had failed to prevent a race for fish and overharvesting, and in which economic efficiency, safety, and product quality suffered. For example, in Alaska's halibut fishery prior to implementation of the IFQ programs, the season was progressively reduced in an attempt to maintain the annual catch of halibut within the TAC. In response, fishermen increased the number of vessels in their fleets and used larger and larger vessels, with more and more gear. The frenzied *derbies*[1] sometimes forced the fishing fleet to operate in dangerous weather, exacerbated *ghost fishing* from gear lost in the race for fish, and created incen-

---

[1] See glossary (Appendix F) for definition of terms.

tives to waste other species caught in the process. The cyclical nature of the fishery left consumers facing gluts of fresh halibut for a few weeks each year and buying frozen fish for the remainder of the year.

The Alaskan IFQ programs for halibut and sablefish addressed and reduced these problems. Evidence from the Alaskan IFQ programs suggests that the derby has been eliminated, safety has improved, and ghost fishing has been reduced. At the same time, these IFQ programs have left the halibut and sablefish fisheries with fewer fishermen (as intended) and have enriched many of those whose catch history qualified them for quota shares.

The capacity of IFQs for transferability, consolidation, and leasing has led to a general concern that independent owner-operators of fishing vessels or crew members will be led into economic dependence on absentee owners as quota shares increase in value and small investors are excluded from the field. Consequently, some programs (e.g., Alaskan halibut and sablefish) have adopted owner-on-board and other provisions intended to prevent absentee ownership.

Other fisheries in which IFQ programs have been used—the Atlantic surf clam and ocean quahog one, for example—were of somewhat different nature. However, even though that fishery did not have open access, the management situation through the 1980s created the equivalent of "derby" fishing, when boats were allowed to fish for very short periods of time in order to make the TAC stretch out over the year. A more striking difference from the Alaska IFQ programs is that the surf clam/ocean quahog IFQ program, the first in the United States, was based on free-market principles, with few restraints on ownership, transfer, or consolidation of shares. This program was extremely effective in eliminating economically excessive effort, but in so doing highlighted the trade-offs involved in terms of the loss of jobs, and decreased opportunities for young people and hired captains to become vessel owners and for independent harvesters to find markets for their clams.

CONCERNS ABOUT THE USE OF INDIVIDUAL FISHING QUOTAS

The National Marine Fisheries Service and other agencies routinely estimate the size of marine fish stocks to determine the amount of fish that can be harvested in a given year so that fisheries can be sustained; this amount is the *allowable biological catch*. The catch level that fishermen are allowed to take is the TAC, which must be equal to or lower than the allowable biological catch. TACs are set for most fisheries. Most other fishery management measures are designed to help fisheries meet, and not exceed, the TAC. Reliance solely on TAC-based controls can induce fishermen to apply excessive inputs of labor and capital to a fishery as they compete for their share. Thus, arguments have arisen in recent years for controlling fishing activity, restricting access to fisheries, and relying on input controls, such as gear restrictions, and output controls, such as quotas and trip limits. Without controls on the amount of fishing, many fisheries are plagued by overcapitalization, waste, and pressures for management measures that place fish stocks at risk. Different methods of limited entry have been developed to control access to fisheries.

The IFQ is one means to limit entry in order to reduce overcapitalization and the wasteful practices that occur under other systems. A major intended effect of IFQs is to create economic incentives for owners of vessels to decrease their inputs of labor and capital to a fishery. Thus, in fisheries with excess harvesting or processing capacity, vessels may be laid up and some crew members may lose their jobs, although others may increase their employment from a few days to several months per year. Processing plants may require fewer workers when processing is spread across a longer period of time. On the other hand, with IFQs, economic resources are no longer wasted through overinvestment in capital and labor. Changes in the harvesting and processing patterns resulting from IFQs could be beneficial to consumers favoring year-round fresh product. Decreased costs and increased profitability can benefit consumers and the nation.

Although Congress requested a review of IFQs at a *national* level, it is difficult to discuss the implementation of these programs without consideration of the specific nature of each fishery and the social and economic communities associated with it, as the cases of the existing U.S. IFQ fisheries demonstrate. Each region is unique in terms of its biologic, social, and economic characteristics. To accommodate this regional uniqueness, Congress has delegated the development of fishery management plans to regional councils.

A number of advantages and concerns were identified from the range of IFQ programs implemented in U.S. fisheries, through comments in favor of and against IFQs at the committee's public meetings, examination of published information, and the committee's knowledge of IFQs and other management techniques:

• *Advantages*—IFQ programs are widely identified as being a highly effective way of dealing with overcapitalization in the fishing industry. Removing the race for fish has reduced the incentive to buy ever-larger vessels and more equipment and to fish during unsafe conditions. Consumers have been able to purchase fresh fish during longer periods of the year. Many fishermen testified that IFQs provided the opportunity to utilize better fishing and handling methods, reducing bycatch of nontargeted species and maintaining higher product quality. Gear conflicts may also be reduced by IFQs.

• *Concerns*—A number of problems were identified in operative IFQ programs during the committee's work. Prominent among them are concerns about the fairness of the initial allocations, effects of IFQs on processors, increased costs for new fishermen to gain entry, consolidation of quota shares (and thus economic power), effects of leasing, confusion about the nature of the privilege involved, elimination of vessels and reductions in crew, and the equity of gifting a public trust resource.

SUMMARY OF RECOMMENDATIONS

IFQs can be used to address a number of social, economic, and biologic issues in fisheries management. Alternative management approaches can achieve some, but not all, of the objectives that can be achieved with IFQs. There are no general threshold criteria for deciding when IFQs are appropriate; the use of IFQs should be considered on a fishery-by-fishery basis. IFQs can be used to remedy the effects of overcapitalization and overfishing or to prevent the development of these negative effects. As discussed in greater detail later, decisions to develop IFQs or to use alternative methods of fishery management should be the responsibility of the regional councils. The following recommendations are directed separately to Congress, the Secretary of Commerce and the National Marine Fisheries Service, the regional fishery management councils, and states and others. However, some of the following recommendations overlap because different institutions share responsibilities related to the specific issues of fisheries management.

IFQs should be allowed as an option in fisheries management if a regional council finds them to be warranted by conditions within a particular fishery and appropriate measures are imposed to avoid potential adverse effects. The issues of initial allocation, transferability, and accumulation of shares should be given careful consideration when IFQ programs are considered and developed by regional councils and reviewed by the Secretary of Commerce.

What Should Congress Do?

Because the committee believes that most decisions about IFQs are most appropriately made at the regional level, rather than the national level, the committee's recommendations to Congress relate primarily to changes that should be made to the Magnuson-Stevens Fishery Conservation and Management Act to govern the use of IFQs by regional councils. Congress should recognize that the design of any limited entry system in relation to concentration limits, transferability, and distribution of shares will depend on the objectives of each specific fishery management plan. This underscores the importance of providing flexibility for regional councils in developing IFQ and other limited entry programs. Congress should do the following:

Lift the Moratorium. **Congress should lift the moratorium on the development and implementation of IFQ programs established by the Sustainable Fisheries Act of 1996.**

Encourage Cost Recovery and Some Extraction of Profits. Congress should permit (1) assessment of fees on initial allocations of quota and first sale and leasing of it; (2) imposition of an annual tax on quota shares; and (3) zero-revenue auctions (see Box 5.1). The Magnuson-Stevens Act presently imposes limits on various fees that may be used to recover the cost of IFQ management and enforcement, but Congress should increase these limits so that costs of IFQ management and other forms of limited entry can be recovered fully. Additionally, revenues extracted from IFQ fisheries could be used to mitigate some of the potential negative impacts of IFQs and to support research to improve fishery management. Two forms of new value can be created by IFQs: windfall gain available immediately and rents[2] generated later. The committee recommends that the Magnuson-Stevens Act be amended to

• Allow the public to capture some of the windfall gain sometimes generated from the initial allocation of quotas in new IFQ programs;
• Recover the incremental costs of IFQ management by authorizing the collection of fees from the transfer and/or holding of IFQs, even if these costs are greater than the existing limits; and
• Authorize the extraction of some of the fishery profits (rents) in excess of cost recovery. Priority should be given to dedicating such revenues to improving the fisheries rather than to the general treasury.

Support the Council Process. The Magnuson-Stevens Act gives responsibility for developing fishery management plans to the regional councils. The Secretary of Commerce bears the burden of implementing fishery management plans. Councils must consider conflicting interests and weigh competing considerations. In many cases, councils have spent years developing management plans, including those involving IFQs. **Congress should recognize that the design of an IFQ or other limited entry system in relation to concentration limits, transferability, distribution of quota shares, and other design questions will depend on the objectives of a specific plan, requiring flexibility for regional councils in designing IFQ programs. Regional councils should have flexibility to adjust existing IFQ programs and develop new ones.**

Require Accumulation Limits. Congress should require any council considering an IFQ program to define "excessive share" for the program and use limits on accumulation of quota share or other measures to prevent excessive shares from developing. These limits should be fishery specific and may also be specific to areas and classes of vessel.

Support Additional Study and Routine Data Collection. All fishery management systems, particularly those that limit entry, require social and economic data for

---

[2] See Chapter 1 for an explanation of resource rent.

both planning and evaluation. In addition to analyzing the impacts of regulatory actions, the data should be used to monitor the health of fisheries. Monitoring the status of the industry should be as routine and systematic as monitoring the status of the stocks. To date, the regional councils and NMFS have not had access to the data and studies required. **Congress should ensure that funding is available to NMFS and the states for the routine and nationwide collection of social and economic information on U.S. marine fisheries in state and federal waters.** Where possible, these efforts should be coordinated with cooperative statistics programs being carried out by the states and specific local studies funded through the National Sea Grant College Program and NMFS. It is crucial that all data collection and social and economic research be subject to objective, peer-reviewed selection processes.

Determine Rules for Foreign Ownership. Although foreign ownership was an issue on which comment was specifically requested by Congress, little concern was expressed over it at the committee's hearings. This may have resulted because extensive restrictions on foreign ownership in U.S. waters already exist (by virtue of limits on vessel registration) or because other legislative remedies are being sought to reduce foreign participation in U.S. fisheries (e.g., passage of the American Fisheries Act [S. 1221] in 1998, increasing the minimum ownership requirements for U.S. fishing vessels). It appears that the imposition of further limits on foreign ownership would have profound implications on the holding of quota by processors and harvesters in fisheries where significant levels of foreign ownership already exist. Assessing the extent to which profits from U.S. fisheries are expropriated by foreign nations is beyond the scope of this evaluation of IFQs and limited access systems. If Congress were to decide to control foreign ownership, criteria could be established for IFQ-based and other fisheries. Enforcement would require careful analysis of financial and corporate records and the economic conditions of the fishery, and improved access by regulators to certain types of proprietary data.

Delegate Decisions About the Transferability of Quota Shares. The decision about whether quota shares should be transferable, one of the most critical elements in the design of an IFQ program, should be delegated to the regional councils because it depends entirely on the specific goals and objectives of the management regime.

Define the Nature of the Privilege. Other amendments to the Magnuson-Stevens Act should include provisions to

• Make it clear that the nature of the interest embodied in an IFQ encompasses the right of a quota holder to protect the long-term value of quota shares through civil action against the private individuals or entities whose unlawful actions might adversely affect the marine resource or environment. However, the Magnuson-Stevens Act should be clear that the IFQ privilege does not authorize actions by quota holders against government agencies for decisions designed to protect marine resources and the environment through TAC reductions, area closures, or other restrictions that could affect the amount of fish available for capture. Actions should be available to councils to discourage behavior that degrades the productivity of resources and to reward exemplary behavior without disrupting the security of the harvesting privilege.

• Authorize regional councils to decide on a case-by-case basis whether to limit the duration of IFQ programs through the inclusion of sunset provisions.

What Should the Secretary of Commerce and National Marine Fisheries Service Do?

The committee encourages NMFS to implement the central registry system for limited access system permits (as required by the Sustainable Fisheries Act of 1996) as soon as possible to increase the confidence of lenders in the security of loans for purchase of IFQs and provide opportunities for individuals to obtain financing to enter or increase their stake in IFQ-managed fisheries. NMFS should establish adequate monitoring and enforcement programs once limited entry systems are in place.

Limited entry is becoming more standard in marine fisheries management and NMFS and the regional councils seem ill-prepared to meet the requirements of the Magnuson-Stevens Act for limited entry programs. Funds should be made available through NMFS to strengthen research on the design and impacts of IFQ programs and limited entry systems of all types. NMFS should review its priorities and practices to give greater weight to the social and economic data collection and studies mentioned earlier.

The Secretary of Commerce should consider the following issues in reviewing proposed IFQ programs before implementation:

• *Delegated management authority*—In considering the range of potential management options, regional councils should not be precluded from considering proposals to delegate management authority to other entities within a region that would operate within the framework of the Magnuson-Stevens Act's national standards and NMFS regulatory guidelines.

• *Long-term, routine data collection*—The regional councils and the Secretary of Commerce should ensure that data collection and studies are undertaken as part of long-term, routine activities separate from the consideration of specific management alternatives for a fishery. It is significant that the committee was unable to analyze the full set of costs and benefits of any U.S. IFQ program because of the unavailability of the necessary information (see Appendix H).

• *Regular review and evaluation*—The Secretary of Commerce should ensure that each fishery management plan that incorporates IFQs includes enforceable provisions for regular review and evaluation of the performance of IFQ programs, including a clear timetable, criteria to be used in evaluation, and steps to be taken if the programs do not meet these criteria. Provisions should be made for the collection and evaluation of data required for such assessments. This process could include review by external, independent groups.

• *Inclusion of fishing communities in initial allocations*—Councils should consider including fishing communities in the initial allocation of IFQs (as community fishing quotas), where appropriate. The Secretary of Commerce should interpret the clause in the Magnuson-Stevens Act pertaining to fishing communities (National Standard 8) to support this approach to limited entry management.

What Should Regional Fishery Management Councils Do?

The committee directs most of its recommendations to the regional fishery management councils because they are in the best position to involve regional stakeholders and design management programs appropriate to the species they manage. The committee proposes several mechanisms, including IFQs, that could be useful in considering choices among the range of alternatives available to deal with problems such as overcapitalization and costly races for fish.

Regional fishery management councils should address the following issues or perform these actions in developing and implementing IFQ programs:

• Many individuals and groups have a stake in the development, implementation, and management of IFQ programs. Such stakeholders include vessel owners, hired skippers, crew members, processors, communities, fishery managers, environmental groups, and others. Councils should review the adequacy of stakeholder representation on advisory panels and other bodies and take steps to broaden representation, if necessary, to include representatives of stakeholders potentially affected by limited entry programs.

• The biologic, social, and economic objectives of each fishery management plan and the means for achieving these objectives through IFQs (if they are deemed appropriate), should be specified clearly through a process that encourages broad participation by stakeholders.

• Priority should be given to the question of social, economic, and biologic consequences of a proposed IFQ program and alternatives to it. The councils and NMFS must allocate more resources and attention to impact assessments, which are now required by law but often are given inadequate attention.

• IFQ programs should include a commitment to monitor both (1) short- and long-term impacts and (2) the political, financial, and administrative ability to make changes as required to meet program objectives.

- Control dates[3] should be set early in the development of an IFQ program and be strictly adhered to throughout the development of the program, with a minimum amount of time between the control dates and the initial allocation of quota.
- Councils should demonstrate that a wide range of initial allocation criteria and allocation mechanisms has been considered in the design of IFQ programs. Councils could avoid some of the allocation controversies encountered in the past by giving more consideration to (1) who should receive initial allocation, including crew members, skippers, communities, and other stakeholders; (2) how much they should receive; and (3) how much the potential recipients should be required to pay for the initial receipt of quota (e.g., auctions, windfall taxes).
- Councils should avoid taking for granted the "gifting" of quota shares to the present participants in a fishery, just as they should avoid taking for granted that vessel owners should be the only recipients of quota and historical participation should be the only measure for determining initial allocations.
- When designing IFQ programs, councils should be allowed to allocate quota shares to communities or other groups, as distinct from vessel owners or fishermen. For existing IFQ programs, councils should be permitted to authorize the purchase, holding, management, and sale of IFQs by communities. Such quota shares could be used for community development purposes, treated as a resource allowing local fishermen to fish, or reallocated to member fishermen by a variety of means, including loans.
- Leasing of quota shares should generally be permitted but, if necessary, with restrictions to avoid creation of an absentee owner class. Making shares freely transferable is generally desirable to accomplish the economic goals of an IFQ program. However, if it is desired to promote an owner-operated fishery or to preserve geographic or other structural features of the industry, it may be necessary to restrict long-term transfers of quota shares to bona fide fishermen or to prohibit transfers away from certain regions or among different vessel categories.
- Issues such as shifting distributions of quota share holdings among firms or communities can be addressed through setting upper limits on accumulation of quota shares. If important objectives include maintaining owner-operated fisheries and fishery-dependent coastal communities, greater attention may have to be

---

[3] The date established for defining the pool of potential participants in a given management program. For example, in preparing to establish a limited entry program, a council might decide to establish a date that would serve as a cutoff for eligibility. With such a control date established, the council could proceed to assess alternative limited entry systems and other program design characteristics without the fear of stimulating speculative entry into the fishery. Unfortunately, because councils may be influenced by industry or required by NMFS to change the control date, there is often some speculative entry even when the control date is widely publicized. In the case of the Alaskan halibut and sablefish IFQ programs, delays in program implementation led to speculative entry by a sizable group that actively participated in the fishery between 1990 and 1994 but was left out of the initial allocation of quota shares.

given to equity considerations in setting upper limits on ownership, limiting transfer of quota shares outside communities, and similar measures.

• In any fishery for which an IFQ program is being considered, attention should be given to the implications of recreational participation in the fishery and, where appropriate, to potential application of the IFQ program to both commercial and recreational sectors.

• Councils should design IFQ programs in such a way as to enhance enforcement by (1) ensuring the fairness of program design and (2) using design principles to reduce the incentives to cheat. Programs that are considered fair and desirable by participants are most likely to be respected. Such programs produce higher compliance rates with less necessity for increased enforcement. IFQ programs are more likely to be perceived as fair and desirable if affected stakeholders participate in their creation.

• Councils should proceed cautiously in changing existing IFQ programs. Many individuals have made substantial investments in IFQ programs, even if they received little or no quota initially. Changes should be designed in a way that maintains the positive benefits of IFQs that result from their stability and predictability.

• Councils should explore the use of individual and pooled bycatch quotas to control overall bycatch and encourage fishermen to minimize their bycatch rates.

What Should States and Others Do?

Fish populations often cross boundaries between federal and state waters. States should coordinate with the federal government in designing state fishery management programs that are compatible with federal limited entry systems. Regional councils should—at the earliest opportunity—officially inform affected state fishery agencies that they are considering adoption of an IFQ program for fisheries that occur in both federal and state waters. Proposed regulations implementing a federal IFQ program should specify the manner in which relevant state fishery policies and regulations would be made consistent with the federal system. Conversely, if states in a region have developed a coordinated and effective limited entry program in state waters, including IFQs, the regional councils should, where consistent with the national standards, complement these programs in federal fishery management plans. States should cooperate in the collection of social and economic data through regional cooperative fisheries statistics efforts. In particular, states should contribute to the collection of employment data and information about processing activities.

CONCLUSIONS

Although the IFQ is no panacea, it deserves a place in the array of techniques that may be needed in any particular fishery management plan. Its value in matching harvesting and processing capacities to the resource, slowing the race for fish, providing consumers with a better product, and reducing wasteful and dangerous fishing has been demonstrated repeatedly.

If the regional councils choose to consider IFQs, they must recognize and respect the interests of all those involved in the fishery—crew members, skippers, their families onshore, prospective fishermen, and all related entities. Fairness *and* efficiency are mandated by the Magnuson-Stevens Act.

In allocating harvest privileges to a national resource, managers must recognize that fisheries are held in trust for the nation and that the nation's stewardship as trustee cannot be abrogated. The allocation of permits to harvest a portion of the TAC is a management tool with high potential for efficiency and stewardship in a given fishery. At the same time, it cannot substitute for the federal government's responsibility to exercise stewardship in the national interest.

Finally, it must be recognized that a system that confers harvest privileges in a fishery can be difficult to reverse once expectations have been created. The committee is by no means suggesting that IFQs be considered compensable rights. Rather, the committee recognizes the political and economic forces that are resistant to regulatory change once investments have been made. Care must be exercised balancing between the certainty needed by recipients of these privileges and the trust responsibility on behalf of the people for whom a fishery is managed.

1 | Introduction

Implementation of the nation's most important fisheries law, the Magnuson-Stevens Fishery Conservation and Management Act, has reached a critical stage. Most U.S. fish stocks are in a state of full exploitation or overutilization (NMFS, 1996). The relative proportions of fish species have been drastically altered in some regions (e.g., Georges Bank: Solow, 1994; NMFS, 1996; Fogarty and Murawski, 1998), with populations of previously dominant species collapsing and less abundant groups becoming dominant. The harvesting and processing capacity in many U.S. fisheries far exceeds levels that are consistent with sustainable fisheries.

Although there have been several successes in U.S. fisheries management in terms of maintaining or restoring stocks—examples include Atlantic striped bass, Pacific halibut, Atlantic surf clam, and North Pacific pollock—there have been many serious failures (Parsons, 1996; Botsford et al., 1997; Roberts, 1997a). Reasons include noncompliance with management regulations, the lack of sufficient data and appropriate models for stock assessments (NRC, 1998a,b), a complex interplay between fluctuating marine populations and a political economy that tends to subsidize or overinvest in fishing capacity (Caddy and Gulland, 1983; Ludwig et al., 1993), and an inclination to make risk-prone rather than risk-averse decisions in the presence of uncertainty (Rosenberg et al., 1993; Sissenwine and Rosenberg, 1993). Some analyses of fishery systems suggest that overinvestment and the inclination toward risk-prone decisionmaking result from the "common-pool" nature of most fishery resources and the "open-access" nature of the rules regarding how these resources can be used (e.g., Gordon, 1954). Many of the investment incentives and decisions that created overinvestment were due to processes associated with the maturing of fisheries and misunderstanding of opportunities that would be available to U.S. fishermen as foreign fleets were pushed out. And, in many cases, overinvestment can be linked to declining natural resources, with some of the resource decline possibly caused by environmental degradation and natural climatic shifts. A vivid example is the overinvestment in Pacific salmon fisheries, where the overwhelming source of overinvestment must be attributed to the problems with the resource (NRC, 1996).

The stressed nature of many fisheries is apparent from scientific reports of decreasing numbers of spawning fish, reduced overall biomass and population levels, and lower catch per unit effort (CPUE) in commercial fisheries. Because some management measures have not been very effective in reducing these stresses, fishermen,[1] communities, and policymakers have been seeking ways to manage fisheries that will maintain biologic resources in the long term, avoid misusing capital, preserve employment, and maintain fishing communities. A relatively new policy instrument, the individual fishing quota (IFQ), is among the alternatives being considered as a possible solution to excess harvesting and processing capacity, stock depletion, and possible ecological disruptions that characterize many managed U.S. fisheries, including those that operate under some form of restricted access. Broadly speaking, IFQs are exclusive individual privileges to harvest portions of an overall quota of marine fish or shellfish.

The Magnuson-Stevens Act authorizes the use of a variety of approaches for controlling fishing effort and protecting fish stocks and their environments, including systems for limiting access. In fact, most U.S. fisheries now operate under some form of limited access or limited harvests. The IFQ as a tool for limiting access has evoked considerable controversy, however, because of its potential for creating windfall benefits to the initial recipients, the privileges that IFQs create, and the potential for decreasing employment and changing social and economic relationships among individuals and communities. Through the Sustainable Fisheries Act of 1996,[2] Congress placed a moratorium on the ability of the regional fishery management councils to develop or submit any fishery management plan using IFQs until October 1, 2000. Furthermore, it directed the Secretary of Commerce, acting through the National Marine Fisheries Service (NMFS), not to approve any new fishery management plan that includes an IFQ program. In the meantime, the National Academy of Sciences, acting through the National Research Council (NRC), was requested to prepare a comprehensive report on IFQs. This report is intended to fulfill the congressional mandate.

---

[1] The committee uses the term "fisherman" throughout the report because this is how the practitioners of fishing (both male and female) tend to refer to themselves in the United States.

[2] The Sustainable Fisheries Act of 1996 amended the Magnuson Fishery Conservation and Management Act.

MAGNUSON-STEVENS FISHERY CONSERVATION
AND MANAGEMENT ACT

To understand the potential role of IFQs in the management of U.S. fisheries under the Magnuson-Stevens Act, it is useful to review briefly the major policy objectives of the act that IFQ-based management might address.

Background on the Magnuson-Stevens Act

When the Fishery Conservation and Management Act (FCMA)[3] became law in 1976, its principal policy goal was to assert U.S. federal authority over non-U.S.-flagged vessels operating within a zone extending 200 nautical miles from the U.S. coastline, coincident with ongoing negotiations of the UN Convention on the Law of the Sea for the same extension of jurisdiction for all coastal nations. This assertion of jurisdiction was a stark break with the past and a rejection of international organizations as the forum for managing fish stocks in U.S. coastal waters. The act also established a comprehensive system for regulating the domestic fishing industry in federal waters. This system was based on a promising, but untested, approach to fishery management, the development of regional conservation and management plans by joint federal-state consultative bodies with significant participation of the resource users, the fishing industry. Eight regional fishery management councils were formed and given responsibility for the development of fishery management plans for fish stocks of significance to commercial and recreational fisheries in each region. The benchmark for federal approval and implementation of these plans was a set of seven national standards (expanded to ten in 1996, including the requirement that the plans prevent overfishing and achieve "optimum yield"; see Appendix D).

Since 1976, the Magnuson-Stevens Act has been amended more than a dozen times, and several sets of amendments have marked significant changes in its course and emphasis (Greenberg, 1993; see Appendix E). The early amendments focused on the process of "Americanizing" U.S. fisheries. The Processor Preference Act in 1978 was designed to foster growth of the American processing sector by requiring the denial of permits to foreign processing vessels for fisheries in which U.S. fish processors have adequate capacity. This act was followed in 1980 by the American Fisheries Promotion Act and its "fish-and-chips" policy requiring allocation of foreign fishing privileges on the basis of a nation's reduction of trade barriers against U.S. fish products.

In the second phase of amendments, the focus was on domestic management institutions, particularly the regional councils, and the process for implementing

---

[3] The FCMA was renamed the Magnuson Fishery Conservation and Management Act in 1980 and the Magnuson-Stevens Fishery Conservation and Management Act in 1996.

regulations promulgated in fishery management plans. Amendments enacted in the 1980s were aimed at accelerating the Americanization process, for both the harvesting and the processing sectors, and strengthening the input of the fishing industry to management policymaking. The 1990 amendments returned to international issues, addressing U.S. jurisdiction over tuna species, driftnet fishing on the high seas, and the high-seas fishery for Bering Sea pollock (Greenberg, 1993).

The 1996 amendments (the Sustainable Fisheries Act [SFA]), by contrast, can be viewed as a response to the overwhelming success (and perhaps excess) of the Americanization and industry empowerment policies. The 1996 amendments were a reaction to the substantial depletion of U.S. fishery resources that resulted (at least in part) from the expansion of U.S. fishing power. The rapid growth in domestic fisheries had led to most of the problems often associated with open-access fisheries, including overcapacity, reduced profits, short and dangerous fishing seasons (Figure 1.1), and continuous pressure on the management system to relax conservation and management measures (Huppert, 1991). The amendments emphasized the goal of biological conservation of fish stocks and protection of habitats and emphasized other significant resource management objectives. The 1996 amendments reflect the increased influence of conservation and environmental groups in federal fisheries legislation.

From an economic perspective, these renewed objectives include greater consideration of economic efficiency in light of the overcapacity in many of the newly Americanized fisheries. The amended act, with its definition of "fishing community" and other provisions, also mandates greater attention to the distribution of economic benefits from U.S. fisheries and the effects of management on fishing communities. Because the issues of overcapitalization and distributional effects were largely subsumed in the congressional debate over IFQs, the relative importance of efficiency versus distributional considerations in U.S. fisheries policy remains undefined under the act.

For the first time, the amendments make the duty to prevent overfishing an enforceable obligation; they also require attention to marine resources used for noncommercial purposes and the broader ecological context of fisheries. These conservation objectives include the need to avoid or minimize the biological waste associated with fisheries, including reducing bycatch, minimizing the discarding of fish, and reducing adverse impacts of fishing on critical fish habitats. The 1994 amendments to the Marine Mammal Protection Act had already significantly changed the federal fisheries regime by requiring an explicit system to control marine mammal mortality in commercial fisheries. The 1996 amendments to the Magnuson-Stevens Act may thus be characterized as signaling a reemphasis on those biological and resource conservation objectives of the original act that had been overshadowed by the policy of Americanizing the industry and a response to the lack of clear conservation objectives in some of the council regions.

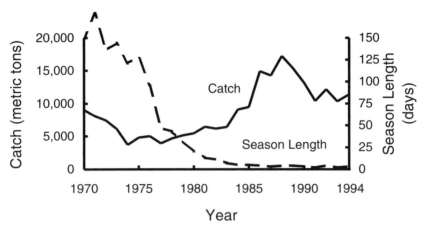

FIGURE 1.1A Changes in catch and season length in the Area 3A (central Gulf of Alaska) halibut fishery from 1970 to 1994, before the introduction of IFQs.

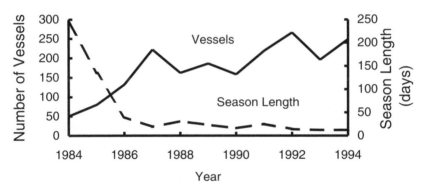

FIGURE 1.1B Participation and season length in the sablefish fishery in the West Yakutat sablefish fishery from 1984 to 1994, before the introduction of IFQs.

Provisions Related to Individual Fishing Quotas (see Appendix A)

The 1996 amendments did not resolve the debate over IFQs, but they did add a statutory definition of IFQs and a statement of policies concerning any IFQ programs designed or implemented after the end of the moratorium.

The act now defines an IFQ as "a Federal permit under a limited access system to harvest a quantity of fish, expressed by a unit or units representing a percentage of the total allowable catch of a fishery that may be received or held for exclusive use by a person" (Sec. 2[21]). The IFQ is defined as a permit for purposes of the act's provisions on prohibited actions, civil penalties, permit sanctions, and criminal offenses. This permit may be revoked or limited at any time in accordance with these provisions, and if it is revoked or limited, the holder is not entitled to any compensation from the government. The act also states that the IFQ shall not create, or be construed to create, any right, title, or interest in or to any fish before the fish is harvested (Sec. 303[d][3]).

After the expiration of the moratorium, Congress requires any council submitting an IFQ program, and the Secretary of Commerce in reviewing that program for approval, to consider this NRC report and to ensure that the program includes a process for review and evaluation; provides for effective enforcement and management; and utilizes a fair and equitable initial allocation, with provisions to prevent excess concentrations of IFQs and to facilitate new entry, especially of those not favored by the initial allocation (Sec. 303[d][5]). The Magnuson-Stevens Act also mandates the development of a central registry system for limited access system permits, including IFQs. The Secretary may collect 0.5% of the value of the permit upon registration or transfer of the permit to fund the operation of the registry (Sec. 305[h]). The act also mandates the collection of a fee of up to 3% of the exvessel value of landed fish to "recover the actual costs directly related to the management and enforcement of IFQ and CDQ programs" (Sec. 304[d][2][A-B]). In IFQ fisheries, up to 25% of the funds collected through such fees can be allocated at a council's discretion to help finance purchase of IFQs by small-vessel fishermen and new entrants to the fishery (Sec. 303 [d][4]). The North Pacific Fishery Management Council was required to recommend to the Secretary of Commerce by October 1, 1997, that the full amount of these fees be used to guarantee loans for small-vessel fishermen and new entrants to the Alaskan halibut and sablefish IFQ fisheries (Sec. 108[g]).

DEFINITIONS

The evaluations presented in this report rely on definitions of a fishery, fishing community, and individual fishing quota that are based on those given in the Magnuson-Stevens Act but differ from the act's definitions in some aspects. The concepts of resource rent and externalities are also discussed in this section.

Fishery

The Magnuson-Stevens Act defines a fishery as one or more stocks of fish that can be treated as a unit for the purposes of conservation and management and that are identified on the basis of geographic, scientific, technical, recreational, and economic characteristics; as well as any fishing for such stocks (Sec. 3[13]). A fishery consists of a suite of biological phenomena and economic, social, and political actors and institutions. A more comprehensive definition, to which the committee subscribes, includes the people engaged in the harvest and use of the fish, whether for commercial, subsistence, recreational, or ceremonial purposes. A fishery includes those who harvest, process, market, and even manage the activity. The dynamics of each fishery can differ based on the uses of harvested fish, ranging from subsistence to harvest for world markets.

Fishing Community

The Magnuson-Stevens Act defines "fishing community" as a community that is substantially dependent on or substantially engaged in the harvest and processing of fishery resources to meet its social and economic needs; vessel owners, operators, crew members, and processors based in such a community are included (Sec. 3[16]). There are two broad aspects to this definition. The first is that of specific, contiguous geographic locations where fishermen or those associated with the fishing industry live and work. In this sense, Gloucester, Massachusetts; Atlantic, North Carolina; Golden Meadow, Louisiana; Fort Bragg, California; Ballard, Washington; and Kodiak, Alaska, might be considered fishing communities, even though some are embedded in larger metropolitan areas.

The second aspect is that of a "community of interest" or "virtual community" (NRC, 1999b), a group of people who share common interests and activities that may or may not be associated with specific, contiguous geographic locations. In this sense, recreational king mackerel fishermen, East Coast longliners, the Gulf of Mexico shrimp fleet, California salmon trollers, and members of organizations such as the Coastal Conservation Association and the Pacific Coast Federation of Fishermen's Associations might be considered fishing communities. The implementation of IFQs creates a new community of interest: those who hold IFQs. The existence of such a community of interest is important in the discussion of co-management and involvement of stakeholders in the management process.

Both definitions of fishing community are relevant to the potential achievement of objectives or assessment of impacts for specific fishery management programs.

A related subject is the issue of "community dependence" as it applies to these two different aspects of the definition of community. For small, isolated geographic communities such as many of those in Alaska, Hawaii, U.S. island territories, and rural areas of the contiguous United States, the notion of dependence may include geographic isolation; lack of employment alternatives; social, economic, and cultural systems that have developed in these locations; and their dependence on fishing as a source of nutrition, livelihood, and life-style. For the more general communities of interest, however, such as a migratory fishing fleet with bases in several different locations or a constituency such as recreational fishermen, the notion of dependence may include the ability to move flexibly among different locations or to use fishing as part of a diverse set of life-styles.

These communities of interest may also have social, cultural, and economic systems that have developed specific to the interests and activities of their members. The fact that these types of communities differ does not mean that either type of community is more or less "dependent" on fishing; rather, they are dependent on fishing in different ways that should be taken into account in the development of all fishery management programs, including those using IFQs.

Individual Fishing Quota

For most fisheries, the most effective mechanisms to ensure that a fish stock can continue to be productive are limits on the amount of fish that is harvested and removed from the breeding population and protection of critical habitat. Two general types of techniques can be used to control the level of harvest: input and output controls.

Input controls attempt to limit catch indirectly through limits on the amount of labor or capital that can be applied to a fishery, for example, by limiting the amount of time fish can be harvested or the amount or design characteristics of gear that can be used. Output controls attempt to directly limit the number or weight of fish that can be harvested. Output controls usually establish a total allowable catch (TAC) for a given fish species and close the fishery once this level is reached. IFQs are a form of output control. Frequently, combinations of input and output controls are used to manage the amount of fishery harvests, the timing of the harvest, and the distribution of harvest activities.

IFQs are allocations of fish harvesting quotas to individuals or firms, specifying that a certain amount of fish or shellfish of a certain species may be caught in a specific area within a given time frame (usually a year, although not necessarily a calendar year). IFQs are not necessarily a replacement for other management tools and are actually complementary to other management measures. IFQs are best suited to fisheries managed by setting a TAC. Indeed, IFQs are usually expressed as shares of the TAC, so that the amount of fish that can be harvested for a given share of quota fluctuates with changes in the level of the TAC.[4] The magnitude of a TAC is usually derived on an annual basis by applying a target exploitation rate to an estimate of the current stock size. Determining the target exploitation rate and measuring the stock size are both subject to considerable uncertainty, because of large variability in the relationship between stock size and the generation of subsequent offspring and to general difficulty of accurately counting and measuring fish in the wild (NRC, 1998a).

IFQs are defined in the Magnuson-Stevens Act as limited access permits to harvest quantities of fish. They represent quasiprivatization of the fisheries, in

---

[4] Although the TAC in IFQ fisheries is usually expressed in weight (biomass), the Wisconsin lake trout IFQ is expressed in number of fish.

that permittees hold exclusive privileges with some of the attributes of private property—such as the privilege to decide when and how to use the quota shares—but not others, including ownership of the resource itself and the ability to decide how much of the resource can be harvested. The latter remains the domain of state and federal governments, which have public trust responsibilities to manage fishery resources for the public.

The term "IFQ" is peculiar to recent U.S. history. Other terms are more widely used. If the quota shares are tied to specific fishing vessels, rather than persons, they are usually known as individual vessel quotas (IVQs). If the quota is more or less freely transferable, it may be known as an individual transferable quota (ITQ). Sometimes the term individual quota (IQ) is used to denote an IFQ program that does not allow transferability. A quota that is allocated for the retention of bycatch rather than for a target species is known as an individual bycatch quota (IBQ) or an individual vessel bycatch quota (IVBQ). Throughout this report, the committee uses the term IFQs to define the general concept of allocating individual privileges to harvest fish. This term includes the concepts of ITQs, IVQs, IQs, and IBQs or IVBQs. When the committee refers to specific types of individual quota systems the appropriate term is used.

Resource Rent[5]

An important economic concept that forms the basis of the economic efficiency arguments in favor of IFQs is the concept of *resource rent:*

- Rents are economic profits that can be earned and can persist in certain natural resource cases due to the fixed supply of the resource. In most industries, expanding supply in the presence of constant demand reduces prices. Rent can arise in a natural resource situation because supply cannot be expanded indefinitely, which can create a scarcity and keep prices higher than necessary to keep the inputs to production in their use in fisheries, instead of some other use.
- In situations in which fixed supply creates scarcity, rent is the difference between total revenue and all necessary costs of production, including a normal return on invested capital (see Panel C in Box 1.1). In normal industries, extranormal profits (those due to a scarcity of supply of the industry's products) are dissipated by an expansion in manufacturing, but this is not always possible with natural resources such as fisheries. In regular production, advantage is gained by producers who adopt cost-minimizing technological changes because total production can be increased. In natural systems, total production is exogenously fixed and technological changes simply serve to redistribute catch shares

---

[5] Anderson (1980) discusses different sources of economic rent from fisheries, including resource rent.

BOX 1.1
A Simple Bioeconomic Model of a Fishery

The Gordon-Schaefer model for a fishery (Gordon, 1954; Schaefer, 1957; Clark, 1985a), illustrates many of the features and problems observed in fisheries. The model deals only with a single fish species in isolation and ignores the substantial and unpredictable environmental fluctuations that give rise to variations in the abundance of fish stocks. Underlying the model is the simplifying assumption that the natural growth is "logistic," with the maximum per capita growth rate (the intrinsic growth rate r) occurring at small stock biomass (Panel A). The model also assumes that all fishing operations are identical and it ignores temporal variability in the price for fish and the costs of fishing. Throughout the report, these shortcomings are identified and addressed at pertinent points. Despite these shortcomings, the model highlights how open access can lead to overexploitation, a feature observed repeatedly in real-world fisheries. In the absence of fishing, the stock will have a growth rate of zero and be at equilibrium when its biomass is equal to the carrying capacity (K). The overall growth rate (as opposed to the per capita rate) is a parabolic function of the stock biomass (Panel B). If the harvest rate is proportional to both the amount of fishing and the stock biomass, it can be represented as a straight line through the origin, with slope equal to the catch rate per fishing operation (q) times the number of operations (f).

Panel A   Panel B

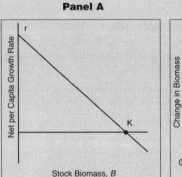 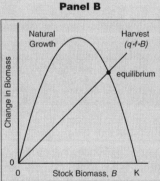

The intersection of the harvest line and the growth curve represents the equilibrium biomass and the equilibrium harvest rate. If the biomass is less than the equilibrium level, natural growth exceeds the harvest and the stock grows toward the equilibrium; if the biomass is greater than the equilibrium level, harvest exceeds natural growth and the stock shrinks. The peak of the growth curve represents the so-called maximum sustainable yield (MSY), which occurs at biomass equal to B_{MSY} (Panel C). Stocks that have been reduced by fishing to below the B_{MSY} level are usually considered overfished.

If the price for fish is constant, the total revenue flow from this fishery can be represented as a parabolic function of either the equilibrium stock biomass or the equilibrium number of fishing operations (Panel C). If the flows of fixed and operating costs per fishing operation are constant, the total cost flow can be represented as a straight line through the origin, with slope proportional to the cost flow per operation. (Note: This assumes a long-term analysis in which all costs are variable. The total cost line includes a normal profit equal to the amount necessary to keep the factors of production in their existing use.) The difference between the revenue curve and the cost line represents the rent that potentially could be derived from the stock. The intersection of the cost line and the revenue curve represents the open-access equilibrium (OAE) associated with f_{OAE} fishing operations and a stock biomass of B_{OAE}. If the number of fishing operations is less than f_{OAE}, the participating operations will make abnormally large profits that will entice additional operations into the fishery, provided entry is not limited. If the number of fishing operations is greater than f_{OAE}, net losses will cause some operations to exit the fishery. Rent from the fishery is zero at the open-access point. The location of B_{OAE} relative to B_{MSY} depends on the magnitude of the fishing costs (the slope of the cost line) relative to the price for fish. Fish that are expensive to harvest will not be biologically overfished unless they are high priced.

The diagonal total cost line in Panel C includes the fixed costs associated with fishing vessel ownership, to cover debt servicing if the vessel was purchased with borrowed money or to provide a reasonable return on investment otherwise. If the government subsidizes vessel purchases, the vessel owner will experience lower costs and the total cost line will be less steep, which will produce an open access equilibrium point with a smaller stock of fish but more fishing vessels making low or negative returns on the investment. Because investments in vessels and permits are fixed in the long run, the decision to fish during any given season will depend on the short-run variable costs, which are less than the total costs. As a consequence, the vessels may make unsustainably large harvests in the short run that lead to long-run losses.

Panel C

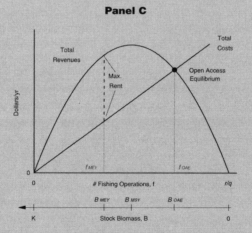

If the fish stock was privately owned, the "static maximum economic yield" (MEY) that produces the maximum stream of profit flows (the maximum rent) given a zero discount rate, would be achieved by limiting the number of fishing operations to f_{MEY}, resulting in a stock biomass of B_{MEY}. Unless fishing costs are zero, the biomass that produces MSY will always be less than the biomass that produces the static MEY ($B_{MSY} < B_{MEY}$). With a nonzero discount rate the stock biomass that produces the maximum stream of discounted profit flows, the so-called dynamic MEY, will depend on the ratio of the discount rate over the stock's intrinsic growth rate (Clark, 1985a). With an infinite discount rate, future harvests have no value and the stock biomass that produces the dynamic MEY is B_{OAE}, coinciding with the open access equilibrium.

among participants, increasing costs and decreasing profits. Rent can exist in an efficient fishery (see Bjorndal and Munro, 1998), although rents may be small or nonexistent if supply is high in relation to demand or if some other product can substitute for the product from the specific fishery (e.g., the same species of fish caught or cultured elsewhere: Box 1.2).

• Rent can be dissipated. Rent in a fishery can be dissipated in several different ways. Attracted by the prospect of appropriating some of the rent, people enter the fishery. Initially, when the harvesting capacity is small, new arrivals do not cause problems; the resource is plentiful enough to supply all harvesters with a good livelihood without depleting the stocks. As capacity continues to grow, however, a point in the race for fish is reached eventually at which both the rent is reduced and the biological viability of the resource is threatened. A particularly striking example of the pace and consequences of the race for fish is reported in Gunderson (1984). The rapid development and subsequent decline of the Pacific coast fishery for widow rockfish (*Sebastes entomelas*) began in 1979 when a trawler, returning late from his fishing grounds, observed a large concentration of fish on his sonar. When he ran a test tow through the school, he was surprised to find that the haul consisted of widow rockfish, a species that was not previously known to be available to trawl gear. In the early nights of this fishery, he and others were able to make hauls that averaged 31 metric tons per hour. Gunderson (1984) reports that the fishery grew from 3,291 metric tons in 1979 (9 vessels) to 20,158 metric tons in 1980 (52 vessels) and 28,419 metric tons in 1981 (70 vessels). Efforts to control exploitation rates were

---

BOX 1.2
Effects of Substitutability on Rents

One might expect that those with access to the few sources of a scarce resource could ask for exceptionally high prices and expect not to be bid down. Newfoundland might provide a case in point because the northern cod fishery has been closed due to stock collapse since July 1992 and only a few cod stocks have been reopened to fishing since then. However, fish—especially cod—have been an international commodity for at least 500 years and the prices for cod in the few places where the fishery is allowed in Newfoundland and elsewhere in the Canadian maritime provinces are low because the Russian, Norwegian, and Icelandic fisheries for Barents Sea and other cod are doing quite well, more than satisfying the markets. Rents can be low even if a resource is scarce when a fishing area is less productive than others or when the same species from a different area or a different species from the same area are substitutes for the scarce species. Rents also can be low when a species is so scarce that it is not economical to harvest it.

IFQs and the steps toward them (e.g., getting rid of small-boat "marginal" fishers; using trip limits and vessel quotas) are ways to increase the chances of making a living from a scarce resource, but rent may be small or nonexistent.

hampered by limited information on biological parameters and an unwillingness to impose restrictive harvest limits without better information on the stock. The Pacific Fishery Management Council set the allowable biological catch for 1983 at 10,500 metric tons.

New entrants to a fishery may dissipate the economic rent both by flooding the market with harvested resource during the most productive harvesting times (thereby lowering the price received by harvesters) and by escalating the costs of harvesting. Escalation of costs is driven by the race for fish induced by the "ownership by capture" rule. According to this practice, the right to claim the rent associated with fugitive property (such as fish in the ocean) is established by the successful harvest of the resource. Fish landed in a boat become the property of the harvester, thereby entitling the harvester to normal economic returns and any rent associated with the landed fish. However, when access to the resources is not limited, people will compete for the greatest possible share of the rent. This competition will raise the costs of extraction until the rent has been absorbed by excessive costs (i.e., wasted inputs of capital and labor). Individual fishermen, by being the first to adopt an effective new harvesting technology, achieve a temporary advantage over other fishermen such that they obtain catch increases that more than offset the added cost of the new technology. But since the total catch is fixed by resource conservation and stewardship requirements, as expressed through the TAC-setting process, gains to the initial adopter of the new technology are offset by equal losses to other fishermen (i.e., it is a "zero-sum game"). The other fishermen quickly recognize their comparative disadvantage and also adopt the new technology. The outcome is that the total harvest is unchanged while the costs of harvesting are increased.

The race for fish is but one source of rent dissipation. Another source emerges from variations in productivity across fishing grounds (Gardner et al., 1990). Certain areas within fishing grounds may be substantially more productive than surrounding areas. As fishermen race for places to fish that are near their home port or for locations that have high catch rates, they expend resources that would not have been used if they had coordinated the use of fishing spots. This form of rent dissipation quickly accumulates as more fishermen enter a fishery and increasingly find their preferred spots taken (Higgs, 1982). Rent dissipation also arises from the management process, as potential harvesters expend resources in an attempt to influence allocation decisions (called *rent seeking*) (Krueger, 1974); Edwards (1994) showed that rent seeking occurring in association with IFQ programs could decrease their economic benefits.

Upon implementation of IFQs, a portion of the rent develops rapidly through a reduction in redundant vessels, reduced gear loss, and reduced catch discards. Rent sometimes also develops in response to increases in exvessel prices, such as occurred with halibut as more product could be sold fresh. See Box 1.1 for a simple bioeconomic model of a fishery.

Externalities

Externalities occur when the costs or benefits of a resource user's actions are not borne fully by the individual user; other resource users share the costs or benefits. Because of the common-pool nature of fisheries (see Box 2.1), fishermen impose externalities on one another. Such externalities occur through highgrading, as well as when fishermen race to maximize their share of the total catch.

HISTORY OF INDIVIDUAL FISHING QUOTAS[6]

The concept of IFQs arose in the context of long histories of conflict over limiting access to marine resources and, until recently, an apparent social and legal commitment to the principle of open access. They have been depicted as " . . . a part of one of the great institutional changes of our times: the enclosure and privatization of the common resources of the ocean" (Neher et al., 1989; p. 3), following as they did the expansion of national jurisdiction over the seas in the late 1970s. However, their appearance in fisheries management in the 1980s and 1990s is more a product of the historical process of widening and deepening the role of markets (Helgason and Pálsson, 1997) and increased recognition of economic factors in protecting environments and managing natural resources. "Thus, ITQs are part of the current global expansion and integration of markets, extended to fisheries" (Squires et al., 1995; p. 143).

Although their history is short—no more than approximately two decades—IFQs are firmly rooted in the long tradition of Western thought and policy, where markets are the source of efficiency and, ultimately, of economic growth and social welfare; exclusive, transferable, and well-defined property rights are essential to markets. From this perspective, cases of overuse and abuse of the "commons" are caused by the failure of markets to give proper signals due to the lack of appropriately specified property rights. These arguments were articulated early in the eighteenth century in the work of economists such as Adam Smith (1957 [1772]) and less well-known people such as William Forster Lloyd (1968 [1837]), the inspiration for Garrett Hardin's essay on the "tragedy of the commons" (Hardin, 1968). They express a particular political philosophy and social psychology, one that has come to define the "modern" era in the history of Western culture: that people and their societies are driven by individuals seeking maximization of their own welfare.

Accordingly, it was only a matter of time, technological and other changes that reduce costs, and increased pressure on shared resources before exclusive property rights were proposed to manage marine fisheries (e.g., Anderson and Hill, 1975; Johnson and Libecap, 1982; Libecap, 1989). Two questions can be

---

[6] This section summarizes how the literature has characterized fisheries management; the committee does not necessarily endorse the terms and concepts used in this section.

posed: Why did it take so long? How, exactly, did it come about? One reason exclusive harvest privileges were not implemented until recently in fisheries is the practical difficulty of erecting boundaries around fish. In common law, such "fugitive" creatures as fish and wild birds or game are not "owned" until they are captured. This rule of law poses an obstacle to privatization.

Yet privatization of aquatic animals is not impossible, nor is it absent from history. Sedentary shellfish can be fenced and points of access to marine resources can be treated as private property. Farmers, fishermen, lords of the manor, monasteries, enterprising fish traders, Native Americans in the Pacific Northwest and elsewhere, and others have often sought to secure their interests by claiming exclusive rights to shellfish beds, places to use fishing weirs and seines, and particular fishing grounds. During the Tokugawa period in Japan (from 1603 to 1868), fishing territories were owned by feudal rulers who extracted labor and taxes in exchange for giving rights to fish to communities and individuals (Kalland, 1988). There is also a historical record of exclusive rights of access to highly valued sports and recreational species for both inland and anadromous stocks, many of which continue (Netboy, 1968; Brubaker, 1996).

An even more significant obstacle to exclusive access to marine resources has been social resistance to it, based on very different conceptions of property rights, configurations of interest, and in some cases, world views, which to some extent continue today and influence the debate about IFQs. Fish weirs were outlawed on the rivers of England by the Magna Carta of 1215, which became the political and legal source of the liberties of British subjects. Over time, in many Western nations, legal principles were established that supported the inalienability of public as opposed to private rights of commerce, navigation, and fisheries. The public trust doctrine, as it came to be known in the United States, has persisted despite society's embrace of the market economy imperatives of privatization (Sax, 1970; Rose, 1986; McCay, 1998). This legal doctrine protected the interests and rights of the public to free access to marine and riparian places for fish and navigation by asserting state ownership of the beds of tidal and navigable waters, the waters themselves, and fish and shellfish, on behalf of the public. The public trust doctrine, discussed in Chapter 2, protects public rights in tide-washed lands, navigable rivers, lakes, and their resources. In some areas, it also has been interpreted as justifying the right of the state, or the Crown, to ignore or replace local customary and formal systems of allocating rights. Accordingly, the development of this argument can make it difficult for local communities to integrate the interests of all users of the resource, as opposed to the public at large.

In the Pacific islands, traditional cultures often treated fishing grounds, rights to take certain species, and even rights to use specific gear types as the exclusive property of individuals or family corporations. Thus, in the codification of customary law relating to land and sea resources in the Gilbert Islands in the early 1940s, British colonial officials encountered a conflict between British law, in which tidal and submerged lands belonged to the Crown, with attendant public rights of navigation and fishing, and Gilbertese law, in which every stretch of reef and every coral head and shoal that was a feeding ground for fish was privately owned (Goodenough, 1963). A similar system characterized customary law in Chuuk (Truk) (Goodenough, 1951), as well as islands of Papua New Guinea (Carrier, 1987a,b), where even the right to use certain types of fishing technology for certain species of fish might be privately owned by individuals or family corporations such as lineages. However, colonial and postcolonial impositions of Western law, as well as profound political, economic, and cultural change, have weakened and destroyed many of these systems of sea tenure (Johannes, 1978; Cordell, 1989). Accordingly, the issues of open-access fisheries are widespread, although not universal.

In the United States, competition between those who claimed the freedom to fish and those who wanted the benefits of exclusive property rights in order to develop fisheries was particularly evident in the estuarine shellfisheries for oysters and clams. Similarly, exclusive use rights to fishing sites and species were held by clans, tribes, and family groups in the Pacific Northwest and Canada (e.g., McEvoy, 1986; Bay-Hansen, 1991; Newell, 1993). Higgs (1982) provides a thoughtful discussion of how the collapse of traditional (Native American) fishing rights on the Columbia River precipitated an economically wasteful race for fish in the mid-nineteenth century. By the end of the nineteenth century, the arguments had come closer to those of economists today: the dangers of open access or "public rights" and the need for some sort of privatization to provide incentives to people to take care of and invest in the future of shellfish resources (Brooks, 1891; McCay, 1998). Leaseholds or private property were created in many states for shellfish enhancement and aquaculture. Otherwise, government-based fisheries management proceeded in the directions of hatchery-based enhancement and regulation of fish harvesting through total allowable catches, closed seasons, gear limits, and other tools, while maintaining more or less open access (Nielsen, 1976).

Admonitions to recognize the role of economics in fisheries management reappeared in the early twentieth century (e.g., Warming, 1911; Graham, 1943) but were largely ignored until the 1960s and 1970s. Scheiber and Carr (1997) found the germ of the IFQ concept in a plan proposed for the State of Maryland to limit licenses in a fishery on the Chesapeake Bay during World War II. This region featured a long-standing concern about the problems of managing fisheries through gear restrictions, which meant managing for inefficiency, especially in the shellfisheries (Brooks, 1891; McHugh, 1972). The new proposal included ideas about "economic rent," basing rights on historical performance, and finding ways to retire excess vessels (Scheiber and Carr, 1997).

At the federal level, fisheries management was being based on the goals of maximum sustainable yield (MSY) and taking an ecosystem approach to achieve

MSY (Scheiber and Carr, 1997). The International Pacific Halibut Commission (IPHC) was an important focus of these debates. The success of the IPHC in restoring the depleted Pacific halibut stock by using annual quotas was invoked as evidence for the importance of focusing on conservation, rather than allocation and economic issues (Scheiber and Carr, 1997).

The Maryland license limitation proposal influenced a Canadian economist, H. Scott Gordon, who in 1953 published "the first major paper setting forth a theory of property rights in fisheries" (Scheiber and Carr, 1997; p. 238). Gordon (1953, 1954) argued that as long as fisheries were considered a "common property" (i.e., open-access) resource, economic rent from nature could not be captured by society. It would be dissipated in a process of escalating harvest costs that results from the race for fish. Gordon demonstrated that a "sole owner" would harvest the stock in such a way that he or she would capture the available rent. In the work of economists who followed Gordon, the otherwise esteemed work of organizations such as the IPHC was accused of creating economic inefficiency (Crutchfield and Zellner, 1962), and MSY itself was construed as a "socially meaningless objective" (Christy and Scott, 1965).

Soon after Gordon's paper was published, the Food and Agriculture Organization (FAO) held a conference on the economics of fisheries management (FAO, 1956; see Scott, 1993), and university conferences and working groups followed (e.g., Crutchfield, 1959). An international community of fishery economists developed around the topic of achieving economic objectives and using economic incentives as tools in fisheries management.

By the mid-1970s, the concerns of economists had become part of broader assessments of fisheries management, which criticized the dominant goal of achieving MSY (Nielsen, 1976; Larkin, 1977). Economists and their concerns also played important roles in changes in the international law of the sea, as Iceland, the United States, Canada, and eventually all other coastal nations claimed 200-nautical-mile exclusive economic zones (EEZs) and revised their domestic policies and management structures to handle the new responsibilities of extended jurisdiction. The economists' argument for "limited entry" to counter the problems of open access had been well developed in works such as Christy and Scott's (1965) general treatise on the problems of open access and the need for limited entry and Crutchfield and Pontecorvo's (1969) book about the economic losses that resulted from open-access management of salmon fishing. However, the concept of limited entry became an easily understood and more politically acceptable reality as applied to foreign fleets than to domestic fishermen.

In Canada and some other countries, domestic management moved in the direction of limited entry as well, guided by economists and others trying to bring the goal of economic efficiency into fisheries management. Limited entry was established in 1968 for salmon and halibut in the province of British Columbia, Canada (Fraser, 1979). Soon after establishing its EEZ in 1977, Canada's Ministry of Fisheries and Oceans began an eventually successful campaign to limit entry to all commercial fisheries (Parsons, 1993). In sharp contrast, the U.S. fishing industry showed major resistance to limited entry in fisheries, although limited entry programs were adopted for salmon and roe herring fisheries in Alaska as well as several fisheries in the Great Lakes (Cicin-Sain et al., 1978), and the State of Washington had a ballot initiative to limit entry as early as 1934 (Benson and Longman, 1978). Limited entry often appeared as an attempt to protect local fishermen from competition from outsiders moving into local waters; in the United States, it was therefore subject to court review in relation to constitutional protections of interstate commerce.

In the course of discussions by fisheries economists from the 1950s through the 1970s, several important ideas arose that contributed eventually to the idea of IFQs. At an FAO conference held in Ottawa, Canada, in 1961, the economist James Crutchfield suggested the idea of creating a property right in licenses, not fish stocks; he called these "limited property rights" (FAO, 1962). This might be considered a predecessor to the idea of creating property rights in shares of a quota. At the same meeting, Anthony Scott, who had argued for some time for sole ownership as opposed to limited entry (e.g., Scott, 1955),[7] suggested that if a sole-owning authority of some kind assigned each vessel a right to fish, it would work toward greater efficiency and reduce overcapitalization (FAO, 1962). This idea is a precursor to that of allocating to individual vessels shares of a quota, which may become transferable. In 1963, a group of scholars from the University of Washington brought these and other ideas together in the first major report on limited entry in U.S. fisheries (Royce et al., 1963).

At numerous meetings and in book chapters and journal articles, economists continued their work on an economic theory of fisheries management, focusing on the neoclassical principle of the "nexus between efficiency and the vesting of property rights in marine resources" (Scheiber and Carr, 1997; p. 251). In 1969, Francis Christy wrote an article in which he argued that the right to exclude others was "fundamental to the achievement of economic efficiency" (Christy, 1969, cited by Scheiber and Carr, 1997; p. 251). By 1973, this idea was formulated as a tradable quota assigned to a fisherman, vessel, or firm (Christy, 1973; Elliot, 1973). Christy (1973) argued for allocating to each fisherman—whether owner or captain—a share of a quota or percentage of catch. These might be leased, although there would be limits to prevent excess accumulation, and they could be sold only to an administering agency. However, fishermen would have as much freedom as possible in determining how to fish and hence achieving some efficiencies. The premise was that the set of co-owners of the fishery might

---

[7] Scott (1955) formalized the arguments developed by Gordon; in a later paper (Scott, 1979), he extended the comparative static approach of Gordon (1953, 1954) and Scott (1955) to a dynamic optimization context.

achieve some of the rent-conserving behaviors available to the hypothetical sole owner.[8]

In footnotes and asides, Christy acknowledged the difficulty of initial allocations as well as several possible drawbacks of limited entry licensing schemes such as this in contrast to open access, including their attraction for "big business" and hence the possibility of the loss of valued attributes of fisheries, including the independent, individualist, risk-taking way of life made possible by open access (Christy, 1973, 1977). As noted earlier in this chapter and discussed in later chapters, these problems are indeed real ones in existing IFQ programs. However, the momentum was in the direction of testing individual quotas in actual fisheries. Beginning in 1976, the herring fishery of the Bay of Fundy region of Atlantic Canada was managed by allocating shares of the annual herring quota to vessels in the industry, the owners of which set up a co-management scheme with the support of the government fisheries agency (Kearney, 1984; Stephenson et al., 1993). The key papers from a conference on economics and fisheries at Powell River, British Columbia, were published in a special issue of the 1979 volume of the influential *Journal of the Fisheries Research Board of Canada* (now the *Canadian Journal of Fisheries and Aquatic Sciences*), the culmination of a year of meetings in Seattle, Denver (Rettig and Ginter, 1978), and Lake Wilderness, Washington. They included reviews of existing limited entry systems that highlighted the problems created by simply limiting entry without having other means of preventing overcapitalization and inefficiencies (Fraser, 1979). Several articles focused on the need to go further than limiting entry, adding limits on the amounts each licensee could catch and perhaps making these transferable (Copes, 1979; Moloney and Pearse, 1979; Scott, 1979), that is, ITQs.

Fundamental to these discussions was increased concern about technological advances and increased effort in many fisheries of the world, rising disenchantment with the idea of managing for MSY, and by the mid-1970s, enclosure of EEZs around coastal nations. With the Fishery Conservation and Management Act of 1976, U.S. fishermen and fishery managers viewed limited entry largely as a matter of excluding foreign vessels from newly claimed waters. Otherwise, the principle of open access was defended as it long had been in U.S. political and legal history, often with reference to the Commerce Clause of the U.S. Constitution (e.g., *McCready v. Virginia*;[9] Libecap, 1989, pp. 79-80; McCay, 1998).

---

[8] The linkage from Gordon through Scott to IFQs followed the logic that multiple users would compete away the rents in a race for fish, whereas a sole owner would, acting as a monopolist, capture the rents. Scott extended the work of Gordon to show the same result in a dynamic context. The next leap was that perhaps individuals with secure access to predetermined quantities would behave in a manner analogous to the sole owner (Keen [1988] argues against this, asserting that sole ownership is required to resolve property issues in fisheries). The assumptions that drive this conclusion are that the harvest right is secure, that cheating does not occur, that there are no unique spatial or temporal concentrations that could lead to a race for fish, and that any returns to scale are captured.

[9] 94 U.S. 391 (1887).

Consequently, in the United States, any interest in individual quotas was probably stymied by widespread resistance to limited entry, as noted above. The surf clam fishery of the Mid-Atlantic region was the first federal fishery to have limited entry, beginning in 1978; it remained an exception for most of the next decade.[10] The principle of open access remained dominant. Management bodies such as the regional councils concentrated on Americanizing fisheries that had been dominated by foreign vessels prior to the establishment of the EEZ. In some regions, particularly the northeastern United States, there was also strong resistance to the use of TACs in fisheries management after their brief and contentious use in New England groundfish management (Miller and van Maanen, 1979). TACs are still not being used in this region, although a moratorium on entry to the groundfish fishery was established in 1992 and effort is controlled through allocation of days-at-sea limits to individual vessels.

From the late 1970s and particularly the 1980s, leading scientists and officials in NMFS favored limited entry and IFQs. However, the only case of limited entry in federal waters into the early 1980s was the surf clam fishery, which was managed with a moratorium on entry as well as TACs and time limitations. The surf clam fishery quickly became a model of how a limited entry program could create an exclusive group engaged in rampant overcapitalization (see Chapter 3). In Canada, the limited entry salmon and halibut fisheries had shown similar problems (Fraser, 1979; Pinkerton, 1987), and the transferable vessel allocation program for the herring fishery created in 1976 in Atlantic Canada seemed doomed because of inadequate enforcement (Stephenson et al., 1993). In the early 1980s—like many other nations in the context of expanding fishing jurisdictions—the fishing industries and ministries of Australia and New Zealand held exchanges of fishery managers and economists who were developing the idea of IFQs. Similar exchanges among managers, industry people, and economists helped shape the early IFQ programs of Iceland, Norway, and Canada. Chapter 3 and Appendix G provide additional details about the development and characteristics of IFQ programs in the United States and abroad.

Certainly, the design of IFQ programs has been an experimental process. For example, lessons about the costly risks of allocating pounds of fish rather than a percentage of a quota share came from the New Zealand IFQ program. The Canadian experiments, starting with Bay of Fundy herring, showed the critical importance of investing in monitoring and enforcement, as well as the potential for industry involvement in "co-management" and cost sharing. The U.S. surf clam/ocean quahog (SCOQ) IFQ program, as well as the wreckfish system, the first two in U.S. federal waters, both benefited from these experiences. The

---

[10] Limited entry within state waters was much more common, used not only by Alaska and the Pacific states but also by the Great Lakes states and New Jersey. During the early period of the Magnuson-Stevens Act, there was concern about states rights being swamped by federal authority, which may have played a role in the lack of federal limited licensing programs.

SCOQ experience, in turn, has led to greater attention to problems of excessive accumulation and concentration in industries under IFQs, a question that was taken up in earnest in the Alaska halibut and sablefish IFQ planning process. In addition, failures to take into account the claims of hired captains and crew to initial allocation in the Alaskan programs probably influenced the decision in the Gulf of Mexico red snapper IFQ plan (approved but not implemented) to provide for allocation to captains under specific forms of contract. The red snapper planning experience (like that of New Zealand), in turn, has highlighted the challenges of using IFQ tools in fisheries with large and growing recreational participation.

GENERAL RATIONALES AND ISSUES FOR IMPLEMENTING INDIVIDUAL FISHING QUOTAS

The reasons for using IFQs can vary widely. The most general reason is to counteract negative consequences of open or limited access management systems, particularly where TACs are used. A TAC without any limitation on fishing by the individual fisherman provides incentives for all participants in the fishery to harvest the TAC as quickly as possible before the fishery is closed. This typically leads to excessive fleet capacity and fishing effort and increasingly shorter fishing seasons (see Figure 1.1). A central objective of many fisheries managed by IFQs is to avoid the undesirable consequences of this race for fish.

Three more specific rationales that have been offered for implementing IFQs are (1) improving economic efficiency by providing incentives to reduce any excess harvesting and processing capacity; (2) improving conservation by creating incentives to reduce bycatch and lost gear and engaging in other activities that conserve the resource; and (3) improving safety by reducing incentives to fish in dangerous conditions. Although many of the benefits and costs derived from IFQ management might be based on economic principles, the potential social effects are also likely to be central concerns in the design of any IFQ program. A wide variety of motives may influence the development of any specific IFQ program. The following discussion of the three principal rationales for implementing IFQ management provides an overview of the potential benefits and costs of using this form of management.

Economic Efficiency

In terms of the national standards contained in the Magnuson-Stevens Act, IFQs could be used as part of a strategy to satisfy the requirement that "conservation and management measures shall, where practicable, consider efficiency in the utilization of fishery resources; except that no such measure shall have economic allocation as its sole purpose" (National Standard 5, Sec. 301[a][5]). By dividing the TAC into shares that are allocated to individuals who can then determine when and how to use them, economic efficiency can be increased, particularly if the quotas are allowed to be transferred, as discussed in later chapters.

The race for fish described above has serious economic consequences. It can lead to more intensive fishing, more gear being deployed, and increased capital expenditures to catch the same TAC. This amounts to wasting productive resources, because the fish could otherwise be taken at a lower cost, and perhaps transformed into a more valuable product, if the landings were spread over a longer period of time and fishermen had more time to handle fish more carefully. In a number of IFQ fisheries that the committee reviewed, improved product yield and/or value of the product has occurred. The race for fish also leads to costly (and otherwise unnecessary) modifications of fishing vessels to make them more effective in catching fish as quickly as possible (e.g., more powerful engines, expensive fish-finding equipment, larger size). Growth in fish-processing capacity both stimulates and is stimulated by the race for fish as processors expand their facilities and develop distribution chains to handle large pulses of fish and compete with each other to attract landings. These pulses also directly affect the price and quality (fresh versus frozen) of fish available to consumers. TAC-based management alone will not promote efficiency if more boats and people enter the fishery without controls. All of these developments make the race for fish more acute over time. IFQ management promotes efficiency by eliminating incentives for fishermen to apply excessive capital and labor inputs to a fishery.

Nevertheless, improving economic efficiency can dramatically alter the characteristics of a fishery and can have significant social implications. If harvesting and processing capacity are removed from the fishery, communities that were once dependent on the race for fish can lose employment and revenues that were generated formerly. (However, such communities will eventually lose employment and revenues anyway if the race for fish is not controlled.) Testimony received by the committee indicated that these changes have reduced employment in regions with limited opportunities. In particular, two features of IFQ program management are controversial and can result in profound socioeconomic changes in a fishery: (1) the initial allocation of quota and (2) the transferability of quota (see Chapter 5). The IFQ programs evaluated by the committee vary with respect to these features.

A confounding factor complicates the economic efficiency arguments: not all components of commercial fishing industries operate according to a common economic logic of firms. Abundant empirical evidence exists to demonstrate that these assumptions are not always true. In their study of fisheries in the U.S. Northeast, Doeringer et al. (1986) differentiate between what they call a kinship sector and a capitalist sector and indicate that the kinship sector thrives and expands under conditions that are detrimental to the capitalist sector. Apostle and Barrett (1992) make a similar distinction not only between fishing operations but processors as well in Atlantic Canada, and Durrenberger (1996) found the same to be true in the U.S. Gulf Coast. The existence of the kinship sector means that features of fisheries management that assume that individuals will make decisions on strictly economic grounds may be invalid and that management measures such as IFQs and other limited entry systems may have economic effects different from those that might be predicted on purely theoretical grounds. Thus, fishery managers should take into account the kinship sector in designing new management schemes, particularly in fisheries and areas characterized by small-boat fisheries with a long history.

Conservation

Another rationale given for implementing an IFQ program is the promotion of conservation. IFQs may promote conservation, if properly monitored and enforced, by keeping the catch within the TAC by making fishing more orderly, limiting the race for fish, and creating a penalty for individuals who exceed their individual portion of the TAC. In fact, most IFQ-managed fisheries are successful in maintaining the cumulative catch for the fishery (at least the recorded part) below the TAC, (see Figure G.4) whereas the same fishery managed without IFQs often exceed their TACs. Under IFQs, fishing time and area can be chosen more carefully by fishermen and less gear may be set (and lost), reducing both ghost fishing and reducing the potential damage that lost gear may cause to the marine environment. The added time available to the IFQ fisherman may also reduce the bycatch of non-target species since operations can be moved to target more favorable harvesting conditions, or it might allow the opportunity to develop practices that could reduce bycatch. Because IFQs allow more time to harvest and process fish, the amount of product recovered from the individual fish can be higher, reducing discarded product.

Additionally, the holder of the quota has an incentive to ensure that the fishery continues to be productive and that the quota continues to be valuable. It is argued by some that this incentive will encourage behavior to conserve the resource, conduct needed research, and assist the enforcement and monitoring of the fishery so that the health of the stock and the future value of the quota are preserved (Neher et al., 1989). Similar assumptions are implicit in much discussion of fisheries management and were explicit in testimony to the committee. Much of the political support for IFQs is similarly driven by faith in the assumption that privatization will foster ecological sensibility. This argument is based on the premise that the community of IFQ holders will behave in a manner analogous to the sole owner, as described in, for example, Gordon (1954) and Scott (1955). Another aspect of this argument is that an IFQ program that limits access to the resource will accumulate value that becomes capitalized in the value of the individual quota. The better the fishery is managed, the higher will be the value of the individual quota share.

However, quota shares are not rights to particular fish. Consequently, quota holders have no assurance that other quota holders will refrain from practices that prevent the sustainable use of fish stocks. Some argue that precisely because IFQ management provides an opportunity to conduct fisheries more slowly, selective harvesting of higher-value fish (*highgrading*) may occur. Highgrading is most likely to occur when catch rates are high and there is a significant price advantage to fish of a particular size, gender, or spawning condition. The incentive to highgrade is also increased when the TAC is expressed in the total number of fish rather than their total weight. Some also believe that because monitoring and enforcing harvests is difficult, the incentive to misreport catches, also known as *quota busting*, is sufficiently high to outweigh the potential risk of being caught. Individuals practicing stewardship may incur the full marginal cost of forgone catches and receive only the average of the increased future benefits. This illustrates the phenomenon of *externalities*, situations in which the costs or benefits are not fully borne by the individual user. Therefore, IFQ holders may have an incentive to conserve at less than the socially optimal level, especially when there are large numbers of them (and hence a smaller average benefit). This rationale demonstrates that IFQ fisheries require effective monitoring, enforcement, and penalties to achieve their benefits.

The net effect of IFQs on conservation will depend on the relative strength of the stewardship effect balanced with enforcement and the incentives for each individual quota holder to cheat. Sorting and discarding fish to highgrade costs money and will occur only if the expected benefits exceed the cost of sorting and the cost of catching replacement fish (including the opportunity cost of time and the expected cost of penalties and sanctions). The general conclusion of a 1990 workshop on the effects of different fishery management schemes on bycatch, "joint catch," and discards was that IFQ programs are no better or worse than other fishery management schemes in relation to these factors (Dewees and Ueber, 1990). Beyond the theories, few data exist regarding the positive or negative stewardship effects of IFQs, although there are some indications in the Pacific halibut fishery (Gilroy et al., 1996) that IFQs decrease regulatory discards and ghost fishing.

Safety

The third rationale for implementing an IFQ program is to improve safety in a fishery, a goal of the new National Standard 10 (Sec. 301[a][10]). It is argued that because an IFQ program allows greater freedom for the individual to choose when to fish, weather conditions, the condition of the vessel, or other safety factors can be considered and hazardous conditions can be avoided. Although empirical evidence suggests that safety has improved in some IFQ-managed fisheries, it is not clear that safety has improved in all fisheries managed using IFQs.

Other Rationales

A variety of other rationales have been used to justify the development and implementation of IFQs. For example, the surf clam/ocean quahog IFQ program was developed (in part) to reduce administrative and enforcement burdens. The wreckfish IFQ program was developed to try to prevent overcapacity from developing when the fishery was new and seemed to be in the midst of unchecked expansion.

OUTLINE OF THE REPORT

The overall goal of this report is to provide Congress with a comprehensive review and analysis of the use of individual fishing quotas and to recommend national policies on the implementation and use of IFQs, addressing the issues that Congress identified in the Magnuson-Stevens Act (see Appendix A). In Chapter 2, the committee reviews some of the theories and practices of common-pool resource management and the use of the public trust doctrine in managing natural resources. In Chapter 3, the committee evaluates the experiences of IFQ management in federal waters of the United States and abroad. In Chapter 4, the committee examines the alternatives and complements to IFQs used in fisheries management. In Chapter 5, based on the analyses undertaken the committee discusses issues to be considered in developing a national policy on IFQs. Finally, in Chapter 6, the committee presents its findings and recommendations.

2 | Fisheries Compared With Other Natural Resources

Debates about the use of individual fishing quotas (IFQs) and other limited access measures in fisheries management are grounded in more general debates about how natural resources with common-pool characteristics should be owned, allocated, and managed. Common-pool resources are those having features that make it difficult to exclude others from them (hence the notion of "common") in which one person's use can affect what is available to another person (Box 2.1). Where common-pool resources are scarce, the problem of allocating access or rights to them is the difficult but central problem (Edney, 1981). Societies have addressed this problem in several general ways (Edney, 1981; Fiske, 1991). One is to do nothing. In fisheries this is known as "open access," and typically leads to overuse (Gordon, 1954; Scott, 1955).

In the sustainable, productive management of fisheries, catch is limited to surplus production of the stock, by taking into account natural population dynamics. This leaves the stock undiminished through time, providing a perpetual stream of production and harvest. However, nothing in an unregulated open-access situation confines catches to the surplus production. As catches exceed surplus production, the population declines (NRC, 1998b). Under certain circumstances, it is even possible for harvesting under open access to reduce the stock below the critical minimum stock size, resulting in commercial extinction of the species. For example, the formerly substantial fishery for Atlantic halibut has been so severely depleted that it cannot support a directed commercial fishery.

This situation is exemplified by the open-access problem in fisheries, in which each person gains the incremental benefit of their action while sharing the costs of that action with all other users. Consequently, individually rational actions may be contrary to the collective interest.

BOX 2.1
Fish as Common-Pool Resources

Fish populations are often thought of as common-pool resources (V. Ostrom and E. Ostrom, 1977). Common-pool resources have two major characteristics:

1. It is difficult and costly to exclude potential users of the resource because of the resource's physical characteristics. For example, it may be difficult to identify and monitor boundaries for some resources, such as migratory fish, that range far from land, and such resources may cross multiple jurisdictions. Laws and customs protecting public or communal rights can also make exclusion difficult.

2. The resource is finite, and extraction by one user diminishes the amount available to other potential users. This is known as subtractability or rivalry in consumption (Plott and Meyer, 1975; V. Ostrom and E. Ostrom, 1977). Fish are owned by individuals for their own benefit only when captured. Once captured, they are not available for others to capture and use, or to contribute to growth and perpetuation of the stock.

These characteristics create numerous challenging problems when the supply of a resource is limited in relation to its demand. When exclusion and subtractability are costly, many individuals can access and use a common-pool resource.

The following sections discuss how the public trust doctrine applies to fisheries, how other common-pool resources are managed, and similarities and differences in the management of different resources.

THE PUBLIC TRUST DOCTRINE AND FISHERIES

All fisheries management in the United States takes place within the context of a cultural and legal framework that strongly influences what is and can be done and also how various management measures are implemented. The public trust doctrine is a significant component of this framework. It is a common law doctrine (i.e., judicially developed, rather than statutory) that reflects popular and general political and cultural concepts, in particular the idea that the resources of the seas within U.S. jurisdiction belong to the public and that the government holds them in trust for the public.

Applicability of the public trust doctrine to U.S. fisheries has several ramifications for this study. First, in light of the essential inalienability of public trust resources, it reinforces concerns about the "giveaway" of public resources to private interests. Second, it confers on government a continuing duty of supervision and a responsibility to choose courses of action least destructive to trust resources. Third, it strengthens the principle set forth in the Magnuson-Stevens Act that individual quotas are privileges, creating no property rights and therefore subject to modification or revocation without compensation to their holders. Finally, it suggests that conferring exclusive rights of use should be accompanied by some form of compensation to the public.

Background

The public trust has its roots in the principle of Roman law that certain things, such as "the air, running water, the sea and consequently the shores of the sea,"[1] are incapable of private ownership. As it developed, this principle was extended to fisheries. As the medieval scholar Bracton put it: "By natural law, these are common to all: running water, the air, the sea, and the shores of the sea . . . *hence the right of fishing in a port or in rivers is common*"[2] (italics added).

In England, the Crown held trust resources for the nation. When the 13 U.S. colonies gained their independence, they assumed the trust prerogatives of the Crown over tidal and submerged lands, and other navigable waterways. Their interest in these waters, including the right to fish, was described as an essential attribute of their sovereignty and included a responsibility to manage public trust resources for the benefit of the nation.[3]

In one of the first major public trust cases, the New Jersey court in 1821 reiterated the nature of "the air, the running water, the sea, *the fish, and the wild beasts*" (italics added) as "common property," to be held and regulated for the common use and benefit by the sovereign. Its reasoning was subsequently adopted by the U.S. Supreme Court.[4]

As it developed in this country, the public trust has three important attributes:

1. *The public trust is inalienable.* It was a principle of medieval law that public rights in such common resources as the sea, its shore, and its resources were inalienable. Consequently, Bracton concluded that "all things which relate peculiarly to the public good cannot be given over or transferred . . . to another person, or separated from the Crown."[5] More cogently, the U.S. Supreme Court held in *Illinois Central Railroad v. Illinois*, one of the cornerstones of U.S. public

---

[1] The Institutes of Justinian 2.2.2 (T. Cooper trans. & ed., 1841).

[2] H. Bracton, *On the Laws and Customs of England*, 39-40 (S. Thorne trans., 1968).

[3] *Martin v. Waddell*, 41 U.S. (16 Pet.) 367 (1842).

[4] *Arnold v. Mundy*, 6 N.J.L. 1(1821); *Martin v. Waddell*, 41 U.S. (16 Pet.) 367 (1842); see generally McCay (1998).

[5] H. Bracton, *On the Laws and Customs of England*, fn. 2 above.

trust doctrine, that a legislative grant of the Chicago waterfront to a private railroad was necessarily revocable because:

> The state can no more abdicate its trust over property in which the whole people are interested, like navigable waters and soils under them…than it can abdicate its police powers in the administration of government and the preservation of peace…. So with trusts connected with public property, or property of a special character, like lands under navigable waters, they cannot be placed entirely beyond the direction and control of the State.[6]

As a public trust principle, this applies to fish, with the qualification that individuals can acquire title to fish once they have been reduced to possession.[7]

2. *The public trust gives the government, as trustee, continuing authority and responsibility for stewardship.* The scope of the public trust as a constraint on the powers of government to deal with public resources has long been noted (Sax, 1970). In 1983, the California Supreme Court emphasized the scope of the trust and explained the responsibility of government as trustee with respect to trust resources. Relying on the principles of *Illinois Central*, the court held that the state has the power and duty "to exercise a continuous supervision and control over the navigable waters of a state and the lands underlying those waters" and that no one can claim a vested right to divert waters "once it becomes clear that such diversions harm the interests protected by the public trust" (in this case the shrimp and brine flies of Mono Lake in California).[8] Subsequent litigation over nonnavigable tributaries into Mono Lake proceeded under statutes requiring that the fish in such streams be kept in good condition, which the appellate court held were legislative exercises of the public trust.[9]

---

[6] 146 U.S. 384, at 453-454. In *Illinois Central*, the Supreme Court laid down a two-part test for determining the validity of grants of the beds and banks of navigable waters: (1) Does the disposition affirmatively aid or improve the public interest in navigation or other public use of the particular area of the waterway? (2) If the legislative grant does not affirmatively aid or improve the public trust, does it substantially impair the public interest in the remaining lands and waters of the particular area of the waterway? Thus, while the state on behalf of the people have leased out the right to extract public resources, such as oil, gas and other minerals, such grants have had to comply with overall trust inhibitions. The State, as trustee for the people, cannot abdicate its ultimate responsibility for the waters involved; see *Boone v Kingsbury*, 206 Cal. 148, 273 P. 797 (1928). Although it may incur liability for the breach of contracts validly made with grantees, it may not place them "entirely beyond the direction and control of the State" *Illinois Central*, supra, 146 U.S. at 453-454.

[7] *Pierson v. Post*, 3 Cal. T.R. 177 (N.Y. Sup. Ct., 1805).

[8] *National Audubon Society v. Superior Court*, 658 P.2d 709 (Cal. 1983). The public trust doctrine calls for continuing supervision and permits the revocation of the right if the public trust is impaired (see *Boone v. Kingsbury*, 487 F.Supp. 443 [1980] and *National Audubon*), although revocation must not violate any vested contract rights without compensation (*Union Oil* 146 U.S. at 453-54 [1892]).

[9] *California Trout, Inc. v. State Water Resources Control Board*, 207 Cal.App.3d 585 (1989).

3. *The public trust applies to fisheries.* Although this doctrine has been traditionally associated with the state's title to the beds of waterways, it also applies to wild creatures. A traditional view of the ownership of wild animals, such as fish, is that they are ownerless until reduced to possession.[10] In the late nineteenth century, some courts confused the Roman concept of ownerless things with the idea that such resources were owned by the state in trust for the people. Thus, a well-known scholar in water law noted the tendency to substitute the positive expression that such things belong to the state in trust for the people, for the older concept that they were "common" property, part of a "negative community."[11] In *Geer v. Connecticut,*[12] the Supreme Court accepted the state's argument that the title to wild animals "so far as they are capable of ownership, is in the State, not as a proprietor but in its sovereign capacity as the representative of and for the benefit of all its people in common" (Id. at 529). However, the Court subsequently repudiated the concept that states "own" wild creatures in the conventional sense, describing it as "pure fantasy,"[13] when state ownership of fish was claimed to shield state regulations from constitutional scrutiny. Nevertheless, the Court did not consider the applicability of the public trust to *ferae naturae.* A number of state court decisions, like *National Audubon,* hold that protection of the public trust justifies measures such as limiting diversions of water, *in the interest of protecting a fishery* as a public trust resource (Johnson, 1989). These decisions are based on the concept that fish are public trust resources, subject to the same protections as other trust assets.

Although the Court subsequently repudiated the concept that states "own" wild creatures in the conventional sense when state ownership of fish was claimed to shield state regulation from commerce clause constraints, the applicability of the public trust to *ferae naturae* was never questioned. The contemporary view is that the state has no title to fish as personal property but they are nevertheless "owned" by government in its sovereign capacity as trustee for the benefit of its

---

[10] Shellfish are anomalous. Unlike finfish, most shellfish species are sessile and found very close to land; some can be "cultivated," making it feasible and possible to claim private property in them. On the other hand, they are wild creatures—*ferae naturae.* Even though a shellfish released in a river will not bound away as a deer will, its larval offspring will move afar. Consequently, American courts and agencies have settled on rules such as that shellfish found in places where they grow naturally, whether planted there or not, belong to the people as their "common property" whereas those found in places where shellfish do not naturally grow may be claimed as private property (McCay, 1998; see also *U.S. v. Long Cove Seafood, Inc.* [2d Cir. 1978] 582 F.2d. 159, 165). This has been a powerful constraint on the privatization of shellfish beds.

[11] S. Wiel, *Water Rights in the Western States,* sec. 6, pp. 11-12 (1911), citing *Geer v. Connecticut,* 161 U.S. 519 (1896).

[12] 161 U.S. 519.

[13] *Douglas v. Seacoast Products, Inc.* 431 U.S. 265, 284 (1977) quoted in *Hughes v. Oklahoma,* 491 U.S. 322, 338 (1979).

citizens. Thus, a recent court quoted as the "contemporary view" the following language from the California Supreme Court:

> The wild game within a state belongs to the people in their collective, sovereign capacity; it is not the subject of private ownership, except insofar as the people may elect to make it so; and they may, if they see fit, absolutely prohibit the taking of it, or any traffic or commerce in it, for the public good.[14] We conclude that like other wild game, the abalone caught in the state's coastal waters belong to the people of the State of California in their collective, sovereign capacity. No individual property right exists in these shellfish. Rather the state acts as trustee to protect and regulate them for the common good. In this representative capacity, the state does not have a proprietary interest in these abalone that can be equated with the personal property of the state. Ex parte Maier (1894) 103 Cal. 476, 483, quoted with approval in *People v. Brady*, (1991) 234 Cal.App.3d 934, 286 Cal.Rptr 19.

Historically, fisheries have been at the core of trust protections. One of the earliest judicial declarations of the public trust in the United States dealt with the inalienability of tidal oysters beds. Thus, the government holds fisheries in trust for its people and has a continuous supervisory responsibility to manage them to achieve trust objectives (i.e., conservation, management in the public interest, and continuing oversight so that changes in fishery management can be made when the need arises).

Does the Public Trust Doctrine Apply to the Federal Government?

The relevance of the public trust doctrine to issues surrounding fisheries in federal waters hinges on its applicability to the ownership of wild animals in federal waters, beyond state jurisdiction. Relatively few courts have considered whether the public trust applies to the federal government in the same way that it applies to the states. This difference may result from a greater dependence of federal wildlife protection and management on laws and statutes, rather than case law. Although there is little, if any, appellate authority directly concerning this issue, the U.S. Supreme Court has inferred that the trust is applicable to both federal and state sovereigns in several cases. In what is probably the leading case, *Illinois Central Railroad v. Illinois*,[15] the court describes the trust in general terms, suggesting that it is a doctrine of federal law applicable nationwide. In

---

[14] *Takahashi v. Fish & Game Commission*, 30 Cal.2d at 728-729 cited as in accord. The court cites several more recent federal cases as supporting its contemporary view analysis, such as *U.S. v. Long Cove Seafood, Inc.* (2d. Cir. 1978), 582 F.2d 159, 163-164. and *U.S. v. Tomlinson* (D. Wyo. 1983), 574 F.Supp. 1531, 1535, Perkins Boyce, Criminal Law, Ch. 4, Sec.1 at 292-295.

[15] 146 U.S. 387 (1892).

Shively v. Bowlby,[16] the court, in describing the historic trust obligations of the Crown that passed to the original 13 states, stated that as to trust resources *not* within the 13 colonies, "the *same* title and dominion passed to the United States (emphasis added)" Federal trial court decisions can be found on both sides, although the better view supports a federal trust. In the matter of *Steuart Transportation Co.,* a federal court held that both the states and the United States are trustees, with the "right and duty to protect and preserve the public's interest in national wildlife resources"[17] (see Archer et al., 1994) and the weight of scholarly comment appears to suggest that the public trust is a federal as well as a state rule, applicable to resources within the jurisdiction of the United States as well as the states. For example, see Archer et al. (1994) and Wilkinson (1980).

In the Magnuson-Stevens Act, the United States has undertaken to impose sovereign rights and exclusive management authority on the fisheries of the exclusive economic zone (EEZ), surrounding them with statutory protections and restrictions tantamount to exercising dominion and control over them, just short of reducing them to possession. Under similar circumstances, a federal court has held the United States had acquired a property right in wild and free-roaming horses and burros (*U.S. v. Tomlinson,* 574 F.Supp. 1531 [D. Wyo. 1983]).

Moreover, although the public trust traditionally has been applied to state waters as an attribute of sovereignty,[18] its historic roots in English common law provide a basis for its application to waters managed by the federal government (Jarman, 1986; Jarman and Archer, 1992). The states are the constitutional repositories of sovereign power in the United States.[19] However, within the EEZ and beyond state waters the United States has "paramount power," asserts "national dominion," and exercises sovereign authority.[20] With sovereign authority comes sovereign responsibility, including that imposed by the public trust.

---

[16] 152 U.S. 1 (1894).

[17] In re *Steuart Transportation Co.,* 495 F.Supp. 38, 40 (E.D. Va. 1980). In several decisions dealing with the Redwood National Park, a federal court held that the Secretary of the Interior has trust and statutory obligations to protect park resources from outside threats. *Sierra Club v. Dept. of the Interior,* 376 F.Supp. 90 (N.D. Cal. 1974); same case, 398 F.Supp. 284 (N.D. Cal. 1975), 424 F.Supp. 172 (N.D. Cal. 1976). See also *United States v. 1.58 Acres of Land,* 523 F.Supp. 120, 122 (D. Mass. 1981) suggesting that the public trust, which it held survives federal condemnation, is "administered by both the federal and state sovereigns." In accord with this ruling, the federal district court in California ruled that the public trust survives acquisition or condemnation by the United States, and the federal government acquires trust obligations as a result. *City of Alameda v. Todd Shipyards,* 632 F.Supp. 333 (N.D. Cal. 1986) and 635 F. Supp. 1447 (N.D. Cal. 1986). But, see *United States v. 11.037 Acres,* 685 F.Supp. 214 (N.D. Cal. 1988), reaching the opposite result; *Sierra Club v. Andrus,* 487 F.Supp. 443 (D.D.C. 1980) (Congress intended to eliminate public trust from statutes governing national parks and public land).

[18] E.g., *Shively v. Bowlby,* 152 U.S. 1, 24 (1894).

[19] U.S. Const. Art. X; *Pollard's Trustee v. Hagan,* 44 U.S. (3 How.) 212 (1845).

[20] *United States v. California,* 332 U.S. 19, 34, 38 (1947); see Magnuson-Stevens Act, Sec. 101, 16 U.S.C. 1811.

Conclusion

Fisheries within federal waters are held in public trust for the people of the United States. Public trust principles are thus applicable to any allocation of fishing rights. The government has an affirmative duty to take the public trust into account in conferring IFQs. Such allocations cannot be irrevocable, but remain subject to the government's continuing supervisory responsibility over them, to hold and manage them on behalf of the people. Although fishing privileges can be granted, they remain subject to modification in light of current knowledge and current needs.

LESSONS FROM OTHER COMMON-POOL RESOURCES[21]

Fisheries are not unique among common-pool resources in generating conflicts among users. A variety of approaches have been used to manage and reduce such conflicts in other resources; these approaches may provide insights for fisheries management.

Air

Another resource besides fish that has been managed with the use of quotas is the air we breathe. In this case, the quotas limit the amount of pollutants that can be emitted into the air.

Nature of the Resource

Managing air pollution is similar to managing fisheries in several ways. Both have historically been treated as open-access resources, and as a result, both have experienced unsustainable levels of exploitation. Whereas the problem with a fishery involves excessive catch, for the airshed the problem is emissions levels that exceed human health standards. Both resources are considered part of the public domain, creating resistance to transfers of the resource to private owners.

Spatial aspects are important for both resources. For some pollutants, the location of emissions is very important. (Emissions that are concentrated in space, for example, cause more damage than dispersed emissions.) In some fisheries, *where* the fish are caught matters as well. Different areas may have much higher bycatch rates of juveniles or non-target species, or may reduce the foraging success of at-risk marine mammals and seabirds, and the concentration of fishing activities (e.g., bottom trawling) in certain areas may cause more significant environmental impacts than if these activities were more dispersed.

---

[21] An interesting source of information about comparisons of fishery management with eight other forms of management is Bigford and Bribitzer (1985).

Both resources typically involve multiple control targets (species in the case of fisheries and different pollutants in the case of the air). In practice, this means that controlling any one species or pollutant will inevitably produce consequences for others.

Perhaps the strongest contrast between the issues faced by managers of these two resources involves attitudes toward the resource. Whereas fishing is considered an important part of the culture of communities affected by it, activities that generate pollution sometimes get little respect in terms of the importance of the industry in the community. In addition, whereas most holders of emission quotas are commercial and industrial firms, IFQ holders range from owners of large corporate factory trawlers to individual fishermen operating out of small boats.

The Path to Quotas

The initial approach to air pollution control relied on a rather traditional form of legal regulation in which the government took upon itself the responsibility for defining management goals, choosing the best approaches for meeting these goals, and monitoring and enforcing compliance with its mandates.

Ambient standards, which establish the highest allowable concentrations of the specified pollutants in the ambient air or water, represent the targets of this approach. To reach these targets, emission or effluent standards (legal discharge ceilings) were imposed on a large number of specific discharge points such as stacks, vents, outfalls, or storage tanks. Following a survey of the technological options of control, the control authority selected a control technology and calculated the amount of discharge reduction achievable by this technology as the basis for setting the emission or effluent standard. The responsibility for defining and enforcing these standards has been shared in legislatively specified ways between the national government and various state governments.

In 1975, the government introduced a form of tradable permit to increase the flexibility with which emission targets could be met. Evidence available at that time suggested that the traditional approach was much more expensive than necessary. Providing greater flexibility was seen as one way to reduce these costs and this expectation has been validated by experience with the lead phase-out and sulfur allowance programs (Tietenberg, 1985, 1995; Hahn and Hester, 1989). Estimates suggest that the flexibility provided by the sulfur allowance program has saved about $225-$375 million (Ellerman et. al., 1997, p. 62), whereas the lead phase-out program saved about $265 million (Nussbaum, 1992, p. 35).

Since these initial efforts, tradable quota programs to control air pollution have been used specifically to eliminate ozone-depleting gases, to eliminate the lead in gasoline, and to control acid rain and tropospheric ozone. The adoption of this approach has also spread to other countries; Canada, Chile, and Germany all have air pollution control policies that involve some form of tradable permits (Tietenberg, 1995, 1998).

Comparing Management Systems

Tradable quota programs for pollution control share several similarities with IFQs. For example,

- Both involve limiting access. Fishing quotas limit catch or effort. For acid rain control, quotas limit the total amount of emissions allowed from utilities.
- Both typically involve an initial allocation of the quota that is based on previous activity (harvesting or emitting).
- Both involve a market process for transferring quota and may involve some restrictions on transferability.
- Both involve significant monitoring and enforcement components.
- Both have attempted to resolve the tension between the desire to protect the investment of quota holders and the desire to ensure that the public trust nature of the resource is adequately respected by defining the "rights" as privileges of access rather than ownership rights.

These two types of quota systems also have some rather interesting differences. For example:

- Although auctions typically play no role in IFQs, the acid rain program has an auction component that serves to ensure both the availability of quota and good price information to potential buyers. This "zero-revenue" auction (see Box 5.1) does not raise money for the government; rather it returns auction proceeds to the quota holders who are required to place a small percentage of their quotas up for auction each year (Hausker, 1990, 1992).
- Under the acid rain program, the costs of continuous monitoring are borne by the quota holders, whereas in U.S. fisheries, most monitoring and enforcement costs have been borne by taxpayers. In New Zealand fisheries, monitoring and enforcement costs are borne by quota holders.
- IFQs usually limit the purchase of quota shares to narrowly defined eligible populations of those engaged in fishing. Several of the air programs allow anyone, including environmental groups, to purchase and retire quota shares. Since retired quotas are not used to authorize emissions, they directly result in better air quality (Tietenberg, 1998).
- While for IFQs the total allowable catch (TAC) may be specified annually and based on the latest data on stock dynamics, for air pollution programs the emissions limit is typically specified several years in advance. In some cases,

such as the RECLAIM system in Los Angeles, the limit declines by a fixed percentage over time (Fromm and Hansjurgens, 1996).

• Whereas IFQs are normally defined in terms of a percentage share of the TAC, for air pollution the quota is typically defined in terms of an authorization to emit a specific number of tons in a given year.

• Air quota programs share with some IFQ programs the characteristic that some of the rent created by the quota program is transferred to the larger community, but the sharing may occur in rather different ways. In the Alaskan IFQ programs, this sharing is accomplished by community development quotas. In the ozone-depleting gas program, rent is transferred by means of a tax on the activity authorized by the quota; the revenue goes to the general treasury (Tietenberg, 1990).

Transferable quotas to control air pollution have a sufficiently different purpose than IFQs that experience with them is certainly not automatically relevant for evaluating IFQs. On the other hand, such experience does provide a potentially useful source of ideas on some of the issues with which this report must grapple.

Surface Water

Property rights can be acquired in surface water, but unlike those in land and tangible things, they are characterized as usufructory[22] in nature, more limited to begin with and subject to a greater reach of the police power. In *United States v. Gerlach Live Stock Co.*,[23] the U.S. Supreme Court characterized these rights as follows:

> As long ago as the Institutes of Justinian, running waters, like the air and the sea, were *res communes*—things common to all and property of none. Such was the doctrine spread by civil-law commentators and embroidered in the Napoleonic Code and in Spanish law. This conception passed into the common law. From these sources, but largely from civil-law sources, the inquisitive and powerful minds of Chancellor Kent and Mr. Justice Story drew in generating the basic doctrines of American water law.

These principles are essentially the same for both riparian and appropriative rights.[24] The principal difference between the two systems, of course, is that *riparian* rights arise from ownership of the land adjacent to the water source,

---

[22] A usufructory right is the right to use something in which one has no property, that is, the right to take the fruits of property owned by another. The owner of surface waters is the public.

[23] 339 U.S. 725, 744-45 (1950).

[24] Cf. *Tyler v. Wilkinson*, 24 Fed. Cas. 472 (C.C.D.R.I. 1827), *Vernon Irrigation Co. v. Los Angeles*, 106 Cal. 237, 39 P. 762 (1895).

whereas *appropriative* rights are based on beneficial use and the "first-in-time" principle.[25] Interestingly enough, Chancellor Kent analogized the acquisition of water rights to the *ferae naturae* theory—a property right based on capture similar to the manner in which ownership of wild game and fish is acquired.[26]

Most significant to a study of IFQs are these points:

• Water rights are transferable and may be used as collateral for loans.[27] However, such things as changes in source, use, and point of diversion are subject to review under most systems.

• Appropriative water rights—those almost universally employed in the arid West—are based on Locke's concept of labor. They are rewards for the diversion of water "by costly artificial works . . . for miles over mountains and ravines."[28] They rest on the first-in-time principle, and water acquired pursuant to the appropriative theory must be dedicated to beneficial use.

• Virtually all water rights in the United States are now subject to a permit system imposing numerous conditions to protect the public interest.

• In an increasing number of states, water rights are subject to modification at any time based on reasonable or beneficial use principles or on public trust needs.

The scope of property rights in water has been characterized as becoming progressively smaller in the face of increasing regulatory demands and the growth of the public trust doctrine. Perhaps the most dramatic trend in this context is the application of public trust principles. In *United Plainsmen's Assn. v. North Dakota Water Conservation Commission,*[29] the North Dakota court held that the state has a trust responsibility to consider and plan comprehensively for the overall impacts of major water diversions on trust uses. Then, in *National Audubon Society v. Superior Court,*[30] the California court held that the state has a continuing supervisory authority over water rights under the public trust, a responsibility to consider adverse effects on trust uses by diversions, and an obligation to avoid such adverse effects whenever feasible. This approach has been adopted by a number of other state courts, including those of Alaska, Idaho, Montana, and Washington.

The law of water rights furnishes a number of useful analogies to the use of IFQs. Both systems involve the problems inherent in allocating a common re-

---

[25] See Tarlock, *Law of Water Rights and Resources,* Secs. 2.05 et seq (1997).

[26] Kent, Commentaries 347 (First Ed. 1828); cf. *Pierson v. Post*, 2 Am. Dec. 264 (N.Y. 1805).

[27] Tarlock, supra, Sec. 5.12, p. 5-59.

[28] *Irwin v. Phillips*, 5 Cal. 140, 146 (1855).

[29] 247 N.W. 2d 457 (N.D. 1976).

[30] 33 Cal. 3d 415, 658 P.2d 709, cert. denied sub. nom. *Los Angeles Dept. of Water & Power v. National Audubon Society,* 464 U.S. 977 (1983).

source. Both involve questions of state regulation and conditions and the extent to which vested rights are created.

1. The allocation of water rights, particularly under the appropriative system, recognizes the past experience and labor of the permittee. Appropriative rights are based on the application of water to a beneficial use. There is a rough analogy between the allocation of water rights based on the work and experience of the diverter and the distribution of IFQs based on the previous activities of the fisherman.

2. The nature of IFQs as property is conditioned on the public trust nature of the fishery, just as water rights are held subject to public trust. Under the *National Audubon* rule, once an appropriation has been approved,

> [T]he public trust imposes a duty of continuing supervision over the taking and use of the appropriated water . . . the state is not confined by past allocation decisions which may be incorrect in light of current knowledge or inconsistent with current needs. The state accordingly has the power to *reconsider* allocation decisions. . . . No vested rights bar such reconsideration. (33 Cal.3d at 447 [italics added]).

Fish and game have historically been considered as held by the state in trust until reduced to capture.[31,32] In *California Trout, Inc. v. State Water Resources Control Board*,[33] the court states, "Wild fish have always been recognized as a species of property the general right and ownership of which is in the people of the state. And fisheries are one of the oldest trust uses."[34]

The significance of the application of the public trust is heightened by the U.S. Supreme Court's admonition that limitations on the exercise of property rights equivalent to a deprivation of all economic use may be a "taking" that must be compensated by the government, unless justified by background principles of nuisance or property law.[35]   Consideration of a fishery as a common resource

---

[31] *Geer v. Connecticut* 161 U.S. 519 (1896).

[32] *Hughes v. Oklahoma* 441 U.S. 322 (1979).

[33] 207 Cal.App.3d, 585, 630 (1989), citing *People v. Stafford Packing Co.*, 193 Cal. 719, 727, 227 P. 485 (1924), *Geer v. Connecticut,* supra; *People v. Monterey Fish Products Co.*, 195 Cal. 548, 563, 324 P. 398, 38 A.L.R. 1186 (1925); *LeConte v. Dept. of Conservation*, 263 U.S. 545 (1924); *Mountain States Legal Foundation v. Hodel*, 799 F.2d 1423 (10th Cir. 1986) (en banc), cert. denied, 480 U.S. 951 (1987); *Moerman v. State*, 21 Cal.Rptr. 329 (1993), cert. denied, 114 S.Ct. 1539 (1994); *Clajon Product Corp. v. Petera,* 854 F.Supp. 843 (D. Wyo. 1994); *Columbia River Fishermen's Protective Ass'n. v. City of St. Helens,* 87 P.2d 1195 (Ore. 1939); *People v. Truckee River Lumber Co.,* 116 Cal. 397 (1897). See generally, Rieser (1991).

[34] E.g., *Martin v. Waddell,* 41 U.S. (16 Pet.) 367 (1842); *Carson v. Blazer,* 2 Binn. 475, 4 Am.Dec. 463 (Pa. 1810).

[35] *Lucas v. South Carolina Coastal Council,* 505 U.S. 966 (1992).

held by government in trust for the people, albeit allowing harvest privileges, largely eliminates the issue of an unconstitutional takings when the privilege is curtailed.

3. Water rights systems may furnish guidelines to the termination of IFQs. The history of water rights is one of steadily increasing state regulatory authority, whether in the form of standards for reasonable use, beneficial use, or continuing public trust jurisdiction. In addition, a substantial body of law has accumulated governing the manner in which water rights permits may be terminated. This can take place in a number of ways, including forfeiture, abandonment, revocation, and prescriptive rights. Such experience may be applied to future situations in which IFQ privileges must be terminated.

Surface water resources and fisheries also share the characteristic of being stochastic resources. Variability in water flows and fish stocks has led to problems in management of both resources. For example, the initial allocation of Colorado River water rights was based on average flow levels during a sequence of El Niño years in the late nineteenth century. Because the rights were to acre-feet rather than to a percentage of the annual flow, the system was unable to satisfy all the assigned rights during normal and dry years until a long and costly renegotiation of rights had been achieved.

Groundwater

In many states, the hydrologic connections between surface water sources and groundwater basins are not recognized. Consequently, a distinct body of statutes and cases governs surface water and groundwater.

Groundwater is considered a public resource and, until recently, has been subject to few access and use restrictions. For most states, owners of overlying land were allowed to pump and use as much groundwater on their land as they could put to reasonable and beneficial use. Apart from a few jurisdictions that considered the transport of groundwater off overlying land to be an unreasonable use, reasonable and beneficial use failed to restrict groundwater pumping. This situation can result in overdraft (the removal of groundwater in excess of the capacity of the water basin to recharge) and subsidence (the downward movement of the land surface due to collapse or compaction of the underlying aquifer as a consequence of groundwater pumping).

Citizens, particularly in the western United States, have found themselves embroiled in groundwater basin dilemmas, primarily overdraft and subsidence, but also interference among wells. Groundwater pumpers receive the full benefit of all water that they pump, but are able to spread the costs of pumping to all users of a basin. Each well contributes to declining water tables in a basin. As water is removed from the ground, in addition to ground subsidence, trees and plants that rely on groundwater may perish as water tables decline. If substantial declines occur in water tables, water quality problems may emerge because water at greater depths is often of poorer quality. In coastal zones, groundwater pumping may also lead to saltwater intrusion, affecting water quality, soil fertility, and the viability of estuarine systems.

Groundwater pumpers and citizens have been aided in their efforts to address these dilemmas by state governments, which have responded by defining property rights in groundwater. Groundwater pumpers are given exclusive rights of use, but not of ownership, of groundwater. States have taken additional steps to locate alternative sources of water and to create water conservation plans.

A variety of mechanisms can be used to define groundwater rights and to manage groundwater basins to address the multiple dilemmas that groundwater pumpers face (Blomquist, 1992). For instance, in a "home rule" state such as California, the state does not define groundwater rights, nor does it engage in much regulation of groundwater. Instead, the state has made available several different management options from which groundwater users may select. Groundwater users may choose to adjudicate their basin. Hydrologic studies are conducted to determine the extent of the overdraft and the amount of water that can be withdrawn from the basin annually without lowering water levels. Groundwater pumpers devise a sharing scheme among themselves, allocating shares of the groundwater available for pumping each year. Markets have developed to enable groundwater pumpers to buy and sell water rights. Thus, these self-organized systems legitimated by the court system are, in essence, individual transferable quota systems for groundwater. Typically, sharing schemes are based on historical pumping rates. When most groundwater pumpers agree to the allocation scheme, a state court is asked to recognize the contract. The court appoints a water master to monitor the basin and the sharing scheme. Each year, the water master publishes each pumper's pumping record. The state does not enforce the sharing schemes. Instead, it is the responsibility of the pumpers in a basin to take civil action against contract violators. Some of these systems have been operating for more than 40 years at very low transaction costs and very low rates of overpumping, for example, in Los Angeles County (Ostrom, 1990; Blomquist, 1992).

If California groundwater users decline to adjudicate their basins, they may, nevertheless, form water districts. Several different types of districts may be formed, with varying levels of authority. Some districts engage in limited forms of regulation, such as prohibiting the transport of groundwater outside a basin. Other types of districts actively manage and carefully monitor their groundwater basins. For instance, the Orange County Water District controls groundwater pumping through pricing. The price of water increases as the amount of water pumped increases, encouraging groundwater pumpers to use surface water supplies made available through the water district. The water district uses the pumping fees to import surface water and to recharge the groundwater basin.

Thus, California has made available to its citizens a number of different mechanisms by which they may address groundwater basin dilemmas. However, the state does not require groundwater pumpers to adopt one of the alternatives. As a consequence, a number of basins that were once in overdraft are now actively managed, and a number of basins that are experiencing critical problems are not managed at all.

Whereas California is at one end of the spectrum in helping pumpers to devise their own solutions to groundwater dilemmas, Arizona is at the other. The State of Arizona, through its active management areas, directly manages the state's most overdrafted basins. The 1980 Arizona Groundwater Management Act created the Arizona Department of Water Resources and directed it to define and quantify groundwater rights based on type of use. Agricultural, mining and industrial, and water utility rights have been defined and allocated. Agricultural rights, and mining and industrial rights have been allocated strictly on the basis of historical use. For instance, for each acre-foot of water a mine pumped on average each year for the five years prior to 1980, the mine was granted a property right. Groundwater pumpers are required to report their groundwater pumping annually. The Arizona Department of Water Resources monitors and enforces groundwater rights.

Groundwater rights were allocated with little immediate consideration given to limiting the overdraft of basins. Groundwater overdraft and subsidence problems will be addressed over time through the retirement of groundwater rights, the gradual phasing-out of agriculture, the importation of surface water supplies through the Central Arizona Project, and increasingly strict water conservation requirements.

Most states fall along the continuum defined by California and Arizona. For instance, Colorado, which relies on a system of state water courts, water masters, and individual rights holders to define and enforce surface and groundwater rights, comes closest to California. Nevada, on the other hand, has adopted an Arizona-like groundwater management approach. No matter which approach, or combination of approaches, a state has adopted to address groundwater aquifer dilemmas, all states have attempted to define rights of use, while subjecting these rights to consideration of the public's interest.

State programs that define exclusive use rights in groundwater share several similarities with IFQs, for example:

- Both involve limits on harvesting the resource flow.
- Both involve an initial allocation of use rights or quota based on historical harvesting activities.
- Both involve a market process for transferring use rights or quota. In the California adjudicated basins, water rights are freely transferable among pumpers in a basin. In Arizona, groundwater use rights may be transferred among pumpers in a use class, or across uses.
- Both involve restrictions on transferability of "use rights" or "quota."

Most states place restrictions on transferring groundwater outside the basin of origin. Other restrictions include limits on the amount of water that may be transferred. For instance, in Arizona, only a portion of a groundwater right may be transferred from one use to another.

• Both require significant monitoring and enforcement.

• Both involve a tension between the desire to protect quota or use-rights holders and the desire to ensure that the public trust nature of the resource is adequately respected. In many state constitutions, water is specified as being owned by the state, with citizens having the right to use such water.

These two types of systems are also different in some ways:

• Most states impose some type of a pump tax or fee to pay for the costs of monitoring and enforcement. In fisheries, most monitoring and enforcement costs have been borne by taxpayers.

• In many states, groundwater monitoring and enforcement are shared by the state and pumpers. For instance, in California's adjudicated basins, the state monitors pumping, but pumpers are authorized to take action against rule breakers. In Colorado, enforcement of water rights is left to individual water rights holders. In fisheries, users typically do not participate in monitoring and enforcement, which are conducted by one or more public agencies.

• The purchase of groundwater use rights is generally less restricted to users than the purchase of fishing quota shares.

• In many states, groundwater use rights may be banked, or stored, from year to year. IFQ programs differ in terms of whether they let harvesters bank unused quota.

Finally, there is a fundamental difference between fish populations and groundwater basins. A groundwater basin stores water that has percolated into it through the soils from rainfall and surface runoff. With modern technology, accurate estimates of the groundwater resources, which come very close to an average sustainable yield, can be determined and quotas assigned. In bad years, the water levels in the basin may be brought down, but in good years the storage space is reoccupied by water and the total storage space is unaffected by overdrafts (unless overdrafting is accompanied by collapse of aquifer structures). In fisheries, recruitment can vary dramatically from year to year (depending on stock size, water temperature, and other exogenous factors). Fishing of a quota in a bad recruitment year may drastically affect the remaining stock on which future years depend and adversely affect the recovery of the stock. Thus, in fisheries, unlike in groundwater basins, stocks are more difficult to estimate accurately and drawing down the stocks is more risky.

Oil and Gas

Initially, oil and gas were treated like fish, as "minerals ferae naturae."[36] It was recognized, however, that the owner of land under which these minerals lie is their absolute owner. Mineral rights in oil and gas can be, and are, conveyed as real property.

For historical reasons, "hard" and "soft" minerals on public lands were treated very differently. Under the common law, minerals on lands belonging to the Crown (and, by the same token, to the United States or various states of the union) were the property of the Crown. The same general rule applied under the Spanish and Mexican law to the lands acquired by the United States from Mexico. However, the pressures of the Gold Rush resulted in de facto recognition of rules developed by the miners and applied to the public lands on which they found their treasure, without regard to the legalities of title. In effect, an open-access system existed for gold mining, at least until a claim was perfected.[37]

Oil and gas received very different treatment. The enactment of the Mineral Leasing Act of 1920 represented a historic shift from more than a century of privatization. Supporters of this act characterized it as changing "the whole principle of the public land laws . . . by refusing to allow the title to lands to go into private ownership and (adopting) for the future a government leasing policy . . ."[38]

This act has been described as a "bargain," in which states relinquished historic claims to the public lands within their borders in exchange for a federal quid pro quo—assurances that 90% of the revenue from these lands would go to the states—either directly or through the Reclamation Fund. The Bureau of Land Management awards oil and gas leases on public lands, and the proceeds from such leases are subject to state and federal taxes of various kinds.[39]

Where ownership of the land above oil and gas reservoirs is very fragmented, the landowners or those who have leased the mineral rights sometimes compete for the oil underneath, because the oil migrates toward the wells where the pressure is lower. This is a direct analogy with fisheries and has led to similar efficiency problems (excessive drilling of wells). The way the authorities (e.g., the Texas Railroad Commission) have dealt with this situation has sometimes

---

[36] *Westmoreland & Cambria Natural Gas Co. v. De Witt*, 18 A. 724, 725 (Pa.).

[37] The Mining Act of 1872 is recognized as resulting from congressional acquiescence to the rules and customs of local mining districts. Under the Mining Act, simply locating a valuable deposit, staking a claim, and performing minimal work was sufficient to establish title to the minerals and to the land containing them.

[38] 59 *Cong. Rec.* 2711 (1920).

[39] Many states impose severance taxes on oil and gas extracted within their borders; see Grew (1982).

been as detrimental to efficiency as the typical fishing regulations, with the production per well being limited (prorationed), which further encourages the drilling of excessive wells. Alternatively, fractional ownership has often been addressed through *unitization* (e.g., in the North Sea). Under unitized development agreements, resource claimants negotiate sharing agreements and designate one firm to serve as resource manager. All firms share in the cost of drilling and extraction and in the resulting revenues. The field manager identifies optimal well spacing and pumping rates. The share of oil in the reservoir that each license holder can claim is fixed, eliminating the incentives for competitive drilling. The shares are sometimes subject to revisions according to agreed procedures as more becomes known about the reservoir.

Timber and Timberland Resources

With respect to the primary definition of common-pool resources having characteristics of "difficulty of exclusion and subtractability," timber and land do not qualify as common-pool resources. However, it is interesting to compare and contrast timber and fish resources in light of property rights, renewable characteristics, public ownership, and changing public attitudes.

Timber and land rights have evolved over time through government action to encourage settlement and development of both public and private ownership. Government actions in the form of land grants, homestead acts, and sale of public lands have encouraged private development throughout the United States, particularly during the nineteenth century. With the set-aside of large areas as Yosemite National Park in 1864 and Yellowstone National Park in 1872, the federal government set the pattern for long-term public ownership and created the foundation for the U.S. national park system. In 1891, the General Revision Act authorized the President to "set aside and reserve in any State or Territory having public lands bearing forests, any part of the public lands wholly or in part covered with timber or undergrowth." This act created the basis for our national forest system.

National parks have from the outset been oriented toward preservation. National forests, in contrast, have historically been managed under the concept of conservation and multiple use. Under the leadership of Gifford Pinchot and the "scientific approach," timber was managed as a renewable resource based on sustainable harvest. Economic development of the forest was encouraged to support local economies by providing valuable jobs and products to the citizens of the United States. The conflict created between preservation and conservation has produced ongoing debate and at times caused significant change in long-term forest management policy. In the last decade, a shift in public values resulted in a shifting goal for national forests away from the harvest of timber to preservation of the forests for other public benefits. This has caused considerable economic dislocation for both individuals and communities and a shift in dependence from public to private-sector forest resources.

Under existing programs, most fisheries are managed on a sustainable yield basis through scientific assessment of fish stocks and the regular setting of TACs. The increasing public concern for habitat and wildlife protection has resulted in a shift in public attitudes from support of fish harvest to preservation of marine environments similar to the trends experienced in national forests. This shift is demonstrated by actions being taken by some organizations to promote the establishment of new marine protected areas and by the changes incorporated in the Magnuson-Stevens Act in 1996. Such shifting attitudes are creating economic impacts on fishing communities similar to those experienced in communities that depend on the harvest of forest products. The impacts on fishing communities will be more severe, however, because there are few private marine fishing grounds comparable to private timberlands to cushion the impact of the removal of public trust lands from commercial access. (Note, however, that aquaculture is a significant alternative to some wild harvest fisheries.)

The forest products industry relies on sawmills, plywood mills, and pulp mills to process its harvest. The fishing industry relies on both land-based and at-sea processors. Like most large fish processors, sawmills require large amounts of capital to be competitive and depend on a steady and reliable supply of raw material. When decisions were made to reduce the supply of public timber sharply, federal funds were made available to support retraining and provide other forms of social assistance to individuals, but sawmills and other timber processors were not provided financial compensation for changes in public policy.

In contrast to fishing, timber has a defined market value by species that is reflected in the purchase price paid at auction for standing timber. This value is recognized as economic rent owed to the public (for that part of the value exceeding operational costs and anticipated profit of the successful bidder). Currently, there is no comparable economic rent supplied to the public from fisheries.

Under existing regulations, most timber harvested from federal lands is required to be processed in the United States. Timber from private lands is unrestricted by regulation and can be sold to the highest and best markets, whether foreign or domestic. Existing laws limit foreign ownership and fish harvest in U.S. waters. Processing of fish is unrestricted in whether it occurs in the United States or abroad, but fish must be at least initially processed soon after capture in U.S. waters, because of their perishable nature. Additional processing may occur in other nations, such as surimi production in Japan from pollock caught in U.S. waters.

Grazing Lands

Public grazing lands have been subject to leases for many years, but only in relatively recent times has the amount charged for such leases been subject to serious scrutiny. The U.S. Forest Service began charging minimal grazing fees in 1906, but no fees were charged on public domain grazing lands until the enactment of the Taylor Grazing Act in 1934.[40] The Federal Land Policy Management Act (FLPMA)[41] established the policy of retention of the public lands and attempted to create a uniform approach for their administration. Bureau of Land Management grazing leases, renewed automatically, set at controversially low levels, and treated as adjuncts to privately held ranches, have been subject to periodic scrutiny. Efforts to have them treated as property, protected by the takings and compensation provisions of the Fifth Amendment, have thus far proven unsuccessful. Grazing permits are, by law, revocable licenses. Transferable forage rights have been proposed for federal rangelands (Nelson, 1997). Anderson and Hill (1975) present the evolution of grazing rights as an interplay between resource demand, transactions costs, and technological change. Extending their argument to the fishery, "property rights" to fish could become ever more tightly defined as technological change and increased competition for access lead to the adoption of new institutional structures.

---

[40] Secretary of the Interior and Secretary of Agriculture, Study of Fees for Grazing Livestock on Federal Land 2-1 to 2-3 (1977).

[41] Pub. L. No. 94-579, 90 Stat. 2743 (codified at 43 U.S.C. Secs. 1701-1782 (1976) and sections of 7, 16, 30, and 40 U.S.C.

3 | U.S. and Foreign Experience: Lessons Learned

This chapter reviews the genesis, characteristics, and outcomes of individual fishing quota (IFQ) programs that are currently implemented in the United States and abroad. The core case studies, summaries of which are presented in the text of this chapter, discuss the federal IFQ programs currently implemented under the Magnuson-Stevens Act, selected examples of IFQ programs in other countries, and the available literature on IFQ programs worldwide (e.g., ICES, 1996, 1997; OECD, 1997). The full texts of the selected case studies and associated literature citations are presented in Appendix G.

IFQ programs reviewed by the committee are a subset of a larger set of management alternatives intended to restrict fishing participation or effort (see Chapter 4). This larger set of alternatives includes license limitation and more direct effort controls such as transferable trap certificates. Each IFQ program currently in place was adapted to the particular circumstances of the fishery or fisheries in question. The common characteristics of these programs are summarized below, according to the following categories:

- Prior regulatory conditions in the fishery
- Prior biological and ecological conditions in the fishery
- Prior economic and social conditions in the fishery
- Problems and issues that led to consideration of an IFQ program
- Objectives of the IFQ program
- IFQ program development process and the transition to IFQs
- The IFQ program
- Outcomes of the IFQ program

SURF CLAM/OCEAN QUAHOG (SCOQ) FISHERY[1]

General Description

Surf clams (*Spisula solidissima*) and ocean quahogs (*Arctica islandica*) are bivalve mollusks that occur along the U.S. East Coast, primarily from Maine to Virginia, with commercial concentrations found off the Mid-Atlantic coast. Surf clam fishing began in the 1940s and ocean quahog fishing began in the 1970s. These two closely related fisheries are largely conducted by the same vessels, using hydraulic clam dredges. There are a small number of landing sites and processing facilities, some of which are vertically integrated in that they also own harvesting vessels. Most of the catch is shucked and processed into products such as minced clams, clam strips, juice, sauce, and chowder. In addition, a small fishery for fresh in-shell ocean quahogs in the Gulf of Maine began in the 1980s. Apart from a small bait fishery, recreational fishing is insignificant. The SCOQ fishery was the first to be managed under the Magnuson-Stevens Act and the first individual transferable quota (ITQ) program approved under the act.

Prior Regulatory Conditions in the Fishery

Prior to ITQs, the SCOQ fishery was managed through a combination of size limits, annual and quarterly quotas, and in the case of surf clams, fishing time restrictions intended to spread out the catch and even out product input to processors. All vessels were required to detail their catches in official logbooks. These logbooks yielded a clear record of individual vessel performance. Permits were required, but were not restricted in number or availability.

Prior Biological and Ecological Conditions in the Fishery

The biomass of surf clam and ocean quahog populations is dominated by a few large year classes, and year-to-year recruitment variability is high. Neither species demonstrates a statistically significant relationship between the size of the spawning stock and the number of clams recruited. Consequently, harvesters rely on a few large year classes to buffer interannual variability. Surf clams grow slowly and are long-lived, but are sedentary and thus easy to exploit when found. Surf clams were subject to heavy fishing pressure from the late 1960s to the mid-1970s, localized stocks were depleted, and the fishing fleet moved to new grounds. In 1976, a period of low dissolved oxygen killed a large portion of the surf clam stock off New Jersey, prompting tighter harvest restrictions.

---

[1] See Appendix G for a more thorough review.

Prior Economic and Social Conditions in the Fishery

A moratorium on new entrants into the fishery was begun in 1977. Under the moratorium, which lasted until 1990 when the ITQ program was implemented, the number of permitted vessels remained essentially unchanged at approximately 140. Nevertheless, fleet harvesting capacity increased because of the nature of the vessel replacement policy. In addition, the number of crew members employed declined during the moratorium period as vessel owners adapted to fishing time restrictions by using the same crew members on more than one boat. Thus, the moratorium significantly affected the social and economic character of the industry. Although crew members who continued to work on clam vessels received a greater number of fishing days and higher incomes, they were less likely to see fishing as a challenge or adventure than other types of commercial fishermen, and there was a somewhat lower degree of commitment to and dependence on clam fishing compared to other types of commercial fishing.

As early as 1980, a trend toward concentrated market power became evident in the processing sector, and market concentration continues to characterize the SCOQ processing sector. A few large, vertically integrated firms dominated the industry in their dealings with numerous small processors and independent vessel owners, including a few owners who themselves amassed large fleets during the moratorium. Many of the clam vessels were unionized prior to 1979 and thus captains and crew members had some union representation in their dealings with vessel owners. After that time when vessels were sold, mostly to their captains, unionization ended, and no association arose to represent the interests of captains and crews. However, both vessel owners and processors were very active in the management process, and several organizations appeared from time to time to help galvanize efforts to cooperate with the Mid-Atlantic Fishery Management Council (MAFMC).

Fishing ports and processor locations for the SCOQ industry are spread throughout the Mid-Atlantic region and into New England. Processors are found in both seaport and inland communities. The processing labor force is dominated by ethnic and racial minorities and in some places is dependent on immigrants transported from inner cities. The fishing fleets move around quite a bit over time, following clams or clam buyers; hence many crew members are long-distance commuters (e.g., between New Bedford, Massachusetts, and Cape May, New Jersey). Crew members often come from the hinterlands of port communities. Thus, the Atlantic City fleet has little direct connection with Atlantic City; the owners and crew live primarily in old "baymen" towns such as Absecon and Tuckertown, New Jersey. In ports such as Cape May and Wildwood, New Jersey, where fishing is one of the very few year-round occupations, the clam fleet is part of a much larger fishing fleet embedded in a seasonal tourism economy. Occupational health and safety issues loomed large in this fishery.

Vessels frequently sank and fishermen's lives were lost each year off the Mid-Atlantic coast through the late 1980s.

Problems and Issues That Led to Consideration of an ITQ Program

The moratorium established in 1977 was widely considered a success. In concert with other fishery regulations, it reduced the overharvest of surf clams and fostered development of the ocean quahog fishery. The regulatory system under the moratorium, however, was cumbersome and costly to enforce. The rules restricting fishing time, in particular, were complicated and led to a large "ghost fleet" of mostly unused fishing capacity and to health and safety problems resulting from the fishermen feeling that they had to fish in bad weather. Cheating in the form of ignoring regulations on time, area, and clam sizes was alleged to have been rampant. Excess capacity clearly existed in the fleet, and financial institutions were notably reluctant to support fishing ventures.

Objectives of the ITQ Program

The objectives of the 1977 SCOQ fishery management plan (FMP), as amended in 1987, included the following:

1. "...[C]onserve and rebuild Atlantic surf clam and ocean quahog resources by stabilizing annual harvest rates throughout the management unit in a way that minimizes short-term dislocation";
2. "Simplify...the regulatory requirements of clam and quahog management to minimize the government and private cost of administering and complying....';
3. "...[P]rovide the opportunity for industry to operate efficiently, consistent with the conservation of clam and quahog resources, which will bring harvesting capacity in balance with processing and biological capacity and allow industry participants to achieve economic efficiency including efficient utilization of capital resources by the industry"; and
4. "A management regime and regulatory framework which is flexible and adaptive to unanticipated short-term events or circumstances and consistent with overall plan objectives and long-term industry planning and investment needs" (MAFMC, 1988, p. 1; MAFMC, 1996, p. 30).

ITQ Program Development Process and the Transition to ITQs

The 1977 moratorium was intended to be a temporary measure. Instead, it lasted for 12 years. During this period, a Plan Development Team, advised by the council's SCOQ Committee, worked though several phases of discussion regarding potential long-term management frameworks. Prominent in this period were the alternative of individual vessel allocations, which was eventually rejected, and the issue of potential industry consolidation and the development of oligopsonistic[2] or monopsonistic[3] systems. The final ITQ program was adopted by the council in 1989 and approved by the Secretary of Commerce in 1990.

The ITQ Program

The ITQ Management Units. The ITQ has two components: (1) the "quota share" expressed in percentages of the total allowable catch (TAC) that can be transferred permanently, and (2) the "allocation permit" issued in the form of cage tags[4] that are valid for, and can be transferred only within, a calendar year. Annual individual quotas are calculated by multiplying the individual quota share by the TAC in bushels. Bushel allocations are then divided by 32 to yield the number of cages allotted, for which cage tags are issued.

The Initial Allocation of Quota Shares. The initial allocation of quota shares was to owners of permitted vessels that harvested surf clams or ocean quahogs between January 1, 1970, and December 31, 1988. Different formulas were used for allocations of surf clams in the Mid-Atlantic region, surf clams in New England, and ocean quahogs in both regions. For Mid-Atlantic surf clams, 80% of the allocation was based on the vessel's average historic catch in the qualifying period, and 20% was based on a "cost factor" involving vessel capacity. For ocean quahogs and New England surf clams, the allocation was based solely on average catch during the qualifying years.

Accumulation and Transfer of Quota Shares. The minimum holding of SCOQ ITQ shares is five cage units. There is no maximum holding or limit to accumulation, except as might be determined by U.S. antitrust law. Anyone qualified to own a fishing vessel under U.S. law is entitled to purchase SCOQ ITQs.

Setting of Quotas and Other Biological Parameters. The SCOQ FMP is a framework plan that establishes an allowable range of harvest, but each year the MAFMC, in conjunction with an industry advisory panel, recommends specific TACs. Council policy is to set the quota within a specified range of optimum yield at a level that will allow fishing to continue at this level for a specified period (for surf clams, 10 years; for ocean quahogs, 30 years), "and within the

---

[2] A market situation in which each of a few buyers exerts a disproportionate influence on the market (Merriam-Webster, Inc., 1998. *The WWWebster Dictionary* [Online]. [available: http://www.m-w.com/cgi-bin/dictionary] September 1, 1998).

[3] An oligopsony limited to one buyer. (Merriam-Webster, Inc., 1998; *The WWWebster Dictionary* [Online] [available: http://www.m-w.com/cgi-bin/dictionary] September 1, 1998).

[4] One tag is affixed to each cage, which is a large cubical mesh container holding 32 bushels of clams.

above constraints the quota may be set taking into account economic information to set the quota to consider net economic benefits over time to consumers and producers, within the framework of greatest national benefit."[5]

Monitoring and Enforcement. The harvest is monitored through the cage tag requirement and vessel log and dealer reports. There is heavy emphasis on shoreside monitoring and enforcement, although some air and at-sea surveillance is also conducted.

Administration and Compensation. No resource rents are collected from SCOQ ITQ fisheries. Allocation permit fees are collected to help defray administrative costs.

Evaluation and Adaptation. Evaluation and adaptation take place through the FMP amendment process, as well as through reviews by the National Marine Fisheries Service (NMFS) and outside groups. After the defeat of several lawsuits filed by industry groups challenging various features of the ITQ plan, the general approach of industry appears to be acceptance and desire for consistency and predictability, as opposed to frequent change.

Outcomes of the ITQ Program

General. TACs have not been exceeded since implementation of the ITQ program. Natural growth of major year classes of clams and greater targeting of fishing effort subsequent to ITQ implementation led the MAFMC to suspend the minimum size limit on surf clams. The number of vessels active in the surf clam fishery in federal waters went from 128 in 1990, at the initiation of the ITQ program, to 33 in 1997, a 74% reduction. Active vessels in the ocean quahog fishery had less of a decline: from 52 in 1990 to 31 in 1997 (in 1997 14 boats were used in both fisheries; the total fleet numbered 50). Effects on employment have not been quantified, but reports suggest commensurate reductions in jobs, both at sea and on land, as well as increases in working hours at sea for crew.

Biological and Ecological Outcomes for the Fishery. Considerable uncertainty and contention exist regarding the status of the SCOQ stocks and the effects of clam dredging on seafloor habitats. The ITQ program is alleged to encourage targeting and selection of clam populations that meet industry demand, that is, high catch per unit effort (CPUE), achieved by harvesting relatively large clams from relatively pure aggregations. There has been a decline in the discard of

---

[5] MAFMC, April 1998 meeting, as reported in Memo to Surf Clam and Ocean Quahog Committee, surf clam and ocean quahog advisors, and others, July 30, 1998, p. 17.

small clams under the ITQ program. Incentives for discards decreased when the council abolished minimum size limits because of data showing relatively low proportions of undersized clams. ITQs may have provided some of the incentive for more effort to find locations with large clams, although this has not been documented.

Economic and Social Outcomes for the Fishery. Evaluations of the SCOQ fishery have shown that economic efficiency has increased and excess harvesting capacity has declined since the introduction of ITQs. Although some small firms were resilient in the fishery, purchasing more quota shares, many small firms sold out in the first two years after implementation of ITQs. Medium-sized firms were the most likely to purchase more quota shares, while the largest firms remained essentially constant in their holdings. Many quota share recipients ceased fishing and leased their quota shares to other firms. Ownership became increasingly concentrated for ocean quahogs but did not change significantly for surf clams. Between 1988 and 1994, market share was unrelated to price received for catch, suggesting lack of monopoly power in the seller's market. After ITQs were implemented, a few buyer-processors gained dominance, and the processing sector has begun to move to southern New England. There has been a northward shift in landings, due in part to declining CPUE off Virginia and southern New Jersey and in part to the shift in processing locations. Reliance on a single buyer increased the likelihood of exiting the fishery by the end of 1993, while reliance on multiple buyers decreased the likelihood of exiting the fishery, suggesting the power of buyers in the system. The surf clam fishery tends to have a bimodal distribution of large versus small operators, whereas the ocean quahog fishery is more evenly distributed, with a middle class of quota shareholders as well as large operators.

Economic and Social Outcomes for Fishery-Dependent Communities. Employment in the clam industry has declined due to the reduction of vessels and a concomitant decline in the bargaining power of crew and captains, symbolized and to some degree exacerbated by changes in the share system of returns to owners and crew. No research has been done on the effects of ITQs on local communities. Improved safety was a major selling point for the ITQ program, given frequent losses of boats and lives prior to ITQs. Reducing the size of the fleet, removing older vessels, and replacing time limits with ITQs would remove pressures to fish in unsafe ways and conditions (McCay, 1992). However, between 1990, when ITQs went into effect, and February 1999 nine clam boats and at least fourteen lives have been lost in this fishery, a rate of loss comparable to that of the 1980s. Clearly, sea clamming remains a dangerous occupation. The role of ITQs in either mitigating or enhancing its dangers is not known.

Administrative Outcomes. Enforcement was problematic at the beginning of the program, although problems were mitigated somewhat by the cage tag, logbook, and dealer reporting systems. The issue of monitoring of concentration of ownership has been particularly problematic for two reasons. First, it is practically impossible to ascertain the exact identity of "owning persons" due to the nature of the record-keeping process. Second, the critical term "excessive share" is not defined in the Magnuson-Stevens Act or the SCOQ FMP, and thus far courts have not given attention to the issue of concentration unless it approaches monopoly levels, which does not appear to be the case in the SCOQ fishery.

Current Perceived Issues. The major current issues relating to the SCOQ ITQ program are (1) concern with the security of the program, given the recent attempts by Congress to hinder the existence of such programs; (2) perceived inadequacies in the stock assessment and economic studies used in the quota-setting process; (3) adequate enforcement in both state and federal waters; (4) concentration of quota share, even though it may be short of the official definition of "monopoly"; and (5) the need for a lien registry to improve lender confidence so that ITQs can better function as collateral.

SOUTH ATLANTIC WRECKFISH FISHERY[6]

The fishery for wreckfish (*Polyprion americanus*) takes place in a relatively small area of the U.S. South Atlantic region, in deep water, using specialized gear. The product is sold in specialized market niches. The number of participants is small (<50), and the fishery was put under an IFQ program within five years of its inception.

Prior Regulatory Conditions in the Fishery

The fishery began in 1987 and was regulated by the South Atlantic Fishery Management Council (SAFMC) under the council's Snapper-Grouper FMP beginning in 1990. Prior to implementation of the ITQ program in 1992, the wreckfish fishery was regulated through a TAC, trip limits, a permit system, a spawning closure, restricted offloading hours, and a bottom longline restriction. A control date for establishing eligibility for potential limited entry was established in 1990.

Prior Biological and Ecological Conditions in the Fishery

Catch in the wreckfish fishery increased from 29,000 pounds in 1987 to more

---

[6] Unless otherwise noted, the information in this synopsis is from SAFMC (1991).

than 4,000,000 pounds in 1990. Little was (or is) known about the biology of wreckfish or the dynamics of wreckfish populations due to the newness of the fishery and the lack of research and reliable stock assessments.

Prior Economic and Social Conditions in the Fishery

The number of vessels in the fishery increased from 2 in 1987 (prior to permits) to 80 permitted vessels in 1991. Most of the vessels were larger than 50 feet in length, had hold capacities of 5,000-20,000 pounds, and were used primarily in other fisheries, such as snapper, grouper, or shrimp. The fishery takes place far offshore (120 miles) compared to most other South Atlantic fisheries, and involves five- to eight-day trips (SAFMC, 1991).

Wreckfish is a market substitute for snapper and grouper. Some economic analysis has been done on the fishery and individual fishing operation characteristics, but no sociological analysis has been conducted.

The relatively small number of participants in the wreckfish fishery come from a large and widely dispersed number of fisheries and communities throughout the South Atlantic region (primarily Florida to North Carolina). There is no discernible community that is significantly dependent on the wreckfish fishery.

Problems and Issues That Led to Consideration of an ITQ Program

The most important factor in the decision to consider an ITQ program for wreckfish was the rapid rise in both catch (29,000 to 4,000,000 pounds) and participation (2 to 80 vessels) in a short period of time (1987-1991); wreckfish are known to be long-lived, but information about the population dynamics and life history of this species is lacking. The rapid development of fishing capacity was already leading to shortening of the season due to a "derby" fishery. The development of the wreckfish fishery was viewed by the SAFMC as an opportunity to "rationalize" a fishery at its early stages.

Objectives of the ITQ Program

The ITQ program has a number of important objectives:

• To develop a mechanism to vest fishermen in the wreckfish fishery and create incentives for conservation and regulatory compliance whereby fishermen can realize potential long-run benefits from efforts to conserve and manage the wreckfish resource.

• To provide a management regime that promotes stability and facilitates long-range planning and investment by harvesters and fish dealers while avoiding, where possible, the necessity for more stringent management measures and increasing management costs over time.

- To develop a mechanism that allows the marketplace to drive harvest strategies and product forms to maintain product continuity and increase total producer and consumer benefits from the fishery.
- To promote management regimes that minimize gear and area conflicts among fishermen.
- To minimize the tendency for overcapitalization in the harvesting and processing-distribution sectors.
- To provide a reasonable opportunity for fishermen to make adequate returns from commercial fishing by controlling entry so that returns are not regularly dissipated by open access, while also providing avenues for fishermen not initially included in the limited entry program to enter the program.

ITQ Program Development Process and the Transition to ITQs

Development of the ITQ program occurred within the council process, as an amendment to the Snapper-Grouper FMP. Scoping and other meetings and workshops involved industry in developing the program and amending it. Economic analyses were performed on optimal fleet size and individual vessel economics and those data were used in the development of the IFQ program.

The ITQ Program

ITQ Management Units. The management units are percentage shares in the TAC each year. Specific poundages are calculated annually based on the TAC, and coupons are issued in the amount of this poundage to ITQ holders.

Initial Allocation of Quota Shares. Eligibility to receive initial ITQ shares was restricted to permittees who landed more than 5,000 pounds of wreckfish in either 1989 or 1990. Fifty percent of the shares were distributed in proportion to a permittee's landings in 1987-1990; the other 50 percent was distributed equally to all eligible permittees. No "single business entity" could receive more than 10% of initial shares.

Accumulation and Transfer of Quota Shares. There is no limit on accumulation of ITQ or coupon shares by permittees. Wreckfish ITQ shares are freely transferable; yearly quotas (coupons) are transferable separately, but only among permittees.

Monitoring and Enforcement. Monitoring is conducted by the SAFMC and NMFS; enforcement is by NMFS and the Coast Guard. Dealers must hold permits to buy wreckfish.

Administration and Compensation. The wreckfish ITQ program is administered by NMFS and the SAFMC.

Evaluation and Adaptation. Biological and economic parameters are evaluated each year by NMFS and the SAFMC. The program has not been changed since it was implemented and the TAC has remained constant.

Outcomes of the ITQ Program

Biological and Ecological Outcomes for the Fishery. Biological characteristics of landed fish have remained relatively constant and the TAC has remained constant. Landings have been significantly lower than the TAC every year since the inception of the ITQ program; in 1996, only 396,868 pounds were landed out of a total TAC of 2,000,000 pounds. This is due principally to a reduction in fishing trips. Underharvest of the TAC appears to be due primarily to low market prices of wreckfish compared to other species for which the same vessels can fish.

Economic and Social Outcomes for the Fishery. The number of ITQ shareholders has decreased from 49 in 1992 to 25 in 1996, only 8 of which landed wreckfish in the 1996-1997 season (April to April). Thus, shareholders are truly "holding" ITQ shares and coupons; most are engaged in other fisheries. The price for wreckfish has increased somewhat since the ITQ program went into effect, but no analysis has been done regarding the relationship between the ITQ program and exvessel price for wreckfish.

Economic and Social Outcomes for Fishery-Dependent Communities. Effort from the wreckfish fishery appears to have transferred into other fisheries in the South Atlantic region, particularly into the snapper-grouper and shrimp fisheries, fisheries from which the wreckfish fishermen came in the recent past. As mentioned earlier, the fishermen are based in a dispersed set of communities in the South Atlantic region, so the impact of the ITQ program on communities is difficult to discern. Presumably some flexibility has been lost for other, non-ITQ fishermen who might wish to fish for the unused portion of the quota. The other perspective is that these fish are being "banked" by quota holders and they or their offspring could be caught in later years.

Administrative Outcomes. The program is relatively small (25 ITQ holders), and much easier to administer, enforce, and monitor than the fishery management system in place prior to the ITQ program. The Magnuson-Stevens Act mandates the recovery of up to 3% of the costs for the administration and enforcement of IFQ programs, but NMFS and the SAFMC have not yet begun planning a cost recovery system for wreckfish. It is reported that the pressure to increase the TAC that existed before ITQs has disappeared.

Current Perceived Issues. The most controversial aspect of the wreckfish program is the fact that landings have decreased and are less than 25% of the TAC. This had led to some concern by non-IFQ holders that the fishery is not being fully utilized and that quota holders are unfairly excluding others from responsibly harvesting an available resource. The counterargument is that the wreckfish fishery is one for which the population parameters are largely unknown, wreckfish are a long-lived species subject to potential overexploitation, and any shortfall of actual landings below the TAC benefits the wreckfish population and future harvests.

ALASKAN HALIBUT AND SABLEFISH FISHERIES[7]

Fisheries for Pacific halibut (*Hippoglossus stenolepis*) and sablefish (*Anoplopoma fimbria*) occur off the coast of the U.S. Pacific Northwest, British Columbia, and Alaska. Development of large-scale commercial fisheries for halibut[8] was stimulated by the completion of transcontinental railroads in the late 1880s. The directed fishery for halibut uses longline gear. The directed fishery for sablefish uses longline, pot, and trawl gear. Most vessels engaged in these fisheries are catcher vessels, but there are a few catcher-processor vessels in the halibut fishery and a larger number in the sablefish fishery. Vessels engaged in the U.S. fishery are based primarily in the Pacific Northwest region and Alaska.

Prior Regulatory Conditions in the Fishery

Canada and the United States negotiated the Halibut Treaty of 1923 and established what came to be called the International Pacific Halibut Commission (IPHC) to investigate the halibut resource and recommend conservation measures to be implemented by the signatories. With passage of the Fishery Conservation and Management Act, limited entry and allocation decisions for U.S. waters were delegated to the North Pacific and Pacific Fishery Management Councils. Fishermen from each country have been excluded from the waters of the other since 1978. Annual limits on commercial catches of halibut are set for a number of subareas of the region by the IPHC. Commercial catches have historically been controlled through a combination of area, season, and gear restrictions, with amounts of harvest being allocated to particular gear types in particular areas and times. Halibut landings data are collected by the states of Alaska, Washington, and Oregon and by the Canadian government and forwarded to the IPHC. Sablefish catch data are collected by the individual states and NMFS. Both fisheries have had various logbook requirements.

---

[7] See Appendix G for a more thorough review.

[8] Pacific halibut was an important component of trade among the Native peoples of the Pacific Northwest, with fishing removals comparable to modern commercial harvests and trade routes extending hundreds of miles inland (Bay-Hansen, 1991; Newell, 1993).

Prior Biological and Ecological Conditions in the Fishery

Pacific halibut and sablefish are both long-lived demersal species. Their range includes the continental shelf and slope areas from the Sea of Japan, through the Bering Sea and Gulf of Alaska, and along North America's Pacific coast to central California. The distribution of sablefish extends as far south as Baja California. Each species is considered to be a single stock throughout its range. The coastwide biomass of halibut is currently above the 25-year average, but is declining and is expected to continue to decline in the near future. Sablefish biomass has been declining since 1986 and is currently 30% below the recent average (see Figure G.6).

Prior Economic and Social Conditions in the Fishery

In addition to being the focus of a directed commercial fishery, halibut is caught in treaty Indian fisheries, personal-use fisheries, sport fisheries, and as bycatch in a variety of other commercial fisheries. Sportfishing grew from 3% of the total 1984 catch to 11% of the 1996 catch (see Figure G.9). The treaty and personal-use fisheries account for a much smaller portion of the total catch. Halibut caught in the commercial fishery must be discarded if taken with other than hook-and-line gear, if taken when the fishery is closed, or if taken by a longline vessel that has already filled its available quota share. Similar restrictions apply to sablefish, although pots are a permitted gear in the Bering Sea, and a limited amount of the TAC is set aside for a directed trawl fishery. Analysis of the markets before IFQ implementation is limited for halibut (Herrmann, 1996) and nonexistent for sablefish. None of the models of halibut markets account for demand while simultaneously accounting for Canadian, U.S., and Russian supplies and export markets.

Participants in the halibut fishery were heterogeneous geographically, with home ports throughout the Pacific Northwest and Alaska. Although many vessels were specifically rigged for longlining, others were jury-rigged to fish for halibut for the duration of the short open seasons. Many halibut fishermen were engaged primarily in non-fishing occupations and took leave to participate in the short seasons. Halibut and sablefish have accounted, respectively, for 5% and 4% of the exvessel value of commercial catches off Alaska and are regionally significant (see Figure G.10).

Problems and Issues That Led to Consideration of an IFQ Program

The problems and issues that led to consideration of an IFQ program were allocation conflicts, gear conflicts, ghost fishing due to lost gear, bycatch loss in other fisheries, discard mortality, excess harvesting capacity, product quality as reflected in low real prices, safety, economic stability in the fishery and communities, and the development of a rural, coastal, community-based, small-boat fishery. The most striking evidence of some of these problems was the extremely short annual season for halibut, which averaged two to three days per year from 1980 to 1994 in the management areas responsible for the majority of catches (see Figure G.12).

Objectives of the IFQ Program

The North Pacific Fishery Management Council (NPFMC) defined the purpose of and need for action in the sablefish fishery (NPFMC, 1991a) as:

> The problems associated with open access to fishery resources as well as other resources such as air, timber, and water have been widely discussed in the economic and environmental literature (Gordon, 1954; Hardin, 1968). With the current levels of participation and season lengths, there is an intensive race for fish. The amount of fish that a fisherman harvests is determined by how rapidly he can harvest fish before the sablefish TAC is taken and the race ends. Most of the ways in which a fisherman can increase his rate of catch impose increased current and future costs on himself and on others. The increased costs are not offset by increased landings for the fleet as a whole because the landings are constrained by the fixed gear apportionment of the sablefish TAC. The current costs may include increased harvesting and processing costs and decreased ex-vessel and product prices. The future costs may include higher debt service, additional fishing mortality not reflected in landings, increases in fishing accidents, and increased requests for the Council to resolve allocation problems.

> When the race for fish is the allocation mechanism, additional vessels will enter the fishery and the fishing power of the vessels already in the fishery will increase until the increased fishing costs and decreased prices preclude further entry. At that point, the same level of landings could be taken with lower cost and could result in higher-valued products. This is not to say that some fishermen are not making a profit. Rather, they are making much less profit than they could if they were not racing for the sablefish.

> The Council can use traditional management measures to mitigate most of the problems resulting from the race for fish, excluding the dissipation of profits. However, this amounts to treating the symptoms of the problem rather than eliminating the problem, implying that the treatment would have to be ongoing. The need for additional management measures continues with ever more restrictions on harvesting effort (closures, gear limits, etc.) and concurrent increases in fishing and management costs. The costs are expected to increase with respect to sablefish as harvesting and processing capacities for additional groundfish species exceed their TACs and additional vessels enter the sablefish fishery.

Similar language is used in relation to halibut IFQs (NPFMC, 1991b).

IFQ Program Development Process and the Transition to IFQs

Consideration of some form of limited entry in the North Pacific halibut fishery began as early as 1977. However, implementation delays resulted from various interactions with the IPHC and NMFS. IFQs began to be seriously considered for both the halibut and the sablefish fisheries in 1988. In December 1991, the NPFMC approved an IFQ program for both sablefish and halibut. The final rule creating the IFQ program was published in 1993, for implementation in 1995.

The IFQ Program

IFQ Management Units. The halibut IFQ program applies to all commercial hook-and-line harvests in state and federal waters off Alaska. The sablefish program is limited to longline and pot gear fisheries in federal waters off Alaska. The IFQ is the individual's annual allocation and is determined by dividing each individual's quota share by the sum of all quota shares in an identified region, and multiplying the result by the annual fixed gear portion of the TAC for each species. In general, IFQ owners are required to be on board the vessel when the IFQ is being fished.

Initial Allocation of Quota Shares. Halibut quota shares were allocated to the 5,484 vessel owners and leaseholders that had verifiable commercial landings of halibut during the eligibility years of 1988, 1989, or 1990. Specific allocations were based on the best five years of landings for each individual during the qualifying years of 1984-1990. Area-specific shares were allocated based on the geographic distribution of landings during these years. Sablefish quota shares were allocated to the 1,094 vessel owners and leaseholders that had verifiable landings of sablefish during the same eligibility years of 1988, 1989, or 1990, but specific allocations were based on catches from 1985 to 1990. The allocation of quota shares included an adjustment for implementation of the Community Development Quota program in the western Bering Sea region. An extensive review and appeals function accompanied the initial allocation of quota share.

Accumulation and Transfer of Quota Shares. Rules on the accumulation and transfer of quota share are continually evolving. In general, there are limits on accumulation and transferability. No person may own more than 0.5% of the total halibut quota share in combined areas 2C, 3A, and 3B; more than 0.5% of the total halibut quota share in areas 4A-E; or more than 1% of the total quota share for area 2C. No person may control more than 1% of the total Bering Sea-Aleutian Islands and Gulf of Alaska sablefish quota share or more than 1% of the total sablefish quota share east of 140°W longitude. Individuals whose initial allocation exceeded the ownership limit were not required to sell quota share, but were prohibited from acquiring additional quota share. Transferability is restricted across vessel sizes and categories. Catcher vessel quota share is transferable only to certain qualified buyers, whereas catcher-processor vessel quota share is transferable to any person. Lease restrictions apply to certain quota shares. Quota shares of less than 20,000 pounds are "blocked" so that they cannot be further subdivided.

Setting of TACs and Other Biological Parameters. The setting of TACs continues to be based on the process that existed prior to the adoption of the IFQ program. The IPHC (for halibut) and the NPFMC (for sablefish) determine the allowable biological catch and overfishing limits. The NPFMC is responsible for setting the TAC for the commercial fisheries such that the sum of the commercial, sport, subsistence, treaty, and bycatch mortality is less than the overfishing limit. Once the TAC has been determined, the determination of IFQ for halibut is straightforward. In the case of sablefish, approximately 10% of the TAC is set aside for the trawl fishery, and the IFQs are based on the residual.

Monitoring and Enforcement. Monitoring is accomplished through a combination of real-time and posttransaction auditing. Deliveries can only be made to registered buyers following a six-hour notice to NMFS. The real-time accounting is through IFQ Landing Cards and transaction terminals. Posttransaction accounting compares the records submitted by registered buyers with the fishermen's landings records. Some (larger) vessels carry observers for catch and bycatch estimation. Provisions exist for over- and underharvests, where *limited* amounts of annual quota share can be either deducted or credited to the next year's allocation.

Administration and Compensation. The NMFS Alaska Region Restricted Access Management (RAM) Division was created to oversee the initial allocation of quota shares, approve transfers, and monitor compliance. There were no special taxes or fees to cover the cost of developing and administering the IFQ programs before their inception to the present. In keeping with the new Magnuson-Stevens Act requirements, a cost recovery program is now being developed.

Evaluation and Adaptation. The first amendments to the halibut and sablefish IFQ programs had been submitted to the Secretary of Commerce before the program was implemented in 1995. Virtually every meeting of the NPFMC since January 1995 has addressed one or more refinements to the program.

Outcomes of the IFQ Program

Biological and Economic Outcomes for the Fishery. The IPHC estimates that halibut fishing mortality from lost and abandoned gear decreased from 554.1 metric tons in 1994 to 125.9 metric tons in 1995. The discard of halibut bycatch is estimated to have dropped from 860 metric tons in 1994 to 150 metric tons in 1995. However, there is considerable uncertainty surrounding these estimates. There is no clear difference in sablefish bycatch before and after the IFQ program was implemented. There is no evidence of significant underreporting of catches of either halibut or sablefish. The frequency of exceeding the TAC for the fisheries was significantly reduced after the introduction of IFQs (see Figure G.4). There is no evidence that quota holders have tried to increase the halibut or sablefish TACs. The spatial and temporal distribution of halibut catches has changed, but these variables have not been evaluated for sablefish. The biological and ecological consequences of these changes have not been evaluated for either species. With respect to stock assessment methods, it is not certain whether the relationship between CPUE and stock size has changed in the post-IFQ fisheries.

Although there is anecdotal evidence of highgrading, comparisons of halibut size-composition data from Alaskan and Canadian commercial landings and from IPHC surveys suggest that if highgrading occurs, it is not statistically significant. Moreover, no instances of highgrading have been documented or prosecuted. Preliminary comparisons of the size distribution of sablefish in the commercial landings and catches in the NMFS sablefish longline survey suggest that highgrading, if it occurs, is not widespread.

Economic and Social Outcomes for the Fishery. Due to lack of studies and data it is not possible to quantify the net economic impact of the IFQ programs (see Appendix H). Although season length has increased from less than 5 days to 245 days per year for both species and landings are now broadly distributed throughout the season, it is uncertain how costs and revenues have been affected. There are indications that the IFQ program has had a positive effect on the exvessel price of sablefish, but without a comprehensive model of exvessel price formation, and in the face of declining catches and variations in the dollar-yen exchange rate, it is not possible to assign the exact cause of this price increase. The exvessel price of halibut increased slightly with the implementation of IFQs (see Figure G.7), but it is uncertain whether the price increase was the continuation of an upward trend in price or the shift in marketing from frozen halibut to higher-price fresh product. Exvessel price is not a simple function of product form, however; prices depend primarily on supply (affected by the TAC level, landings, and inventory of frozen fish) and demand. The effect of the IFQ programs on halibut exvessel price and on costs and revenues for processors, communities, and consumers are even less well understood. There is anecdotal evidence that an increasing number of halibut fishermen are bypassing traditional processors and marketing directly to wholesalers and retailers, but the magnitude and impact of this phenomenon has not been documented. Casey et al. (1995) showed that in the Canadian halibut fishery implementation of IFQs resulted in a replacement of many of the larger frozen product processors with more individual buyers who added value to the system by searching out new niches and markets for the increased flow of fresh product.

The top five halibut ports have remained the same, with occasional reordering (see Appendix H). The top sablefish ports have also been generally consistent, but since the primary final market for sablefish is Japan, the opportunities for directly marketing are limited, so no change in ports would be expected. The quota share market has been active, with more than 3,800 permanent transfers in the halibut fishery and more than 1,100 in the sablefish fishery. These transfers have led to some consolidation. The number of quota holders declined by 24% in halibut and 18% in sablefish between January 1995 and August 1997. However, the number of quota shareholders still exceeds the annual maximum number of participants in the pre-IFQ fisheries. In both fisheries, the bulk of the consolidation has taken place in the smaller holdings. There is anecdotal evidence that fishermen have reduced crew size and that quota shareholders are crewing for each other. However, since there are few data on pre-IFQ crewing practices, it is difficult to determine the magnitude of changes or the opportunity costs of crew who are no longer in these fisheries.

Economic and Social Outcomes for Fishery-Dependent Communities. The economic and social outcomes of the halibut and sablefish IFQ programs for dependent communities are largely anecdotal. Continued low prices for salmon have made halibut and sablefish catches increasingly important for regional fishing economies. The regional impacts of reductions in crew size are unknown because information on crew participation in the pre-IFQ fisheries, their residencies, demographics, and opportunity costs is limited and has not been compiled adequately.

Administrative and Enforcement Outcomes. Currently, the increased costs of managing and enforcing the IFQ programs are not being recovered from the quota shareholders. However, a cost recovery program is being developed that will assess up to 3% of the exvessel value, which compares favorably with the budget of the RAM Division (see Appendix H). NMFS has successfully prosecuted one case of a sablefish fisherman exceeding his quota share holdings and falsifying landing records. The case resulted in the fishermen forfeiting part of his quota share, and a fine of $16,320 was assessed.

Current Perceived Issues. Some dissatisfaction over the initial allocation continues. This dissatisfaction is related to the delay between the qualifying years and the implementation of the program, and the exclusion of crew members and processors from the initial allocation. The delays in implementation resulted in the exclusion of some fishermen who were active in the years immediately preceding implementation, but were not active during the qualifying years. Similarly, there was dissatisfaction with the award of quota shares to individuals who were active during the qualifying years but inactive in the years immediately preceding implementation. Crew members and processors are discontented that the initial allocation rewarded vessel owners and changed market power in favor of quota shareholders. There are ongoing concerns about the adequacy of enforcement and about community impacts. IFQ implementation has been accompanied by a heightened awareness of subsistence and sport catches and an effort to define harvest limits on these competing fisheries. This competition has led to concerns about localized depletion and preemption of productive sportfishing grounds by commercial fishermen. Expansion of the fishery for sablefish in Alaska state waters and the possible creation of a Gulf of Alaska community development quota (CDQ) program are also of concern.

The characteristics of the U.S. IFQ programs are summarized in Table 3.1.

ICELAND'S INDIVIDUAL TRANSFERABLE QUOTA PROGRAM[9]

Prior Regulatory Conditions in the Fishery

The waters around Iceland are highly productive, and many nations have harvested fish from these waters for hundreds of years. Being keenly aware of their dependence on the sea, Icelanders attempted to reserve their coastal fish stocks by passing a law in 1948 claiming ownership of the living resources in the waters above Iceland's continental shelf. On the basis of the 1948 law, Iceland extended its fishing limits several times in the following decades.

Iceland embarked on an ambitious vessel construction program in the early 1970s and expanded rapidly into the void created by the displacement of foreign fleets with the establishment of Iceland's exclusive economic zone (EEZ). Only a few years later, overcapacity of the fleet and overexploitation of Icelandic fish stocks, particularly cod, were occurring. Gradually, it was recognized that it would be necessary to reduce fishing effort and the capacity of the fishing fleet in order to build up the stocks and increase the catches and the profitability of the industry. From 1977 onward, attempts were made to limit the size of the fishing fleet. These attempts were not particularly effective; in 1977-1983, the value of the fishing fleet increased by about 17% (2.6% annually) and the TAC for cod was consistently exceeded despite a limitation in the number of fishing days. By 1982, politicians and interest groups increasingly believed that more radical measures would be needed to limit effort.

Prior Biological and Ecological Conditions in the Fishery

Major fisheries in Iceland focus on cod, herring, capelin, haddock, and saithe. Following the establishment of Iceland's EEZ, Icelanders rapidly replaced for-

---

[9] See Appendix G for a more thorough review.

TABLE 3.1 Characteristics of U.S. Fisheries Managed Under Individual Fishing Quotas

	Surf Clam	Ocean Quahog	Wreckfish	Halibut	Sablefish
Prior Management Strategy					
TAC	Quarterly apportioned	Yes, but nonbinding	Yes	Yes	Yes
Size limit	No	No	No	No	No
Season	No	No	Yes	Yes	Yes
Fishing time restrictions	Yes	No	No	No	No
Area restrictions	No	No	No	Yes	Yes
Gear	Hydraulic dredge	Hydraulic dredge	LL, except bottom LL	LL	LL, TR, Trap
Vessel moratorium/license limitation	Yes	No	Yes	No	No
Trip limits	No	No	Catch	During end-of-season mop-up	No
Mandatory logbooks	Yes	Yes	Trip coupon	No	No
Program					
Type	% of TAC	% of TAC	% of TAC	% of TAC	% of TAC
Year implemented	1990	1990	1992	1995	1995
Initial Allocation					
Initial recipient	Vessel owner	Vessel owner	Vessel owner	Vessel owner or lease holder	Vessel owner or lease holder
Basis for allocation	Weighted catch history and "cost factor"	Weighted catch history	50% based on catch history, 50% equiproportionate	Catch history	Catch history
Initial concentration limit	None	None	10%	None	None
Characteristics of IFQ					
Durability	Perpetual	Perpetual	Perpetual	Perpetual	Perpetual
Divisibility	160 bushel minimum	160 bushel minimum	No minimum	Minimum blocks	Minimum blocks
Concentration limits	None	None	None	0.5% of regional TAC	1% of regional TAC

Transferability	Eligible to own U.S. fishing vessel	Eligible to own U.S. fishing vessel	Unrestricted	Initial recipients, qualified crew, limits on transfers between vessel size classes	Initial recipients, qualified crew, limits on transfers between vessel size classes
Leasing	Eligible to own U.S. fishing vessel	Eligible to own U.S. fishing vessel	QS holders may lease from each other	Initial recipients may lease up to 10% of their QS during first 3 years	Initial recipients may lease up to 10% of their QS during first 3 years
QS holder required to be on board	No	No	No	Yes, unless owner is a corporation or partnership	Yes, unless owner is a corporation or partnership
Limits on Exvessel Sales					
Port	No	No	No	Any port with a registered buyer	Any port with a registered buyer
Notification prior to landing	Yes	Yes	No, but must be offloaded between 8 a.m. and 5 p.m.	6 hours	6 hours
Eligible buyers	Permit required	Permit required	Permit required and dealer may only receive fish from a permitted vessel	Permit required	Permit required
Cost Recovery					
Windfall tax	No	No	No	No	No
License fee	Yes	Yes	Yes	Nominal	Nominal
Transfer tax	No	No	No	No	No
Landings tax	No	No	No	2%	2%
Rent recovery tax	No	No	No	No	No
Capacity Reduction					
Buyback	No	No	No	No	No
Uncompensated retirement	No	No	No	No	No

(continued)

TABLE 3.1 *Continued*

	Surf Clam	Ocean Quahog	Wreckfish	Halibut	Sablefish
Related Fisheries					
Commercial fishery in state waters	Yes	Yes	No	Yes	Minor
Related Fisheries, continued					
Bycatch in other commercial fisheries	No	No	No	Yes	Yes
Treaty or subsistence	No	No	No	Small	No
Recreation	No	No	No	Moderate	No
Biological Characteristics					
Stock condition	Restored	Localized overfishing	Uncertain	Moderate decline	Strong decline
Year classes in fishery	Few	Few	Multiple	Multiple	Multiple
Overages and Underages Allowed	No (?)	No (?)	?	Yes	Yes

NOTE: LL = long-line; QS = quota share; TR = trawl.

eigners in the harvest of cod and other demersal fish; foreign fishing around Iceland virtually came to a halt in 1976. Icelandic catches of cod increased from around 250,000 metric tons annually in 1971-1975 to a peak of 461,000 metric tons in 1981. In the late 1960s, the Atlanto-Scandian herring stock collapsed, probably because of lower sea temperatures and excessive fishing pressure by Icelandic and Norwegian vessels allowed by the invention of the power block. Two smaller local Icelandic herring stocks also collapsed, and one is believed to have disappeared altogether. The second herring stock was put under a moratorium in 1972, and after a partial recovery the fishery was reopened on a small scale in 1975.

Prior Economic and Social Conditions in the Fishery

The Icelandic economy is heavily dependent on its fisheries. About 73% of the value of goods exported in 1996 consisted of fish and fish products. In 1995, about 11% of the population was employed in fishing and fish processing, which contributed about 15% of the gross domestic product (GDP).

Approximately 90% of Iceland's population lives in villages and towns with more than 200 inhabitants and 60% lives in the capital city of Reykjavík and its suburbs. The towns and villages are located primarily on the coast and scattered almost all around the island, with fishing being a dominant industry in most of these.

Problems and Issues That Led to Consideration of an ITQ Program

Two primary factors led to the initiation of the ITQ program: a desire to improve conservation and a desire to increase economic efficiency. Traditional controls of fishing effort and fleet capacity had not been very effective, and the TAC for cod was consistently exceeded. The politically influenced system of limiting investment in fishing vessels did not succeed in preventing the expansion of a fleet that was already oversized. The system of limiting the number of fishing days was wasteful, since all vessels would try to catch as much cod (the most valuable species) as possible, when they were permitted to catch it. Existing methods of dealing with overcapacity and overfishing in the cod fishery were seen to be too complex, uneconomical, and ineffective. The fleet continued to grow, and temporary bans on fishing on particular grounds failed to reduce the fishing effort. As a result, it was argued, more radical measures would be needed to limit effort. The demand for IFQs also was partly motivated by a general demand for extending the boundaries of the free market and the role of private property in Iceland. Finally, it was argued that an IFQ program would solve other perennial problems, including the problem of safety at sea and the burden of administration.

Objectives of the ITQ Program

The objectives of ITQs were

• To contribute to the conservation of fish stocks by ensuring that the total catch would stay within the limits set by the TAC;
• To make fishing more efficient and reduce overcapacity; and
• To simplify the management program and make it less political and more efficient.

ITQ Program Development Process and the Transition to ITQs

Herring and Capelin. In 1976, vessel quotas were introduced, but each vessel received a very small allocation, due to the low TAC and the large number of vessels with a catch history. At first, the quotas were not transferable, but because of the small size of the quotas and the difficulty of fishing them profitably, transfers were allowed from 1979 on. In 1980, vessel quotas were introduced in the capelin fishery, and in 1986 they were made transferable.

Groundfish. By 1982, Icelandic politicians and interest groups increasingly believed that radical measures would be needed to prevent collapse of the cod stock and to reduce overcapitalization. An ITQ program was introduced by the Icelandic Parliament in 1983 to deal with the problems of the cod fisheries. A new licensing scheme stipulated that new vessels could be introduced to the fisheries only if one or more existing vessels of equivalent size (in gross registered tons [GRT]) were eliminated in return.

The ITQ Program

The fishing law in 1990 incorporated most fish stocks around Iceland into the quota management program. For groundfish, the main exemption is that vessels less than 6 GRT are subject to limitations in the number of fishing days and an overall limit on how much they can catch. Quota allocations are of an indefinite duration and could be revoked by the Icelandic Parliament at any time.

ITQ Management Units. Quota shares are expressed as a percentage of the TAC in metric tons.

Initial Allocation of Quota Shares. When the groundfish (cod) ITQ program was first implemented, each fishing vessel over 10 tons was allotted a fixed proportion of future TACs for cod and five other demersal species. Catch quotas for each species were allotted annually on the basis of this ITQ share. The ITQ program divided access to the resource among vessel owners on the basis of their fishing record during the three years preceding implementation of the program.

Accumulation and Transfer of Quota Shares. In order to be eligible for holding quota, a person or company must have access to a vessel to which the quota is allocated. Initially, groundfish ITQ shares could only be bought or sold undivided along with the fishing vessel to which they were originally allotted, although they could be leased relatively freely; that is, ITQ shares were not fully divisible or independently tradable. Quota shares now can be leased or permanently sold. Leasing of quotas cannot be repeated indefinitely, however; to retain their quota allocations, quota holders must fish at least half of their quotas every second year. Twenty percent of a year's groundfish quota can be shifted to the subsequent year, and an overage of 5% is permitted in any year, without a penalty.

If a quota is to be leased or sold to a vessel operating from a different place, the consent of the municipal government and the local fishermen's union must be acquired. Trading of quotas appears to be brisk; in the "fishing year" 1993-1994 the trading of cod and saithe quotas amounted to 44% and 96%, respectively, of the total catch. Note, however, that the same quota can be traded more than once.

Monitoring and Enforcement. The ITQ program has made it necessary to strengthen monitoring and enforcement. A new government agency has been set up for this purpose. Its role is to issue fishing permits and quota shares, to record information about catches and landings, and to ensure that rules about weighing and landings are followed. Employees of this agency occasionally monitor fishing operations and take samples of landings. There are registered weighing stations in every harbor, and all fish must be weighed and recorded in one of them. Penalties are issued for the discarding, landing, processing, or trading of illegal catches. The penalties for illegal catches are modest or equivalent to the value of the catch. These penalties form a fund that is earmarked to support research and monitoring. Gross violations of the laws about catches and landings are met with legal action and forfeiture of fishing permits. During the last fishing year (1997-1998), there were 57 cases of forfeiture of fishing permits. These, however, are temporary, from one day to one year depending on the seriousness of the violation. There is anecdotal evidence of highgrading and discarding, and some cases of dumping fish not covered by quota have been discovered.

Administration and Compensation. The ITQ program has changed the administrative requirements of Icelandic fisheries. In particular, it has become necessary to strengthen monitoring of landings and activities at sea. The previous system of controlling investment in fishing vessels, which was highly political and not very effective, has been abandoned. Likewise, the system of limiting the number of fishing days for cod and related enforcement activities have been abolished. The Minister of Fisheries determines the fee required to recover the cost of monitoring. This fee cannot be greater than 0.4% of the expected value of the quota in question. There is no compensation to the public beyond this small fee.

Evaluation and Adaptation. The ITQ program was initially put in place for only one year and was seen by many as a temporary emergency measure, to be abolished when the stocks recovered. It was, however, successively prolonged for two or three years at a time, and in 1990 a program of quotas of indefinite duration was emplaced.

With the fisheries law of 1990 passed by Parliament, the program was reinforced and extended into the distant future. First, the program was extended by allocating ITQ shares to approximately 900 smaller vessels (6-10 GRT) that had been fishing without restrictions. As a result, the number of ITQ holders increased by 156% (from 451 in 1990 to 1,155 in 1991). Second, the ITQ program was extended to include all major fisheries. Finally, and arguably most significantly, ITQs became fully divisible and independently transferable.

The 1990 fisheries law is still controversial, however; on December 3, 1998, the Icelandic Supreme Court unanimously concluded that the clause in existing fisheries laws (Art. 5, 38/1990) which privileges those who derive their fishing rights from ownership of vessels during a specific period (during which their "fishing history" was established) is unconstitutional. This privilege, the Court concludes, violates both the Constitutional rule against discrimination (Art. 65) and the rule about the "right to work" (Art. 75). The Court reasoned that while temporary measures of this kind may have been both necessary and constitutional in the beginning, to prevent the collapse of fish stocks, the indefinite legalization of the discrimination that follows from Art. 5 38/1990 is not justified. That Article, in principle, the Court went on, prevents the majority of the public from enjoying the right to work, and the relative share in the common property represented by the fish stocks, to which they are entitled. The implications of the Court's decision will, no doubt, be far-reaching.

Outcomes of the ITQ Program

Biological and Ecological Outcomes for the Fishery. Since the collapse of Icelandic herring stocks in the late 1960s, management of the herring stock has been very successful. Catches have increased gradually, from less than 20,000 metric tons in 1975 to about 140,000 metric tons in the 1994-1995 season, but they fell in the 1996-1997 season to about 100,000 metric tons. Whether or not ITQs have contributed to the general recovery is difficult to determine. The primary tool for conservation is the TAC. To the extent that ITQs have kept the total catch below the TAC, they have helped promote conservation.

Management of the Icelandic cod stock has been much less successful than management of herring, despite the fact that cod is included in the ITQ program and much more important for the Icelandic economy. The cod stock reached an all time low in 1992, but has recovered somewhat since then. The primary reason for the population decline is probably an excessive TAC and catches that have surpassed this TAC by about 12% annually between 1984 and 1996. Overruns of the cod TAC have resulted because of fisheries exempted from the quota program, such as fishing by vessels less than 6 GRT and the hook-and-line fishery in winter. Discards at sea of bycatch and small and immature fish may also be reducing populations of cod and other species.

Economic and Social Outcomes for the Fishery. The ITQ program appears to have improved the profitability of Icelandic fishing firms considerably. The price that fishing firms are prepared to pay for renting cod quota is a possible measure of this profitability. This price has risen from the equivalent of US$0.05-0.09 per kilogram in 1984 to US$0.90-1.00 per kilogram in 1994, and quotations from the summer of 1997 showed prices of up to US$1.25 per kilogram, which is more than one-half of the normal exvessel price. The increase in quota price is much greater than the rate of inflation, so the real price of quota has undoubtedly risen substantially. It must be noted, however, that these figures reflect not only increased profitability of fishing operations but also increasing scarcity of cod. The total productivity of capital and labor in the fishing industry increased by 67% over 1973-1990, despite the fact that the fish stocks were less plentiful in 1990 than in 1973.

ITQs in the herring fishery have led to a substantial increase in economic efficiency. The number of vessels participating in the herring fishery decreased drastically from more than 200 vessels in 1980 to 29 vessels in 1996, at the same time the total catch increased (from 53,000 metric tons in 1980 to approximately 140,000 metric tons in 1994-1995).

The number of decked vessels in the Icelandic fishing fleet began to decline in 1990 when it had reached a peak of about 1,000 and had fallen to 800 by 1996. The size of the fleet in terms of GRT has increased since 1990, when ITQs were extended indefinitely. Thus, there has been a development towards fewer and larger vessels. The Icelandic government initiated a buy-back program in 1994, aimed at removing vessels from the fisheries. The existence of this program indicates that expectations that the ITQ program and the market approach to management would eliminate or reduce overcapacity have not been fulfilled.

Effects on Equity. There has been a steady decrease in the total number of quota holders, with a gradual increase in the number of firms holding more than 1% of the quota each. Currently, 24 of these large firms own almost half of the total quota (a decade earlier these larger firms owned only a quarter of the total quota), and the share of the largest quota holder is about 6%.

Effects on Remuneration and Relative Power. Vessel owners have been permitted to lease their ITQs from the onset of the program. At first, ITQ leasing did not seem to be a particularly common practice, and it was probably undertaken mainly on a small scale by operators who needed extra ITQs after a particularly successful fishing season using their own ITQs. Over time, however, some ITQ holders came to realize that considerable profits could be earned through leasing ITQs on a larger scale, particularly with many fishing operations suffering from the "devaluation" of ITQ shares resulting from repeated reductions in the TAC for cod after 1988. Recently, new and more formalized modes of ITQ leasing have begun to emerge. These transactions involve long-term contracts between large ITQ holders and smaller operators, where the former provide the latter with ITQs in return for the catch and a proportion of the proceeds. Small-scale operators may pay a lease price of up to one-half of the value of the catch and crew shares may be reduced by a similar amount.

Effects on Property Rights. ITQs remain, according to the first clause of the 1990 fisheries management legislation, the "public property of the nation." The laws that eventually were passed reinforced such a conclusion by stating categorically that the aim of the authorities was not to establish private ownership. The issue of ownership, however, is still contested, and quota shares are gradually acquiring the characteristics of private property, despite legal clauses to the contrary.

Effects on Communities. Some companies that have encountered economic difficulties have sold their quota to companies located elsewhere. Also, when TACs are decreased, some quota holders sell out because their share is not viable anymore. Whatever the reason for movement of quota out of communities, it affects the entire community, causing employment problems and eroding the tax base of some municipalities. Small communities, with fewer than 500 inhabitants, have lost a much larger share of their quota than larger communities. In some cases, rural municipalities have tried to reverse the process of decline by buying or leasing quota or investing in local fishing firms.

Effects on Safety. Between 1966 and 1986, 132 fishermen had fatal accidents at sea (108 died by drowning) (Rafnsson and Gunnarssdóttir, 1992), resulting in a mortality of 89.4 per 100,000 person-years. The mortality rate has not changed appreciably during the ITQ period. It is difficult to evaluate the impact of the ITQ program on safety for a variety of reasons: the structure of the fleet and the number of fishermen at risk have changed, there are new regulations on safety precautions, safety data combine ITQ and other fisheries, and no systematic study of the safety effects of ITQs has been conducted.

Current Perceived Issues. Current discontent with the ITQ program can be summarized in several points:

• Many people oppose the privatization of fishing rights within Iceland's EEZ.
• The initial allocation of quota only to vessel owners is often criticized. Prior to the program, fishing was typically regarded as a "co-venture" of vessel owners and crews and many crew members now feel disenfranchised.

• At the present time, industry pays very little in the way of user fees; a fee of up to 0.4% of the catch value is collected to defray the costs of ITQ regulations. The fishing industry is, not surprisingly, adamantly opposed to any collection of fees beyond what would be needed to cover the cost of fisheries management.

• Many Icelanders are wary of the rapid concentration of ITQs in the hands of large vertically integrated companies. Parliament decided in 1998 to set the limit at 10% for cod and haddock and 20% for other species.

• There is much resistance to profit-oriented exchange of fishing rights. Vessel owners who engage in such transactions are labeled "quota profiteers."

• Fishermen and others are concerned with the emergence of the relations of dependency associated with "fishing for others," prompting at least three strikes by fishermen in the past five years.

• The complexity of bureaucratic practices and regulations related to Iceland's fisheries has not been significantly reduced under its ITQ program.

• There is much concern over the threat of municipal bankruptcy in fishing villages that have lost most or all of their quota, with massive unemployment and dissolution of communities. There are demands for effective limitations on quota transfers between regions and communities, to avoid extreme uncertainty in employment.

NEW ZEALAND'S INDIVIDUAL TRANSFERABLE QUOTA PROGRAM[10]

Prior Regulatory Conditions in the Fishery

Prior to the declaration of the 200-mile New Zealand EEZ in 1978, marine fisheries were small and confined to an inshore domestic industry, fishing mostly in depths less than 200 meters. In 1978, a moratorium was introduced on the issuance of additional permits to fish for rock lobsters and scallops. This was followed in 1982 by a moratorium on the issuance of new permits to fish for finfish. The moratoriums limited entry into the fisheries but did not limit fishing power, which continued to increase. In 1979, a number of separately managed limited entry fisheries were established for rock lobsters. Licenses were non-transferable and entry to and exit from the fisheries were managed by a government licensing authority. This system of limited entry failed to control the increase in effort and investment in these fisheries.

Subsequently, the Fisheries Act 1983 was passed, replacing legislation dating from 1908. The new act consolidated previous fisheries legislation and introduced the concept of fishery management plans. The act, and by extension

---

[10] See Appendix G for a more thorough review.

the management plans, recognized the goal of maximizing the economic returns from fisheries, as well as biological conservation, but did not integrate these goals.

Also in 1983, the government issued a Deepwater Fisheries Policy that introduced a system of enterprise allocations for the deepwater trawl fisheries based on company individual quotas. In 1986, the government passed an amendment to the Fisheries Act 1983 that allowed for the introduction of an ITQ program in the inshore fishery and for its broader application to the deepwater fishery.

Prior Biological and Ecological Conditions in the Fishery

Prior to the introduction of ITQs in 1986, there was a widespread perception within government and industry (based primarily on falling catch rates because few quantitative stock assessments existed at that stage) that the harvest from inshore fisheries could be increased in the long term by a short-term reduction in fishing. Initial TACs for most of the inshore finfish stocks were based on average reported landings during periods when the catches were considered to be sustainable. This was a largely qualitative rather than quantitative assessment. For a number of the prime inshore species, the initial TACs were set at levels up to 75% below the catches reported immediately prior to the introduction of ITQs.

Prior Economic and Social Conditions in the Fishery

Prior to the introduction of ITQs in 1986, there was a widespread perception within government and industry that profits from inshore fisheries could be increased in the long term by a short-term reduction in fishing. Again, there was limited economic information to support this perception. The only published information available was a statement that the harvesting sector was overcapitalized by about NZ$28 million, based on insured value (Anon., 1984).

Problems and Issues That Led to Consideration of an ITQ Program

The problems and issues that led to the introduction of the ITQ program were based on the perception that New Zealand's fishery resources would be more productive, both biologically and economically, if fishing activity were reduced temporarily. The industry was overcapitalized, crippled by excessive government management intervention, and subject to rapidly declining economic performance. Recreational fishermen were also concerned about the decline of their fishery.

Objectives of the ITQ Program

During the development of the proposed ITQ program, the government issued a consultation document titled *Inshore Finfish Fisheries—Proposed Policy for Future Management* (Anon., 1984). This document clearly stated the objectives and aims of the proposed ITQ program:

- To achieve the long-term, continuing, maximum economic benefits from the resources; and
- To preserve a satisfactory recreational fishery.

A proposed management regime was developed and used as the basis for discussion. Within this management regime, ITQs were seen as the best mechanism for maintaining the balance between the harvesting sector and the fish stocks, delivering government restructuring assistance, and maintaining profit and equity within the industry.

The aims of the proposed management policy using ITQs as the main management mechanism were as follows:

- To rebuild fish stocks to their former levels;
- To ensure that catches would be limited to levels that could be sustained over the long term;
- To ensure that these catches would be harvested efficiently with maximum benefits to fishermen and the nation;
- To allocate catch entitlement equitably based on fishermen's commitment to the industry;
- To manage the fishery so that fishermen would retain maximum security of access to fish and flexibility of harvesting;
- To integrate the ITQ programs of the inshore and deepwater fisheries;
- To develop a management framework that could be administered regionally in each fishery management area;
- To assist the harvesting sector financially to restructure its operations to achieve the above aims; and
- To enhance the recreational fishery.

ITQ Program Development Process and the Transition to ITQs

The important steps leading up to the implementation of the ITQ program included the following:

- Between 1983 and 1985, possible solutions to overfishing were explored by government and industry, including (1) regulatory intervention based on input controls and (2) actions to establish long-term economic management principles, followed by the reductions of government interference to allow market forces to operate within biologically sustainable levels. After consultation, ITQs were chosen as the preferred management option, with industry support.

• During 1982, a moratorium on new entrants into the inshore fishery was implemented. During 1983-1984, regulations prohibited the participation of part-time fishermen (those deriving less than 80% or their income or NZ$10,000 per year, or both, from fishing).

• In 1982, an enterprise allocation scheme was introduced for seven important species in the deepwater and offshore trawl fisheries.

• The Fisheries Amendment Act 1986 was passed, making the introduction of ITQs possible.

• TACs were established for the inshore and deepwater finfish species that were included in the program.

• TACs were allocated among fishermen, based on their catch history over a period of qualifying years.

• The government provided adjustment assistance to the fishing industry in the form of a buyback of quota entitlements in certain fisheries.

• A computerized reporting system was implemented in 1986, including monthly reports from fishermen and fish buyers, catch logs for vessels, and reports of all quota transfers.

• The ITQ program was implemented on October 1, 1986, and the tendering process was completed by the end of 1986.

The ITQ Program

ITQ Management Units. As of October 1, 1997, there were 30 species or species groups in the quota management system (QMS). The fishery for each species in the QMS is divided into a number of different management units, officially designated as *Fishstocks*. The number of Fishstocks ranges from 2 to 10 for any given species, with a total of 179 different Fishstocks in the QMS.

Initial Allocation of Quota Shares. The initial allocation of ITQs was made free of charge. ITQs were allocated in perpetuity and authorized the holders to take specified quantities of each species annually in each quota area (as opposed to a percentage share of an annually adjusted TAC). Except for the species included in the enterprise allocation system introduced into the deepwater and offshore fisheries in 1983, initial allocation was made on the basis of catch history, modified by the results of a buyback scheme and administrative reductions used to match effort more closely to the available resource.

Initial allocations to the deepwater and offshore trawl fisheries were made on the basis of investment in catching, onshore capital, and onshore throughput. These allocations were converted to ITQs in 1986. Where the sum of the initial allocation was less than the initial TAC, the balance was allocated by tender.

Accumulation and Transfer of Quota Shares. Maximum and minimum holdings of ITQs have been set. No person or company can hold more than 35% of the total of ITQs (for all areas combined) for each of the seven deepwater and offshore species originally allocated under the enterprise allocation scheme or more than 20% of the total ITQ for any single Fishstock area for any other species. These limits apply to the total of owned and leased quota. A minimum quota holding of 5 metric tons was specified for finfish species and 1 metric ton for shellfish.

ITQs may not be held by persons not ordinarily resident in New Zealand or by companies with overseas control. ITQs may not be allocated to or held by owners of licensed foreign fishing vessels, and the government has the sole right to lease ITQs to such vessels. Except for the restrictions described above, ITQs are freely transferable on the open market.

Monitoring and Enforcement. The New Zealand ITQ monitoring and enforcement system is based on documented product flow control that establishes and tracks a fish "paper trail." Fishermen must sell only to licensed fish receivers. All persons selling, transporting, or storing fish must keep business records establishing that the product has been purchased from a licensed fish receiver. Cost-effective enforcement is enhanced by the use of sophisticated electronic monitoring and surveillance information and analytical systems. The system through which quotas are reported and monitored is based on three documents that can be cross-checked—the Catch Landing Log, the Quota Management Report, and the Licensed Fish Receivers Return. The Ministry of Fisheries obtains information from three other sources that can be compared with the information submitted through the quota monitoring and reporting system: the Catch and Effort Returns system, the Observer Programme, and the Vessel Monitoring System. Offenses against the ITQ program are treated not as traditional fishing violations, but as commercial fraud. Penalties include significant fines and forfeiture of fish, vessel, and quota, and are part of an effective deterrent.

Quota busting is known to occur in some fisheries, especially those for high-value species such as rock lobster, paua, snapper, and orange roughy. The illegal catch of rock lobsters in 1993 was estimated as 715 metric tons, about 25% of the total rock lobster TAC (Annala, 1994). Industry is taking a more active role in helping to reduce illegal fishing, especially in the rock lobster and paua fisheries. The discarding or "dumping" of species in the QMS is illegal, except in very limited circumstances. In the multispecies inshore trawl fisheries, fishermen have been known to dump quantities of non-target QMS species rather than use one of the legal mechanisms for dealing with bycatch. In the deepwater trawl fisheries, vessels carrying observers have reported larger quantities of bycatch than vessels fishing the same area that do not carry observers, indicating that discarding probably occurs on vessels without observers. Highgrading has occurred in both the inshore and the deepwater fisheries when a premium price is paid for fish of a certain size or quality and when small fish are discarded because of their unsuitability for processing.

Administration and Compensation. The New Zealand ITQ program is administered primarily by the Ministry of Fisheries, except for quota trading, which is carried out directly among quota holders or through private brokers. The Ministry of Fisheries is consulting with fisheries stakeholders on the transfer of responsibility to the commercial industry for administering the ITQ program. Some of the major administrative issues encountered during the first 10 years of the New Zealand ITQ program include bycatch problems in multispecies fisheries, TAC overruns, and the complicated nature of the quota management system.

Evaluation and Adaptation. One of the glaring gaps in the New Zealand ITQ program is the lack of any systematic, quantitative evaluation of the benefits and costs of the program either by government agencies or by the fishing industry. There is not much in the way of objective, quantitative information available, but there is a great deal in the way of perceptions. A number of adaptations have been made in the first 10 years of the New Zealand ITQ program. The important ones include reducing bycatch problems in multispecies fisheries, settlement of Maori fisheries claims, the change to proportional ITQs from fixed tonnages, and implementation of strategies for adjusting TACs in situations with limited information.

Outcomes of the ITQ Program

Biological and Ecological Outcomes for the Fishery. The major biological and ecological outcomes of New Zealand's ITQ program include improved biological status of fish stocks and development of an open and transparent stock assessment and TAC-setting process. Of the 179 Fishstocks in the QMS as of October 1, 1997, 30 were created for administrative purposes around an offshore island group that is only lightly fished for a few species. Of the remaining 149 Fishstocks, only 11 (7.4%) were estimated to be below a level of biomass that will sustain a stock's maximum sustainable yield (B_{MSY}). Sixteen (10.7%) Fishstocks were estimated to be above and 27 (18.1%) at or near B_{MSY}. The status of the remaining 95 (63.8%) Fishstocks relative to B_{MSY} was not known.

One of the strengths of the New Zealand QMS is the completely open and transparent stock assessment and TAC-setting process. The process is open to all users of the resource and all groups with interests in the fisheries, including Maori, the commercial industry, recreational fishermen, and environmental or conservation groups. All stock assessment data collected by the Ministry of Fisheries are made available (at cost) to all participants in the process. The data are provided only in an aggregated form so that individual fishermen and/or companies cannot be identified. The foundation of the stock assessment process are the Fishery Assessment Working Groups. The working groups analyze the available fishery and research data and prepare draft reports giving the details of the stock assessments and status of the stocks according to agreed terms of reference for all 179 Fishstocks in the QMS. Fishstocks for which the stock assessments indicate a substantial change in the yield estimates or status of the stocks are referred to the Fishery Assessment Plenary (open to all participants in the process). Advice from the plenary session is provided to the Minister of Fisheries and includes other information relevant to the socioeconomic and environmental aspects of each fishery.

Economic and Social Outcomes for the Fishery. The major economic and social outcomes of New Zealand's ITQ program include secure access to the resource; a market-oriented industry structured by market forces; reduced overcapitalization; greater industry freedom, flexibility, and responsibility; and improved industry efficiency, competitiveness, and profitability.

Administrative Outcomes. Sissenwine and Mace (1992) concluded that the QMS had not reduced government intervention. Indeed, the advent of the QMS saw the introduction of new record keeping and reporting requirements such as the quota-monitoring and reporting system and the bycatch trades system. In addition, most input controls—for example, minimum size restrictions, closed seasons and areas—have remained in place.

Current Perceived Issues. In 1996, a new Fisheries Act was passed by the New Zealand Parliament. The act concluded the review of fisheries legislation that had been ongoing since 1991. It provided a complete revision of the Fisheries Act, building on the strengths of the QMS, refined some aspects of the QMS, and added other fishery management features. The act has the following principal components that address many of the current issues with regard to the ITQ program.

Environmental Principles. The act provides the following general environmental principles:

- Stocks must be maintained at or above defined levels. TACs must be set at a level that will maintain stocks at or above a level or move them toward levels that will produce the MSY.
- The effects of fishing on associated and dependent species must be taken into account.
- The biological diversity of the aquatic environment must be conserved.

Consultation. The act formalizes the processes for consultation with sector-user groups. This replaces the current informal advisory group structure. The creation of a National Fishery Advisory Council, with representation from all the sector-user groups, has been authorized.

Conflict Resolution. The act formalizes the resolution of conflicts concerning access to resources. The process first encourages the various sector-user groups to sort out their differences. If the parties are unable to negotiate a solution, the Minister may appoint a commissioner to hold an inquiry and report back to the Minister. All such disputes will be resolved by the Minister.

Addition of New Species to the Quota Management System. The government intends to move all commercially harvested species into the QMS over the next three years. Twenty percent of all new quota will be allocated to the Maori. For most species, the remainder of the quota will be allocated on the basis of catch history. There will be an appeals process for quota allocations, but the process will be stricter than previously. The process will not result in any increases to TACs, and there will be a time limit for filing appeals.

Simplification of the Quota Management System. The new Fisheries Act separates the property right (ITQ) from the catching right by introducing a system of annual catch entitlements (ACEs). For most species, fishermen will no longer be required to hold ITQs before going fishing but will be required to hold ACEs. At the beginning of each fishing year, every person who holds quota will be allocated an ACE based on the amount of quota held. ACEs are superficially similar to an annual lease of quota and are tradable rights like ITQs. When the catch exceeds the ACE, a deemed value is payable. The existing provision allowing 10% overrun of ITQs (with mitigating remedies required) will be abolished.

Institutional Reform. Another issue is the reform of the delivery of fisheries management services. Recent reforms include the provision of services by agencies outside the Ministry of Fisheries (including fisheries research), the transfer of fisheries stock assessment research into a Crown Research Institute, and the establishment of a stand-alone Ministry of Fisheries. The role of the Ministry of Fisheries is being reduced to one of policy advice; determining the standards and specifications for, and purchasing, monitoring, and auditing of, the contestable services; liaison and facilitating conflict and dispute resolution; and enforcement, compliance, and prosecutions.

GENERAL SUMMARY[11]

Prior Regulatory Conditions in the Fishery

All the programs evaluated here had operated under some combination of traditional management measures prior to creation of the IFQ program. Attempts had often been made to achieve the same objectives as IFQs through such mechanisms as trip or vessel quotas, restricted seasons or areas, or even license limitation systems. The transition from traditional management to IFQ management

---

[11] The committee reviewed the four U.S. IFQ programs, plus the IFQ programs of Iceland and New Zealand. The following summary comments focus on these programs, although additional examples are drawn from quota programs in Canada, Norway, and The Netherlands.

has usually proceeded with some intermediate step involving a moratorium on licenses or some other restriction on new entry into the fishery.

Prior Biologic, Economic, and Social Conditions in the Fishery

Prior to the implementation of IFQ programs in the evaluated fisheries, TACs typically had been established, and these catch limits had led to shortened fishing seasons, intensified competition and conflict, changes in historic distributions of costs and benefits from the fishery, and other effects such as increased dangers from fishing in bad weather due to restricted season openings. These factors were in almost all cases exacerbated by an excess of fishing capital, participation, and effort with respect to the available amount of fish under the quota. Many of the subject fish stocks either were overutilized or showed some signs that the populations were being harvested at a greater level than would be sustainable in the long term.

Problems and Issues That Led to Consideration of an IFQ Program

The most common problem cited in IFQ fisheries prior to the adoption of the IFQ program was an excess of capital, participation, and/or effort with respect to the available amount of fish, often resulting in shortened seasons (see Figures 1.1a and b). This had led variously to increased competition and conflict, undesirable price and market effects, increased physical danger to fishermen, administrative and enforcement problems, and potential for undesirable biological impacts through changes in fishing effort patterns. IFQ programs have sometimes been considered for situations in which administration or enforcement of an existing system was costly or difficult under traditional management mechanisms (e.g., surf clams/ocean quahogs). In many cases, some historical participants in the fishery requested the management entity to implement IFQs or some other form of limited entry to address biological, social, or economic issues in the fishery (e.g., halibut, sablefish, wreckfish).

Objectives of IFQ Programs

Despite the claims by some that IFQs have the sole purpose of economic allocation or are a tool for social engineering, a mix of objectives has most often governed the use of IFQs: some biologic (effective implementation of a TAC); some economic (reducing overcapitalization, increasing overall economic efficiency of the fishery); some social (preserving traditional fishing patterns, allocating benefits among individuals, avoiding conflict); and some administrative (more cost-efficient administration, reduction in gear conflicts, better enforcement). The specific objectives of the programs, however, have not always been clear or adequately communicated.

General Characteristics of Existing IFQ Programs

Although no two IFQ programs are exactly the same, the existing programs do exhibit some common characteristics in terms of the initial allocations, how the programs were developed, the nature and duration of IFQs, limits on transferability and accumulation, monitoring and bycatch, and provisions for cost recovery.

Initial Allocation of, and Qualifications for Holding, IFQs

The most common criteria for the initial allocation of IFQs have been those based on catch history. Without any exceptions of which the committee is aware, IFQs have been allocated initially to some license holder of record, most often the vessel owner or skipper. The issue of the lack of initial allocation to those who are not officially associated with the ownership of fishing vessels, such as hired skippers, crew members, or processors, has been raised prominently in several cases (e.g., halibut, sablefish, surf clams, ocean quahogs). The potential for market-based initial allocations such as auctions has been widely discussed, but the committee is not aware of the use of such mechanisms in existing programs. The initial allocation decision is one of the most controversial aspects of an IFQ program, in part because the act of considering IFQs or other limited entry systems often leads to speculative entry. This speculative entry results in increases in participation and effort and dilution of the initial allocation such that most participants will be allocated less than the average of their historic catches. Thus, attempts to be equitable can be unfair.

In terms of qualification for holding IFQs, some programs require IFQ holders to be licensed, if not actually active, fishermen in the fishery (e.g., halibut, sablefish). Some IFQ fisheries have no ownership qualification, except for administrative and record-keeping requirements (e.g., the surf clam/ocean quahog program requires eligibility to own, but not actual ownership, of a vessel). Many programs allow leasing of IFQs to those other than the "owners" of the IFQ (e.g., surf clams; ocean quahogs; wreckfish; several Canadian, New Zealand, and Icelandic programs). Other programs, such as for Alaskan halibut and sablefish, provide limited opportunities for leasing.

IFQ Program Development Processes

The U.S. federal fishery management system under the Magnuson-Stevens Act gave general authority to the regional councils to develop, and the Secretary of Commerce to approve, limited entry programs under a specified set of criteria (Sec. 303[b][6]) that had included IFQ programs until the moratorium set by the Sustainable Fisheries Act of 1996. The processes through which existing IFQ programs have been developed vary widely worldwide. Some have been essentially "top down," with scientists and managers initiating the process and making major decisions. Others have been initiated by scientists and managers but developed substantially by fishermen (e.g., wreckfish). Most have used some form of collaborative process, usually involving task forces, advisory committees, public hearings, and workshops (through regional fishery management councils for U.S. federal fisheries), to gain input from a wide variety of constituents (e.g., for surf clam/ocean quahog, Alaskan halibut and sablefish, and New Zealand fisheries). Few, if any, programs have been developed under a full "co-management" arrangement as this term is currently used, in the sense that fishermen and other stakeholders are full participants in the development process.

Nature and Duration of the IFQ

Most of the existing IFQ programs define the legal status of an IFQ as a "revocable privilege," not a permanent enfranchisement. The quota management program in New Zealand, however, is a prominent exception, granting rights in perpetuity. The more widespread notion is that as long as the program is meeting its stated objectives, it will continue, but the government reserves the right to revoke the privilege for cause. Because none of the major IFQ programs have been significantly altered or abolished, the power of the revocable privilege argument has not been tested. The closest phenomenon has been the "buyback," where privileges were purchased back from the holder by the public sector (e.g., in New Zealand) and retirement of spiny lobster trap certificates (see Appendix G). Some programs have attempted buybacks with funds generated by the fishery via landings or other taxation, with little success. The issue of "sunset" periods for privileges is often raised, but rarely implemented due to the argument that the market transferability and stewardship features of IFQ programs will not work if they have a limited or unknown duration.

Transferability and Accumulation

The majority of existing programs employ IFQs transferable through market mechanisms, some with qualifiers (e.g., transferable only after a certain time period, or among qualified individuals or certain classes). The concern about the potential monopolization of fisheries through IFQ accumulation or aggregation is prominent, and significant (although not legally monopolistic) accumulation or aggregation of IFQs has clearly occurred in some fisheries subsequent to, and as an artifact of, an IFQ program. Some programs have internal rules governing accumulation or aggregation, typically in the form of the maximum amount of IFQ one individual or entity may own or control (e.g., halibut, sablefish, spiny lobster), and other programs do not (e.g., surf clams, ocean quahogs, wreckfish). The latter fisheries did not include accumulation limits because it was believed that rules and procedures external to the fishery itself, such as antitrust laws, would be adequate to address this issue. It is generally true, however, that limits under antitrust legislation are considered unacceptably liberal for most fisheries (Millikin, 1994). For those programs in which transferability is allowed, the quota shares seem to transfer fairly actively (wreckfish being an exception), although the ability to purchase IFQs may be more difficult for those who do not have adequate access to capital.

Monitoring and Enforcement

Most evaluations of existing IFQ programs have questioned the adequacy of catch monitoring and enforcement (e.g., Matthews, 1997, for halibut and sable-fish). Problems with enforcement increase in direct proportion to the geographic extent of the fishery, the number of fishing units in the fishery, the number of landing or sale points, and the ability to sell the fish in a retail market without processing. The use of dealers who must be registered with the fishery manage-ment program with exclusive ability to purchase IFQ fish is common (e.g., hali-but, sablefish, wreckfish). Few programs have adequate internal, long-term moni-toring built into the program itself, and most rely on periodic, specialized evaluations and assessments. The New Zealand ITQ program is an exception, with ongoing monitoring and enforcement activity built into it. In the Alaskan halibut and sablefish programs, enforcement actions have decreased over time (Table 3.2), although enforcement activities have increased since the implemen-tation of IFQs (Appendix H).

Cost Recovery for Administration of the Program and Payments to the Public

Most of the existing IFQ programs provide for minimal, if any, cost recovery for administration of the program. The New Zealand program is a notable excep-tion, in which the attributable and avoidable costs are fully recovered from quota holders. As noted above, most programs essentially give the originally issued

TABLE 3.2   Enforcement Actions in Relation to the Alaskan Halibut and Sablefish IFQ Programs

Year	# IFQ Cases	Overages > 10%	Other
1994	9	0	9
1995	601	436	165
1996	453	302	151
1997	294	179	115

SOURCES: 1994-1996: Matthews (1997), Table 1; 1997: John Kingeter, NMFS.

IFQs to recipients at the initiation of the program. The U.S. IFQ programs now have the mandate to recover up to 3% of exvessel landings value of IFQ fisheries for administrative and enforcement costs and 0.5% of quota value at transfer for the limited access registry system, but none have implemented cost recovery activities yet.

Outcomes of Current IFQ Programs[12]

There are several generalizable outcomes of currently implemented IFQ programs, as reviewed above.   These outcomes vary in terms of their costs and benefits to individuals socially (Figure 3.1) and economically.

• IFQ programs tend to reduce the number of vessels in an ITQ-managed fishery (see Box 3.1 and Figure 3.2). For example, the number of vessels landing halibut has decreased by 42% and sablefish catcher vessels decreased by 52% in Alaskan fisheries. However, some fisheries have actually experienced increases in fishing effort even after IFQs and other limited entry systems were instituted (e.g., in some New Zealand fisheries). As another measure of economic efficiency, IFQs appear to have improved the profitability of Icelandic firms considerably. Quota holdings by size of vessel have changed in programs that do not limit transfers among size classes (e.g., in Iceland; see Box 3.2 and Figure 3.3).

• Many IFQ programs reviewed have experienced a lengthening of the fishing season and, in some cases, increases in exvessel fish prices. For example, Herrmann (1996) reported a statistically significant increase in Canadian exvessel prices for halibut after implementation of an individual vessel quota program.

• Data on changes in human safety in IFQ fisheries are anecdotal for some fisheries, but positive in that fishermen generally report feeling less constrained to fish in bad weather (e.g., in the halibut and sablefish fisheries). However, others emphasize continued pressures to fish in unsafe conditions due to market demands (e.g., surf clams, ocean quahogs). The annual average number of search and rescue missions conducted by the U.S. Coast Guard in Alaska's halibut and sablefish fisheries decreased significantly (p = 0.009) and substantially (about 63%) following implementation of IFQs in Alaska (Table 3.3). In Iceland, there did not appear to be significant improvements in safety with the introduction of IFQs (Figure G.20; see Rafnsson and Gunnarssdóttir, 1992). In the U.S. SCOQ fishery, losses of life and vessels at sea continued for two years after IFQs were

---

[12] Although there may be 50 or more experiments with IFQs (ICES, 1996, 1997; OECD, 1997), many of them do not offer the proper kinds of data for analysis of the effects of IFQs. In addition, for many of these programs, other factors have changed at the same time as the implementation of IFQs. The best data available to the committee are from the Alaskan IFQ programs for factors measured before and after the IFQ programs were implemented.

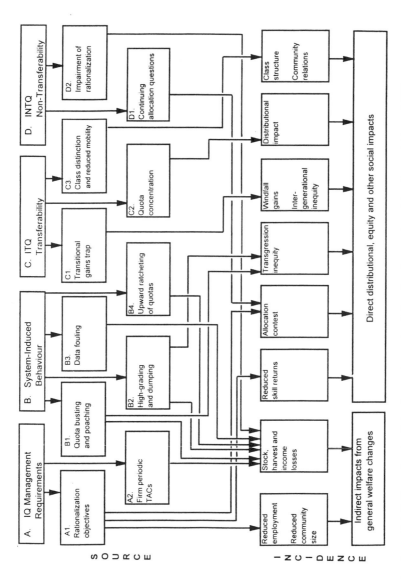

FIGURE 3.1 A partial listing of potential adverse impacts of the development and implementation of IFQ programs, as represented by Copes (1997). Used with permission from the Nordic Council of Ministers.

BOX 3.1
Effect of IFQs on Fleet Capacity

The Netherlands. ITQs have been used in fisheries management in The Netherlands since 1976. The development of fleet capacity and effort since enforcement was tightened is consistent with the effect expected from effective ITQ management. The number of vessels declined from 533 in 1990 to 437 in 1996, and total engine capacity (1,000 units of horsepower) from 544 in 1990 to 467 in 1996 (Smit et al., 1997). It is difficult to ascribe this effect to the workings of the ITQ program alone because other management measures have also been in place, such as licensing of capacity (horsepower) and limits on the number of days at sea. Some of the decline in capacity and effort is due to a stricter enforcement of these measures.

Norway. Figure 3.2 shows the number of licensed vessels and the aggregate licensed cargo capacity of the purse seine fleet in Norway. In the early 1970s, a limited entry system was instituted in the Norwegian purse seine fleet for vessels of more than 1,500-hectoliter hold capacity or longer than 90 feet. Individual vessel quota (IVQs) were introduced also, with the quota allocation of each vessel being determined by the licensed hold capacity through a formula that gave relatively smaller quotas to the larger vessels. Because of economies of scale in the fleet, there was a development toward fewer, larger vessels. In the beginning the total licensed capacity actually increased, due to liberal practice of the rules of capacity replacement when old vessels were replaced by new ones. The reduction in total fleet capacity in the 1980s was due, at least in part, to a buyback program financed by the government.

Iceland. Whether or not IFQs have reduced the excessive capacity of the fleet in the IFQ fisheries is still an open question. The size of the entire Icelandic fishing fleet in terms of gross register tons has increased slightly since 1990, the year when quotas became long-term and could be expected to have an impact on fleet size. However, some of the increase in capacity may be due to increased distant water fishing, which requires large vessels suitable for long trips.

---

implemented, and no lives were lost between 1992 and 1998; however, the sinking of four vessels in early 1999 resulted in the deaths of 10 fishermen.

• Decreases in total harvest-sector employment have been documented in some IFQ fisheries, primarily as a result of decreased numbers of vessels participating and secondarily as a result of less intensive demand for labor compared to "derby" fisheries. However, the length of employment has increased for those who remain employed in some fisheries (e.g., in the Canadian Pacific groundfish fisheries: Bruce Turris, presentation to the committee).

A common perception is that "power" (bargaining for prices or employment; influence over the management structure; economic influence in communities)

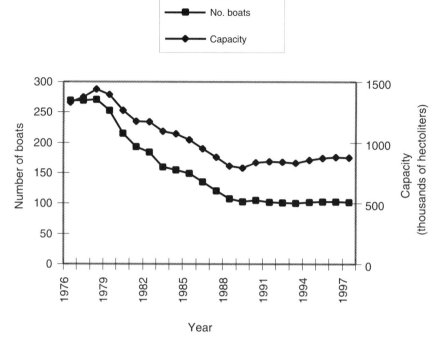

FIGURE 3.2   Number of license purse seiners and total fleet capacity in Norway.

BOX 3.2
Shift in Vessel Size in Icelandic ITQ Fisheries

In the Icelandic case, there has been a significant change in the distribution of quota among vessel size classes from the onset of the ITQ program. Many vessel owners have been dropped out of the program, and a large majority of these were the smallest operators. At the same time, quotas are becoming concentrated in the hands of fewer vessel owners and companies (Figure 3.3; see Appendix G for additional details).

Many Icelanders are wary of the rapid concentration of ITQs in the hands of large vertically integrated companies. A committee appointed by the Ministry of Fisheries recommended that a ceiling for any single quota holder be fixed by law. The Icelandic Parliament decided to set the limit at 10% for cod and haddock and 20% for other species.

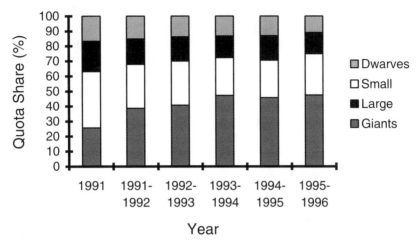

FIGURE 3.3   Changes in holdings of quota share by size of quota holding.

has shifted in many IFQ-managed fisheries. This is usually attributed both to the generation and ownership of new economic value reflected in IFQs and to the fact that ownership of originally issued IFQs is generally concentrated among vessel owners, rather than the crew or processing sectors.

Concern exists in many IFQ-managed fisheries that certain interest groups or communities will become winners or losers due to the shifts in ownership or control of IFQs over time. Thus, some communities fear the loss of an economic base through the exit of IFQs from the community, and some commercial fishermen fear eventual control of IFQs by environmental or recreational interests. Neither of these outcomes has been documented to date, although some IFQ

TABLE 3.3   Search and Rescue Statistics from Alaskan IFQ Fisheries

Year	No. of Search and Rescue Cases	Mortalities
1992	24	5
1993	26	0
1994	33	1
	(IFQs Implemented)	
1995	15	0
1996	7	2
1997	9	1

SOURCE: U.S. Coast Guard.

BOX 3.3
Limitations on Interregional Transfers in Norwegian
Individual Vessel Quota (IVQ) Programs

In 1996, a new Norwegian regulation allowed a person who buys a licensed vessel and scraps it to retain a part of the quota allocation of that vessel for 13 years. How large a part an individual is allowed to retain depends on whether the vessel is being sold from the northern part of the country to the southern part, vice versa, or within each of these areas:

Direction of Sale	% Quota Retained
Northern to southern Norway	50
Within southern Norway	75
Within northern Norway	95
Southern to northern Norway	95

This policy was implemented because most purse seiners previously located in northern Norway have been sold to operators in the southern part of the country, a trend that the government is trying to reverse. The part of the quota that the buyer of a vessel looses is divided among all the licensed vessels in the fleet.

---

programs have been designed to discourage transfer of quota shares among regions (Box 3.3). Few community-based organizations (e.g., municipalities, cooperatives, development associations) have taken the opportunity to serve as lenders for the purchase of IFQs by individuals, although some of the Bering Sea Community Development Quota groups have done so.

IFQs may be used as collateral at commercial lending institutions in some programs, but this option is weakened in the United States by a delay in the implementation of a lien registry.

• Several of the programs reviewed showed clear evidence of the aggregation of IFQs subsequent to the initiation of the program (e.g., surf clams, ocean quahogs). However, other programs, particularly those that had been designed with provisions intended to prevent aggregation, did not show evidence of aggregation beyond the design parameters of the program (e.g., halibut, sablefish). Thus, aggregation appears to have occurred in those programs that were not designed to prevent it, and not to have occurred in those that were designed to prevent it.

Lessons Learned

The following are the general lessons that can be drawn from the above cases and summaries and from the more complete descriptions in Appendix G.

Social, Economic, and Management Issues

• IFQs have had different effects in different fisheries. Within the broad category of "limited entry or access," IFQs are directed toward different objectives and have different effects from other limited entry or access approaches. For example, under IFQs the number of fishing units or participants may vary; under a license limitation program there are generally a fixed number of licenses (if licenses are not transferable). Neither IFQs nor limited entry directly controls fishing effort, although they may create incentives for changes in the amount or distribution of fishing effort.

• Setting clear objectives that are specifically related to the potential effects of IFQ programs is critical. Confusion often exists regarding the mix of biological versus social or economic objectives in the implementation of IFQ programs. The implementation of IFQ programs clearly has the potential to alter, and in some cases the demonstrated record of altering, (1) the distribution of costs and benefits within the fishing community and (2) the management structure. These actual or potential effects may achieve or conflict with the goals and objectives of the Magnuson-Stevens Act and other applicable law.

• The success of IFQ programs in fulfilling their objectives depends on other provisions of the fishing policy and management program. For example, if the TAC is set too high, the program may fail to meet its biologic objectives, and therefore many of the economic and social objectives also.

• All policy and management, in fisheries or any other sector, involves trade-offs. Achieving the goals of increased overall economic efficiency, more effective enforcement or administration, or more effective conservation through the use of IFQs may lead to reduced breadth of participation by fishermen, reduced total employment in the harvesting sector, and other shifts in the distribution of benefits from the fishery. The critical point is that these trade-offs be clearly identified, estimated prior to decisionmaking, and monitored subsequent to program implementation to provide information for adjusting the program over time and for designing subsequent programs.

• For a variety of reasons—from adequate design to increased acceptance of resulting programs—broad involvement of constituents in all phases of program design and implementation is critical.

Stewardship and Biological Conservation Issues

IFQs are not primarily a biological conservation tool; the TAC and other management measures are the main conservation tools in IFQ-managed fisheries. However, IFQs may benefit the resource by addressing either biological conservation or stewardship objectives. Biological conservation can result indirectly from changes in the behavior of fishermen who improve the efficiency of their fishing operations. The effects are largely second-order ones that follow from

IFQ management, such as decreased ghost fishing, decreased bycatch due to improved selection for target species or marketable sizes, and decreased TAC overruns. These behavioral changes usually result from actions on the part of individual quota holders rather than collective actions on the part of all quota holders. Moreover, it is likely that the immediate biological conservation effects of an IFQ program will not necessarily be an indication of the long-term effects (e.g., increased fish size or recruitment); some of the effects may not be measurable for several years (Gilroy et al., 1996; see also Appendix H).

Stewardship. Stewardship objectives are addressed by the direct actions taken by IFQ holders to promote the health of the fisheries resource and the wider ecosystem supporting the resource. In theory, IFQs provide collective incentives for quota holders to undertake actions such as directly funding research to determine biomass and sustainable yields, decreasing bycatch, reducing the effects of fishing on the environment, or voluntarily accepting TAC reductions to promote conservation of the resource because these actions increase the value of the quota· and the potential for increased TACs in future years. These incentives may be stronger with fewer quota holders and the incentive for stewardship may be related directly to the strength of property rights, particularly the length of quota tenure. Alternatively, as mentioned in Chapter 1, stewardship may not be improved by IFQ programs, because like other forms of fisheries management, any individual fisherman reaps the full *benefits* of illegal actions and the much smaller average *costs* of the same action. In Nova Scotia, ITQ holders cooperated with government officials to develop improved conservation measures for their fishery; however, official and anecdotal reports of highgrading and data fouling continued (McCay et al., 1998).

Only limited experience is available regarding whether such theoretical results occur in practice. In a few fisheries in New Zealand, quota holders have formed companies that directly fund research to determine biomass and sustainable yields, to conduct fisheries enhancement projects, and to promote voluntary TAC reductions to enhance conservation of the resource. Likewise, wreckfish IFQ holders have underfished the TAC significantly since implementation of the wreckfish IFQ program. However, the committee also received testimony that IFQs do not promote stewardship.

Biological Conservation. Excess harvesting capacity is a fundamental problem with respect to conservation of fishery resources, and biological conservation is an expressed objective of most IFQ programs. Insofar as an IFQ program contributes to reduction in harvest capacity such that directed effort and catches (fishing mortalities) are reduced and/or the fishery is constrained to its TAC, conservation benefits may be real. In New Zealand, the majority of quota holders perceive biological conservation to be the greatest benefit of their IFQ program (Dewees, 1989; Boyd and Dewees, 1992). In Nova Scotia, McCay et al. (1995, 1998) found evidence of collective efforts to improve conservation through adoption of gear changes and closed areas to protect undersized and spawning fish.

If single-species management, the current practice, is indeed the most effective way to prevent overfishing, IFQ programs as a means to control harvest capacity may promote conservation. Although there is limited experience with the application of IFQs to multispecies fisheries, early experience from New Zealand and British Columbia indicates that appropriately structured IFQs can be an effective management tool for multispecies fisheries (Squires et al., 1998).

In actuality, the TAC, combined with size limits, gear restrictions, protection of spawning areas, and other measures, is the primary conservation measure for many exploited fish species. Thus, biological conservation is best achieved by monitoring, enforcement, compliance, and acceptance of stock assessment findings and the management process. Insofar as an IFQ program contributes to the efficacy of any of the above, real biological conservation benefits may result. The following specific issues have bearing on the efficacy of IFQ programs as conservation measures and have been identified in the refereed literature and public testimony to the committee.

Derby Fishing (the race for fish). IFQ programs have been effective in eliminating the derby nature of fisheries to which they have been applied, thereby decreasing directed effort, stabilizing the supply of fish, and decreasing the potential for quota overruns attributable to the difficulty of monitoring catches during short, frantic fishing seasons. On the other hand, some public testimony, especially by those involved in enforcement, has cautioned that a slower pace and prolonged fishing season place an increased burden on those responsible for monitoring and enforcement, thus making it more difficult to prevent quota overruns. It simply becomes much more difficult to know who should and should not be fishing at any given time and place, increasing the potential for "cheating," especially if exvessel prices are high. This makes at-sea enforcement costs higher for some IFQ programs than under a derby (e.g., halibut and sablefish; see Appendix H).

Data Collection and Data Fouling (underreporting catches, falsifying effort and location data, and making honest mistakes). With the implementation of an IFQ program, the nature of how fish are landed, with respect to both time and space, may change dramatically, thus changing how landings must be monitored. Cheating and data fouling can make the TAC-setting process even more difficult.

Empirical evidence from New Zealand indicates that deliberate underreporting of catches (quota busting) has not increased since implementation of IFQ management, although accurate estimates of fishing effort have been more difficult to obtain because of major changes in fishing operations (Boyd and Dewees, 1992). Similarly, quota busting appears to be minimal in the IFQ-managed Alaska halibut fishery (Gilroy et al., 1996). Some evidence suggests, however, substantial underreporting of total sablefish catches in some years, which may be attributable in part to poor estimates of discarded catch (Gilroy et al., 1996). In New Zealand (Dewees, 1989; Boyd and Dewees, 1992; Annala, 1996) and in the Australian southeast trawl fishery (Squires et al., 1995), the primary resource-related problem identified with IFQ management is the high rate of discarding. This includes both discarding of bycatch for which fishermen do not possess quota (see discussion below) and highgrading to ensure that only the highest-priced portion of the catch is landed and counted against quota. However, fishermen encouraged by high-profile enforcement have learned to modify fishing operations to reduce the amount of illegal discarding as time has progressed (Annala, 1996).

Bycatch and Ghost Fishing. Elimination of the race for fish may provide time for fishermen to search for lower-bycatch fishing grounds (e.g., halibut bycatch in the groundfish fishery in Pacific Canada) and to better care for bycatch species while on deck, thereby decreasing discard mortality. Nevertheless, as Squires et al. (1998) asserted, managing fisheries where several species are caught jointly is especially difficult—part of the mix is likely to be overfished and excessive discards of bycatch can occur.

In New Zealand, IFQs are used in multispecies fisheries and lessons learned there suggest that this form of management can work if sufficient flexibility exists for balancing catches after the fact by acquiring additional quota holdings for bycaught species by the end of some specified time period (Boyd and Dewees, 1992). However, matching the mix of quota held to catches remains a real problem, and excessive bycatch has proven to be a difficulty in certain New Zealand fisheries. In addition, in contrast to U.S. fisheries, in the New Zealand quota management system fishing can continue in multispecies fisheries when either the IFQ or the TAC of a particular species has been filled, if the quota of other associated species has not been caught (Annala, 1996). Thus, many of the overruns in New Zealand TACs have resulted from bycatch in multispecies fisheries (Boyd and Dewees, 1992; Annala, 1996). However, fishermen appear to be adjusting their operations as time passes such that fewer overruns have occurred in recent years (Annala, 1996). Gilroy et al. (1996) estimated that fishing mortality from lost and abandoned gear decreased by 77% in the first year of halibut IFQs. Bycatch discards of halibut in sablefish fisheries decreased by 83%.

Highgrading. In the absence of derby fishing, the incentive for highgrading may be increased as fishermen hunt for fish of the most marketable size and species, but more time for better treatment of discards while on deck may decrease discard mortality of fish caught with some gear types (but not trawls). Empirical evidence from the Alaskan halibut and sablefish fisheries following implementation of the IFQ program indicates that highgrading is not significant in these fisheries (Gilroy et al., 1996; see Appendix G). Indeed, the generalization that highgrading in unlikely to be profitable can be demonstrated (Box 3.4). There is theoretical evidence that the occurrence of highgrading will depend on the unique conditions in each fishery (Anderson, 1994).

Empirical evidence for highgrading in other IFQ-managed fisheries (including some state programs) is mixed. Data from Wisconsin lake trout and Ontario walleye fisheries indicate serious highgrading (Wisconsin lake trout IFQs are

---

BOX 3.4
The High Cost of Highgrading

What's to keep fishermen from highgrading—throwing back all their smaller halibut or sablefish in hopes of catching bigger fish—under an IFQ program? British Columbia IVQ fishermen say they can't afford to highgrade halibut. They plan deliveries, aim for maximum efficiency, and don't want to increase operating costs by highgrading. What's the bottom line on highgrading? Figured at September 1991 prices, highgrading would increase a fisherman's revenue by 3.7%, but he or she would have to catch 24.4% more fish to make up for the discards.

The IPHC sampled a delivery of 2,537 legal-sized halibut totaling 74,514 pounds. They found 38.47% of the fish (19.65% by weight) were 10-20 pound halibut. A fisherman could discard those 14,639 pounds of 10- to 20-pound halibut and try again, but would have to catch 18,217 more pounds of halibut—620 additional fish—to make sure to land at least 14,514 pounds of fish larger than 20 pounds. The additional catch would earn $5,300 more, but would rack up excess operating expenses. In other words, you would forego $30,058 in revenues from the fish discarded to earn an additional $5,300 (see below). The table shows how much more fish a harvester would have to catch to make up for highgrading, and the minimal revenue that highgrading would produce. Results would be similar for sablefish.

Sept. '91

Size	$/lb.	No highgrading		Highgrading: discard 10-20s	
		Lbs. Caught	Revenue	Lbs. Caught	Revenue
10-20 lbs.	$1.65	14,639	$24,153.64	18,217	$0.00
20-40 lbs.	$1.65	28,370	$46,811.31	35,307	$58,255.98
40-60 lbs.	$2.35	11,008	$25,869.02	13,699	$32,193.62
60-80 lbs.	$2.35	10,663	$25,059.13	13,271	$31,185.72
80-100 lbs.	$2.35	5,295	$12,442.71	6,589	$15,484.77
100+ lbs.	$2.35	4,538	$10,664.91	5,648	$13,272.33
Total catch & revenue		74,515	$145,000.72	92,731	$150,392.41
Increase in catch & revenue		0.0%	0.0%	24.4%	3.7%

SOURCE: NPFMC (1992).

---

expressed in numbers of fish), but it appears to be minimal in the Gulf of St. Lawrence trawl, Australian bluefin tuna, and San Francisco Bay herring roe fisheries (Squires et al., 1995). Fisheries in which highgrading is not a serious problem seem to be characterized by minimal price differentials among fish sizes and/or relatively high costs of catching replacement fish (Squires et al., 1995).

Discarding of small and immature fish during fishing operations and highgrading of the catch seem to continue to be a serious problem in the Icelandic fishery, and the problems may have escalated with IFQs. Since quotas are fixed and excessive catch is a violation of the law and subject to prosecution, a quota holder tends to land only the portion of the catch that generates the highest income. It is difficult to estimate the scale of such practices, but the Icelandic Parliament expressed grave concerns and passed strict laws on the treatment of fishing catches in June 1996. Concerns about highgrading, quota busting, and discards are most prominent in fisheries that lack onboard observers. If IFQ-managed fisheries develop bycatch discard problems, it may be necessary to implement or expand observer programs for these fisheries.

Stock Assessment and TAC Setting. Accurate and timely stock assessments to set TACs are an integral part of most IFQ programs because IFQs represent a privilege to harvest part of the TAC. An IFQ program may affect data quality and data collection programs used to set TACs and assess the stocks (Squires et al., 1995). For example, IFQ programs can affect stock assessments due to changes in fishing behavior (Squires et al., 1995). Shifts in fishing location or seasonal patterns may alter catch rates and indices of stock abundance derived from CPUE; there may be changes in the selectivity for different sizes of fish that alter the maximum sustainable yield and the target rates of exploitation on which the TACs are based. TAC setting is invariably a somewhat unpredictable process; these uncertainties affect the expectations of fishermen in their decisions about involvement in IFQ and other limited entry programs.

Underfishing TACs. Clark (1985b) presents a theoretical model that predicts the level that catches will underrun the TAC when managed under an IFQ program. Copes (1986) argued that IFQs hamper reaching the TAC because fishermen are punished for catching more than their quotas. In New Zealand, many TACs have been substantially underfished, even when very large catch reductions were imposed at the time IFQs were introduced (Boyd and Dewees, 1992). The precise reasons for underfishing are unknown, but Boyd and Dewees (1992) suggest that quota busting has been substantially reduced and that fishermen are undercatching many species (especially in multispecies fisheries) because of the limiting effect of possessing sufficient IFQs for other species in their catch mix. Thus, many fish stocks are probably benefiting from lower catch rates, resulting in faster rebuilding of some stocks that were formerly overfished (Boyd and Dewees, 1992; Annala, 1996). Similar reductions in landings of Alaska halibut (a 10% decrease) occurred immediately following the implementation of the halibut IFQ program (Gilroy et al., 1996; Knapp, 1997a, b), but landings have since increased to within a few percentage of the TAC (see Figure G.4). In the wreckfish IFQ program, the TAC has been so substantially underfished that some other factor must be operating. For example, wreckfish fishermen may be maximizing their profit by limiting the supply of wreckfish sold to certain amounts or certain times of the year.

Given this analysis of the strengths and weaknesses of IFQs, what alternative measures might be used to supplement, complement, or perhaps even replace IFQs? The following chapter discusses the range of fishery management measures that have been used to try to sustain marine fisheries.

4 | Alternative Conservation and Management Measures

EFFECTIVENESS OF FISHERY MANAGEMENT MEASURES

The purpose of fisheries management is to control the exploitation of fish populations so that the fisheries they support remain biologically productive, economically valuable, and socially equitable. Maintenance of productive fish populations and associated ecosystems must take into account the variability of ecosystems and to be successful may require caution in setting total allowable catches (TACs). Economic value of fisheries may require creating systems that are economically stable over time and may require minimizing unemployment. Systems to address these goals will fail unless they are administratively feasible and politically acceptable.

A number of biological characteristics of fish stocks impact the effectiveness of management measures: the stock's geographic range, the migration patterns of its members, the usual life span of individuals, the fecundity and spawning potential of the population, the annual variability in recruitment and population size, interactions with other species, and the species' role in the ecosystem.

National Standard 3 specifies that each individual stock should be managed as a unit throughout its geographic range. If individual fishing quotas (IFQs) and other fishery management measures do not encompass the complete stock, they are unlikely to be effective, with the unmanaged portion of the stock becoming overexploited, thereby forcing increased conservation measures in the managed portion of the fishery. This situation is most likely to occur when stocks range across state-federal boundaries, across boundaries between nations, or into the high seas. Close coordination and complementary management systems across jurisdictional boundaries are particularly important.

Many fish stocks undergo predictable migrations on a seasonal or longer time scale, either for feeding or reproduction, or in response to changing environmental conditions or physiological needs. Such stocks can become stratified so that younger and older animals are geographically separated and the fishery may become stratified, with different groups of fishermen harvesting different segments of the stock. If this occurs, a major obstacle to effective fishery management may be resolving the allocation of harvest (and bycatch) among the geographically separated fishermen. In the case of species that migrate across national boundaries (e.g., Atlantic and Pacific salmon, Pacific whiting, and tuna and billfish species), catch allocation conflicts can become especially difficult to resolve, as vividly illustrated by the difficulties surrounding breakdown of the Pacific Salmon Treaty.

Most fish stocks in temperate seas breed seasonally and exhibit high variability in their annual production of offspring. In some stocks, the largest year classes are as much as several hundred times larger than the smallest year classes (Myers et al., 1995). This large variability in recruitment leads to great uncertainty in determining appropriate TACs, which generally are based on the notion of a well-measured, functional relationship between the size of a parent stock and its subsequent production of offspring. Because recruitment is often highly variable and seemingly independent of parental stock size, it is extremely difficult to determine how much of the stock to leave behind (i.e., how to set the TAC). Interannual variability in the growth rates of individuals is another source of uncertainty for some fish stocks. In regions of the Gulf of Alaska in 1980, for example, 12-year-old Pacific halibut were twice the weight of halibut of the same age in 1996 (IPHC, 1997). However, the presence of uncertainty and variability is not adequate grounds for rejecting TAC-based management because TACs can be designed to reflect variability and risk.

Most marine animals are strongly affected by their environment; the influence of the varying biophysical environment on stock size and the condition of individual animals has long concerned fishery scientists. Small (almost unmeasurable) changes in growth and mortality rates during early life that are attributable to environmental variability can lead to significant changes in annual recruitment and persistent impacts on stock size (e.g., Hofmann and Powell, 1998).

Although some recent studies of fish population dynamics emphasize "surprises," discontinuities, and uncertainties (e.g., Botkin, 1990; Ludwig et al., 1993; Wilson et al., 1994), the continuing dominant approach of applied fish population dynamics and bioeconomics that emphasizes deterministic, single-species linear relationships and equilibrium conditions may not account for the realities of many fisheries.

One very difficult fishery management situation, which is sometimes described as the "mixed-stock fishery" (Ricker, 1958; Paulik et al., 1967) or "mixed-species fishery" problem (Clark, 1985a), arises when biologically productive (fast-growing and fecund) and unproductive (slow-growing, late-maturing, slowly reproducing) fish stocks occur on the same fishing grounds and are caught by the same fishing gear. When TACs are based on the more productive stock, the less productive stocks can become seriously depleted or even driven to commercial extinction while the more productive stocks continue to support good harvests. Too many individuals of the less productive species may be taken, reducing the number of fish available to spawn in subsequent years. For example, one rare species of skate in the Irish Sea apparently was harvested to extinction even though it was only caught incidentally (Brander, 1988). To protect the less productive stock, it is necessary to forgo harvesting most of the greater productivity of the more productive stock unless some fishing technique can be developed that allows fishermen to selectively harvest fish from the more productive stock while avoiding fish from the less productive stock. Many salmon fisheries in the Pacific Northwest are managed to protect weak or endangered stocks, with the result that significant harvests of more abundant stocks of hatchery fish must be forgone.

Such characteristics of fish stocks present numerous challenges for designing management systems, especially systems based on setting TACs, such as an IFQ program. Ease in measuring a fish stock, its spatial and temporal distribution, its age-class structure, and the types and numbers of other fish species inhabiting the fishing grounds are among the more prominent characteristics that affect management systems.

Because of such complicating factors, it is not unusual for disputes to emerge among fishery scientists and fishermen over the abundance and robustness of fish stocks, even when scientists are highly confident in their stock assessments. In part, this occurs because fishery scientists and fishermen pay attention to and experience different types of information. Fishery scientists base their models on large-scale characteristics of entire fish stocks, such as fish population recruitment, mortality, and population size over the species' entire range. Fishermen pay attention to small-scale characteristics of fish populations and select fishing areas to maximize catch or net revenue per unit effort. In contrast, fishery scientists base their stock assessments on random samples acquired over the population range. Consequently, each group may develop a very different view of the dynamics of fish stocks. These different viewpoints complicate management by increasing the difficulty of building consensus on problems and solutions and gaining support from fishermen for various management practices, such as setting TACs.

Fishery management under the Magnuson-Stevens Act must be consistent with the National Standards for Fishery Conservation and Management (Sec. 301, Title III, 16 U.S.C. 1851; see Appendix D). These ten standards apply to prevention of overfishing, use of scientific information, equity of allocation, prevention of excessive share, efficiency of utilization, minimization of bycatch, cost-effectiveness of regulations, safety at sea, and importance of fishery resources to coastal communities. As the general guidelines for all U.S. fishery management, the effects of management measures, including IFQs, must be evaluated against these standards.

A number of different management measures address the intent of the national standard guidelines and are discussed in this chapter. Most fisheries, including those with IFQ programs, are managed using a combination of such measures. The degree to which any given measure or combination of measures leads to outcomes that achieve the national standards depends on the interaction between the measures and the biologic, economic, and social attributes of the fishery being managed. Different measures vary in their ability to address the different national standards, and the effectiveness of each management measure depends on the specific management context. Some combinations of measures in some contexts might achieve the same goals as IFQs. For purposes of discussion, fishery management measures are divided into four general types: input controls, output controls, fees and taxes, and technical measures. This discussion follows the general structure of a recent report of the Organization for Economic Cooperation and Development (OECD, 1997) that provides an in-depth review of the performance of various management measures used by OECD member nations. In addition, this chapter discusses alternative management processes based on shared authority between governments and users.

INPUT CONTROLS

Input controls are the oldest type of fishery management tool. Designed to limit either the number of people fishing or the efficiency of fishing, input controls are the type of measure adopted when a fishery is first managed. Input controls include restrictions on gear, vessels, area fished, time fished, or numbers of people fishing. They apply to both commercial and sport fisheries, and may be applied to an entire fishery or to segments of it. Input controls are considered to be an indirect means of limiting the exploitation of fish stocks because they do not directly control the amount of catch (Sissenwine and Kirkley, 1982). Input controls generally lead to inefficient outcomes. They clearly lead to more variable yield than output controls. Fishermen may be able to substitute unrestricted inputs for restricted inputs, thereby maintaining catch levels above those anticipated under the restricted input. Most inputs in most production processes are unrestricted most of the time. For example, there are no legal limits on the numbers of "skates" of longline gear, hook spacing, type of bait, composition of longlines, engine size, or number of crew members that can be used in the halibut and sablefish longline fisheries.

Gear Restrictions

Gear restrictions limit the type, amount, or use of particular fishing gear. Regulations on the *type* of gear include minimum mesh size for trawl codends to allow escape of small fish, excluder devices to minimize bycatch of protected species, specification of trap design and material, limits on the spacing of hooks on longline gear, and minimum mesh size for trawls or gillnets. Regulations on the *amount* of gear include limits on the numbers of traps, the number of longlines in a set, the length and width of a gillnet, and the size of trawl openings. Limits on the *use* of gear include the definition of legal gear for specific fisheries, such as longline gear in the North Pacific halibut fishery, and the prohibition of particular uses of gear, such as trawl gear in the Maine lobster fishery. Sport fisheries are also subject to gear regulations such as limits on the type of hooks or prohibitions against using barbed hooks or live bait.

Gear regulations are widely used in fishery management, often in conjunction with other measures such as time and area closures and license requirements. They often reflect traditional practices in a fishery. Although gear restrictions may be effective in achieving certain limited conservation goals such as decreased mortality of fish returned to the sea or protection of endangered species, they alone are inadequate to control overall exploitation or achieve larger stock protection goals (Rettig, 1991; OECD, 1997). In fisheries that are managed with weak direct controls on exploitation, gear regulations are typically made progressively restrictive over time in an attempt to counter the effects of increasing numbers of participants or intensified fishing. The restrictions decrease the efficiency of fishing, leading fishermen to try to substitute other inputs to make up for constraints on gear. Gear restrictions can, however, be relatively simple to design and enforce (Sissenwine and Kirkley, 1982). If the gear is sufficiently specialized, gear limitations may be effective at controlling effort (e.g., the Florida spiny lobster fishing trap certificates; see Appendix G) (see Hermann et al., 1998 and Greenberg and Hermann, 1994, for discussions of the efficiency and equity effects of pot limits in the Alaskan king and Tanner crab fisheries). Gear restrictions usually do not prevent the race for fish because they do nothing to control the other dimensions of fishing effort, such as vessel size, engine power, or number of crew members. To the extent that substitution of inputs can occur, the race for fish simply switches to a new dimension (Anderson, 1977).

Vessel Restrictions

Vessel restrictions place limits on the type, size, or power of vessels used in a particular fishery. Typical restrictions include limits on vessel design, length, or engine horsepower. Vessel restrictions are frequently used in conjunction with licensing requirements, gear restrictions, and other management measures that attempt to control the amount of fishing. Like gear restrictions, vessel restrictions are an indirect and limited means of controlling fishing and can produce unintended consequences as participants enhance those inputs that are not controlled. For example, limits on the number of vessels allowed in the Bristol Bay sockeye salmon fishery led to the replacement of small, slow vessels with larger, faster vessels (Muse and Schelle, 1989). Subsequent restrictions on vessel length led to increases in vessel width and engine horsepower. Both outcomes hindered the goal of limiting the growth of *fishing power*.[1] In combination with gear regulations, however, vessel restrictions can be somewhat effective in impeding capital stuffing[2] when there is limited ability to substitute unconstrained inputs for constrained inputs (ICES, 1996, 1997; OECD, 1997).

Licenses

Licenses and license endorsements may be used to certify fishermen or vessels, without limitation on the numbers issued, or they may be used as a management measure to limit the number and types of vessels or fishermen that can participate in the fishery. License limitations are intended to limit fishing capacity and effort, but their effect on either is indirect. Limited licenses are used both in federal fisheries, such as the Hawaiian lobster and Pacific groundfish fisheries, and in state fisheries, such as the California sea urchin and Oregon pink shrimp fisheries. Licenses and endorsement limits can also be linked to vessel and gear requirements. In some fisheries, limited licenses are tradable. When licenses are limited and tradable, the value of the license typically varies over time, reflecting changes in expected earnings from the fishery. For example, the price of Alaskan salmon limited entry permits has declined as farmed salmon production volume has increased. Similarly, there was a temporary decline in the value of Prince William Sound, Alaska, salmon limited entry permits that can be attributed to the 1989 *Exxon Valdez* oil spill. (Note, however, that the value of Kenai Peninsula salmon permits averaged $287,222 in 1996.) When licenses are attached to fishing vessels, such vessels typically acquire a value greater than their value as production equipment.

Fleet capacity can be controlled only partially through license limitation. If licenses do not stipulate a maximum vessel size or other limits on fishing power or capacity, the capacity of the fleet can drift upwards as small vessels are re-placed with larger ones. The problem arises because size is only one dimension of fishing power. Also, attempts to control size can lead to adaptations that are inefficient or unseaworthy. To the extent that fishing power can be controlled successfully by license stipulations, such requirements might be impediments to

---

[1] Fishing power measures the ability of a fishing vessel (and its gear and crew) to catch fish, relative to some standard vessel, given that both vessels are fishing under identical conditions (e.g., simultaneously on the same fishing ground).

[2] Investing in gear, technology, engines, processing lines, and other capital components of a fishing operation in order to maximize the ability of a vessel or processing facility to harvest or process fish. These investments are made so that the vessel or processing facility can harvest and process fish as rapidly as possible under a derby fishery or in a race for fish.

technological progress, because new and better vessel designs would not be compatible with the licensed design.

Controlling the fleet capacity by licenses does not encourage economic efficiency to the same extent as controlling fleet capacity indirectly by IFQs. When IFQs would be difficult to monitor or enforce, however, license limitation could be a viable alternative. Nevertheless, license limitation alone is, at best, a short-term approach with short-term benefits. In the long run, the performance of a license limitation program depends on its use in combination with other management measures.

Individual Effort Quotas

Individual effort quotas limit the number of units of effort that a given vessel, license holder, or fisherman can use. In such systems, each participant is allocated a certain number of effort units, such as the number of traps (see Appendix G for a description of the Florida spiny lobster trap certificate program) or days at sea. In the United States, effort quotas have their broadest application in pot fisheries for crustaceans, although they are also used for Atlantic groundfish and scallops through fleet-wide "days-at-sea" limitations (OECD, 1997).

The initial allocation of individual effort quotas can be determined by a variety of mechanisms, including historic catch levels or vessel size. Effort control measures are frequently combined with gear restrictions, license limitations, and vessel configuration limits. The conservation effects of individual effort quotas require limits on entry and are strengthened when combined with a TAC (OECD, 1997). Tradable effort quotas are similar to IFQs, except that as input controls they are only indirectly associated with output. As indirect controls on output, they will be effective in controlling total catch only if there are no other inputs (time, space, gear, behavior) that can reasonably be substituted for the restricted input and if the link between inputs and catch is predictable and relatively stable.

In some fisheries, effort units may be traded among license holders or vessels. If effort quotas are transferable, some efficiencies will be realized as quota shares are fished by fewer vessels. However, effort quotas will not eliminate the incentive to invest in gear innovations to increase catch rates in the race for fish. Evidence collected in OECD member countries supports the expectation of capital stuffing with individual effort quotas and associated increases in operating costs. In addition, individual effort quotas are in many cases difficult and costly to enforce, particularly when strong incentives for compliance are absent (ICES, 1996, 1997; OECD, 1997).

Time limits are one form of effort quota. Time limits seek to reduce the harvest of a given species, or group of species, in an area by reducing the amount of time available for harvesting or by controlling the particular time period over which the species can be caught. The total amount of fishing time allowed at a specific location is often controlled through the specification of a fishing season. Seasonal closures are temporary in nature and are often used in conjunction with area and gear restrictions. Seasonal restrictions have been used extensively throughout the United States and internationally (Rettig, 1991). A single season may be used, or multiple season openings may be set to spread out landings over time. Seasons can vary in length from months to several minutes as in the case of the fishery for herring roe in the North Pacific region (Hourston, 1980). The typical result of time limits is that the length of the season declines over time as fishing effort increases, so without other management measures, time closures lead to a less efficient and more costly race for fish. The use of seasons or time-area closures generally is not effective in meeting either efficiency or conservation goals (OECD, 1997), although time limits can help processors regulate the flow of product more efficiently, as was the case in the surf clam/ocean quahog (SCOQ) fishery management regime prior to IFQs (Appendix G).

Time limit measures may also specifically limit the number of days at sea. In the SCOQ fisheries prior to IFQs (in 1990), each vessel was allocated a number of hours per week or quarter that it was allowed to fish. In the New England and Mid-Atlantic groundfish and scallop fisheries, time limits are now being imposed through limits on days at sea per vessel (NEFMC, 1996).

Time limits are an attempt to control the effect of excess fishing capacity indirectly. They are only indirect controls because, like other input control measures that limit one dimension of fishing effort, they create incentives to develop other dimensions of effort, such as the fishing power of gear and vessels. Time and season controls can be useful, however, to protect spawning stocks, encourage harvesting at times of peak value, and reduce the effects of localized depletion on forage opportunities for marine mammals and seabirds. However, time limits do nothing to prevent overcapitalization; rather, they encourage it.

OUTPUT CONTROLS

Output controls are management techniques that directly limit catch and hence a significant component of fishing mortality (which also includes mortality from bycatch, ghost fishing, and habitat degradation due to fishing). Output controls can be used to set catch limits for an entire fleet or fishery, such as a total allowable catch. They can also be used to set catch limits for specific vessels (trip limits, individual vessel quotas), owners, or operators (individual fishing quotas), so that the sum of the catch limits for individuals or vessels equals the TAC for the entire fishery. Output controls are commonly used in recreational fisheries, taking the form of bag and possession limits that constrain an individual's daily or annual catch.

Output controls rely on the ability to monitor total catch. This can be achieved by either (1) measuring total landed catch with reliable landings records, port-sampling data, and some estimates of discarded or unreported catch; or

(2) measuring the actual total catch with at-sea observer coverage or verifiable logbook data.

Total Allowable Catch

Total allowable catch is a management measure that limits the total output from a fishery by setting the maximum weight or number of fish that can be harvested. TAC-based management requires that landings be monitored and that fishing operations stop when the TAC for the fishery is met. A TAC is based on stock assessments and other indicators of biological productivity, usually derived from both fishery-dependent (catch) and fishery independent (biological survey) data (see NRC, 1998a). Data collected from fishermen, processors, or dockside sampling can be combined with at-sea observations and independent fishery survey cruises to provide information about the total biomass, age distribution, and number of fish harvested. Typically, the TAC is determined on an annual basis, but then partitioned across seasons. To the extent that a TAC is well estimated and enforced, it can control total fishing mortality on a stock (e.g., Pacific halibut). However, experience shows that management by a TAC alone is insufficient to eliminate the race for fish and incentives for capital stuffing. In the long run, without other management controls, management under a TAC leads to dissipation of all fishery rents (Rettig, 1991; OECD, 1997).

The relationship between recruitment and stock size, which is a key part of TAC calculations, is usually difficult to measure reliably because recruitment is often highly variable and, for some species, seemingly independent of parental stock size. Hence, it is extremely difficult to guarantee that conservation objectives will be satisfied by a given numerical TAC or by an IFQ program based on such a TAC. However, stochastiscity and uncertainty about the relationship between recruitment and stock size does not preclude the development of a TAC; it may be possible to develop a risk-compensated TAC based on precautionary principles. The risk of overfishing is greater with no TAC than with a precautionary TAC. Recent National Research Council (NRC) studies have stressed the need to address stochasticity in the development of TAC recommendations (NRC, 1998a,b).

Trip Limits and Bag Limits

Trip limits and bag limits are measures that pace landings by limiting the amount of harvest of a species in a given trip. Trip limits are applied in commercial fisheries when there is interest in spacing out the landings over time or a desire to specify maximum landings sizes, and they are usually accompanied by a limit on the frequency of landings. For example, many Pacific groundfish species are restricted in terms of pounds landed per week or month (PFMC, 1993). The Pacific groundfish trip limit system was adopted for the purpose of maintaining year-round fisheries and provision of fish to markets. Trip limits were an important management tool in the pre-IFQ fishery, in which they were used to minimize TAC overruns during end-of-season mop-up fishing.

Trip limits are usually set to allocate the timing of landings throughout a season when capacity exceeds the TAC. Fleet efficiency declines as vessels make low-capacity trips. Fishing up to the limit during each trip also means that more fish are caught than can be landed, and discards result (Sampson, 1994). Trip limits also heighten the incentive to highgrade catches. As trip limits get more restrictive, the percentage of catch that is discarded increases (Alverson et al., 1994). If trip limits are uniform across the fishery, they may have negative distributional consequences for large vessels, similar to the effect of pot limits on the distribution of benefits described for Alaskan king and Tanner crab fisheries (Greenberg and Hermann, 1994).

A bag limit, used in many recreational fisheries, is similar to a trip limit. Bag limits restrict the number of fish that can be retained in a given day. Bag limits are used in most marine recreational fisheries, including red snapper in the Gulf of Mexico, striped bass in the Mid-Atlantic region, and black rockfish in Oregon and Washington. Trip and bag limits often are combined with license or endorsement requirements, time and area restrictions, and vessel restrictions.

Individual Vessel Quotas

Individual vessel quotas (IVQs) are used in a number of fisheries worldwide, including some Canadian and Norwegian fisheries. IVQs are similar to IFQs, except that they divide the TAC among vessels registered in a fishery, rather than among individuals (Boxes 4.1 and 4.2).

BOX 4.1
Individual Vessel Quotas in Norway

The Norwegian share of the TAC for each fish stock (shared with the European Union or Russia) is partitioned among different vessel groups. For the largest vessels (trawlers, purse seiners) there are individual vessel quotas. For other vessels, there are group quotas and maximum vessel quotas; that is, no vessel can take more than its allotted maximum quota. The maximum quotas are based on vessel size categories, with all vessels in the same size category receiving the same maximum quota. Neither these nor the IVQs are transferable. Individual quota allocations are nevertheless indirectly transferable for the longer term through buying and scrapping a licensed vessel and "stacking" its quota on another vessel.

(Continues)

BOX 4.1
Continued

Vessels fishing under a group quota are regulated by stopping the fishery when the group quota has been taken. In fisheries regulated by maximum vessel quotas, it is easier to ensure that the catch stays below the TAC. The derby effect becomes increasingly forceful, however, when the quota allocations are decreased. In 1990 and 1991, when the TAC for Arcto-Norwegian cod was extremely small, most of the coastal vessels fishing this stock were put under an IVQ regime.

The allocation of annual quotas in the purse seine fishery is regressive; that is, the largest vessels get a proportionally smaller quota than the small vessels. The philosophy behind this is equalization of incomes; there are economies of scale in the purse seine fleet, and larger vessels would obtain a proportionally higher catch value than smaller vessels if all could be used to their full capacity. The formula is determined by the Ministry of Fisheries after consultation with the industry.

BOX 4.2
Individual Vessel Quotas in Canadian Pacific Coast Fisheries

Individual quota programs for the commercial fisheries of western Canada include one established in 1990 for the longline and pot fishery for sablefish; another established in 1991 for the longline fishery for Pacific halibut; and a third established in 1996 for the trawl fishery for groundfish. These IVQ programs are coupled with a variety of additional fishery management measures including limited entry, vessel size limits, gear restrictions, time and area closures, and marine reserves. Prior to IVQ implementation these fisheries used limited entry and were considered to be overcapitalized. Improvement in economic efficiency of the fishing fleet was one of the key reasons for adopting IVQs, from both the industry and the government perspective, and the programs resulted in major reductions in the size of the fleets and the numbers of crew members.

The government and industry in western Canada adopted IVQ programs after having considerable experience with other fishery management tools. The committee heard testimony that the limited entry programs in Pacific Canada had not been effective at limiting fishing effort; instead, the value of the license may have added economic incentives for license holders to fish even more intensely. In some fisheries, trip limits were used to slow down the legal catch but they did not stop the race for fish. Trip limits continued to decrease from one year to the next and resulted in considerable discarding, highgrading, and misreporting. A system of individual transferable effort quotas, which limited the number of fishing days, also failed to stop the race for fish. Instead, it resulted in significant overruns of TACs (>20%), which in turn led to shorter and shorter time allotments in subsequent years.

(Continues)

BOX 4.2
Continued

Implementation of the IVQ programs resulted in significant changes in the fisheries. Geographic and temporal patterns of landings changed over time, particularly in the halibut fishery, where fish were landed at more ports and during a greatly extended season. Halibut were marketed as fresh fish rather than as frozen product, and there was a corresponding increase in landed price (Hermann, 1996). (In fact, this development was one factor influencing the implementation of the Alaskan halibut IFQ program.) In the sablefish fishery, the average license holder's income increased by Can$157,000 annually and income per crew member increased from Can$12,000 to Can$33,000 annually.

TAC overruns had occurred in each of the ten years prior to the implementation of the sablefish IVQ program and in nine of the ten years prior to the implementation of the halibut IVQ program. Subsequent to implementation, overage of the sablefish TAC occurred only in the first year of the program (by 0.5%), and no overages of the halibut TAC occurred. Prior to implementation of the groundfish IVQ program there had been premature closures of the fishery and TAC overages for many of the 55 different component fish stocks that are harvested by this multispecies trawl fishery. However, for 1997, the second year of this IVQ program, it was forecast that the fishery would be under the TACs for all the stocks, on average by 30%. The groundfish trawl fishery also has individual vessel bycatch quotas for halibut that resulted in spectacular reductions, from 1.8 million pounds of bycatch mortality in 1995 to 300,000 pounds in 1996.

In the trawl fishery, there is 100% at-sea observer coverage and full accounting of all catch and bycatch. The considerable expense of this observer program (Can$40,000 per vessel annually) is fully funded by the industry, as are the costs of additional fishery officers and dockside monitoring. In the sablefish fishery, the industry pays all the costs of dockside monitoring, enforcement, stock assessment, and management; in the halibut fishery, the industry pays the management and enforcement costs, but the science activities are funded by the International Pacific Halibut Commission.

Bycatch Quotas

Fishing gear is imperfectly selective. Consequently, most catch includes non-target species, called bycatch, which can be characterized in several ways:

- It may be prohibited to the gear that caught it (e.g., halibut caught by trawl gear);
- It may be too small to sell or undesirable to the market;
- It may be smaller than the legal minimum size;
- It may be a target species for which the quota has already been achieved; or
- It may be an intentional or unintentional component of directed harvest effort with significant market value.

Significant mortality of the discarded fish often results (Chopin and Arimoto, 1995). An important issue of bycatch reduction is how much of the bycatch is behavioral (and amenable to proper incentives) and how much is technological and thus harder (but not impossible) to reduce with incentives.

Some bycatch is retained and marketed. Other bycatch is discarded at sea or after delivery to a processing facility. National Standard 9 of the Magnuson-Stevens Act states that "conservation and management measures shall, to the extent practicable, (A) minimize bycatch and (B) to the extent bycatch cannot be avoided, minimize the mortality of such bycatch" (Sec. 301[a][9]). The Magnuson-Stevens Act also includes provisions to encourage the development of bycatch reduction incentives and the full utilization of unavoidable economic discards (bycatch that has no market). Although the development of individual bycatch quotas (IBQs) is permitted, none of the regional councils have sought review by the Secretary of Commerce of fishery management plan amendments to implement IBQs. An example of a bycatch quota is the individual dolphin mortality limits issued under the 1992 Agreement for the Conservation of Dolphins in the eastern Pacific Ocean tuna fishery.

Bycatch quotas provide incentives for a fishing operation to reduce its bycatch in two ways: (1) reducing *rates* of bycatch (bycatch as a fraction of the targeted catch) means that more of the target species can be caught; (2) reducing total *quantity* of bycatch, so that a portion of the bycatch quota is not needed, means that some bycatch quota can be sold. Tradable IBQs mean that high-bycatch fishermen must pay extra costs to continue fishing, which may encourage them to exit the fishery or change their method of fishing or fishing areas to reduce bycatch. Effective implementation of IBQs requires the ability to account accurately for bycatch through onboard observer and sampling programs. IBQ programs could follow design principles employed in tradable emission quota or lobster trap certificate programs, with a portion on the quota shares attenuating over time as a means of reducing bycatch to a lower target.

Community Development Quotas

Community development quotas (CDQs) are assignments of quota shares to individual communities for the purpose of enhancing fishery-based economic activity. Currently, CDQs are used only in coastal villages bordering the Bering Sea. Many of these communities have substantial Native Alaskan populations and lack opportunities for full employment. The goal of CDQs is to ensure that coastal communities receive a share of fishery benefits. The 1992 Bering Sea CDQ program allocated 7.5% of the walleye pollock quota to Bering Sea communities, requiring that the profits from the allocation of quota be used to improve and advance commercial fishing and related industries (Ginter, 1995; NPFMC, 1998). It is noteworthy that vessels operating in the CDQ pollock fishery achieve higher product recovery rates and lower bycatch rates than they attain while participating in the open-access fishery that immediately precedes the CDQ opening. Subsequent revisions extended the program to Pacific halibut and sablefish and, most recently, to other groundfish and crab fisheries managed under the Eastern Bering Sea Groundfish Management Plan (NPFMC, 1998). The CDQ program requirements stipulate that the proceeds of CDQ fishing be used to enhance fishery-based economic activities. Profits have been used to invest in factory trawlers, port facilities, marine services and cargo handling facilities, and small multipurpose fishing vessels (for crab, salmon, halibut, and cod fishing).

The 1996 amendments to the Magnuson-Stevens Act authorized the development of CDQs for the Western Pacific and mandated an NRC study of CDQs (NRC, 1999a) (see Box 4.3).

BOX 4.3
Findings and Recommendations of the NRC Committee to Review Community Development Quotas (NRC, 1999a)

The Community Development Quota (CDQ) program was implemented in December 1992 by the North Pacific Fishery Management Council. The CDQ program allocates a portion of the annual fish harvest of certain commercial species directly to coalitions of villages, which because of geographic isolation and dependence on subsistence lifestyles have had limited economic opportunities. The program is an innovative attempt to accomplish community development in rural coastal communities in Western Alaska, and in many ways it appears to be succeeding. The CDQ program has fostered greater involvement of the residents of Western Alaska in the fishing industry and has brought both economic and social benefits. The program is not without its problems, but most can be attributed to the newness of the program and the inexperience of participants. Overall the program appears on track to accomplishing the goals set out in the authorizing legislation: to provide the participating communities with the means to develop ongoing commercial fishing activities, create employment opportunities, attract capital, develop infrastructure, and generally promote positive social and economic conditions.

STRENGTHS AND WEAKNESSES OF THE CDQ PROGRAM

Because the program is still relatively new, the data necessary for detailed evaluation are limited and it is not yet possible to detect long-term trends. The six CDQ groups, organized from the 56 eligible communities (later expanded to 57), were of varying sizes and took varying approaches to harvesting their quota and allocating the returns generated. Although not all groups have been equally successful, there were significant examples of real benefits accruing to the communities. All six groups saw creation of jobs as an important goal and stressed employment of local residents on the catcher-processor vessels and shoreside processing plants. All incorporated some kind of education and training component for residents, although to different degrees and with different emphases. Another benefit of the program is that the periodic nature of employment in the fishing industry

(Continues)

BOX 4.3
Continued

preserves options for the local people to continue some elements of their subsistence lifestyles. The CDQ program generates resources that give local communities greater control of their futures. The State of Alaska also has played its part relatively effectively—it was efficient in reviewing the Community Development Plans, monitoring how the communities progressed, and responding to problems. Some of these responses, like reallocating quota share among communities, have been controversial, as might be expected.

Perhaps the greatest weakness of the CDQ program as implemented is a lack of open, consistent communication between the CDQ groups and the communities they represent, particularly a lack of mechanisms for substantial input from the communities into the governance structures. There has also been a lack of outreach by the state to the communities to help ensure that the communities and their residents are aware of the program and how to participate. For the CDQ program to be effective there must be a clear, well-established governance structure that fosters exchange of information among the groups' decisionmakers, the communities they represent, and the state and federal personnel involved in program oversight.

Some debate has centered on uncertainty about the intended beneficiaries of the program. It is unclear whether the program is intended primarily for the Native Alaskan residents of the participating communities or, if not, whether the governance structures should be modified to ensure that non-Native participation is possible. Similarly, there has been dissatisfaction among segments of the fishing industry that are not involved, either directly or as partners of CDQ groups, who believe that the program unfairly targets a particular population for benefits. This conflict is inevitable, given that the CDQ program is designed to provide opportunities for economic and social growth specifically to rural Western Alaska. This policy choice specifically defines those to be included and cannot help but exclude others.

Although it is logical to require initially that all reinvestment of profits be in fishery-related activities because the initial objective of the CDQ program is to help the participating communities to establish a viable presence in this capital intensive industry, over time there should be more flexibility in the rules governing allocation of benefits—perhaps still requiring most benefits to be reinvested in fishing and fisheries-related activities but allowing some portion to go to other community development activities. This will better suit the long-term goal of the program, which is development of opportunities for communities in Western Alaska.

The main goal of the CDQ program—community development—is by definition a long-term goal. Thus there is a need for a set and dependable program duration and the certainty that brings to oversight and management. This will allow CDQ group decisionmakers to develop sound business plans and will reduce pressures to seek only short-term results. However, calling for the program to be long-term does not mean it must go on indefinitely nor that it must never change. Periodic reviews should be conducted, and changes made to adapt rules and procedures as necessary. There can be a balance between certainty and flexibility if the program is assured to exist for some reasonable time (e.g., ten years) and if major changes in requirements are announced in advance with adequate time to phase in new approaches (e.g., five years). The appropriate time scales will of course vary with the nature of the change, with minor changes requiring little notice and major changes requiring enough time for decisionmakers and communities to plan and adjust.

(Continues)

BOX 4.3
Continued

Another long-term issue is environmental stewardship. The CDQ program as currently structured is, in large part, about economic development, but economic sustainability is dependent upon long-term assurance of a sound resource base—the fisheries. Thus, to be successful over the long-term the CDQ program will need to give more emphasis to environmental considerations.

While this report reviews the CDQ program in a broad way, there remains a need for periodic, detailed review of the program over the long term (perhaps every five years), most likely conducted by the State of Alaska. Such a review should look in detail at what each group has accomplished—the nature and extent of the benefits and how all funds were used. For a program like this, care must be taken not to use strictly financial evaluations of success. Annual profits gained from harvest and numbers of local people trained are valuable measures, but they must be seen within the full context of the program. It is a program that addresses far less tangible elements of "sustainability," including a sense of place and optimism for the future.

LESSONS FOR OTHER REGIONS

What emerges from a review of the western Alaska CDQ program is an appreciation that this program is an example of a broad concept adapted to very particular circumstances. Others interested in the application of CDQ-style programs are likely to have different aspirations and different contexts. Wholesale importation of the Alaska CDQ program to other locales is likely to be unsuccessful unless the local context and goals are similar.

One region where the expansion of the CDQ concept has been considered is in the Western Pacific, but such an expansion would need to be approached cautiously because the setting and communities are very different. The major differences between the fisheries and communities of the two regions are: the general lack of management by quota or total allowable catch (TAC) in the Western Pacific; the pelagic nature of the valuable fisheries in the region; and the lack of clear, geographically definable "native" communities in most parts of the region. Application of the CDQ program to the Western Pacific would require the Western Pacific council to define realistic goals that fit within Council purposes and plans. Definitions of eligible communities would need to be crafted carefully so the potential benefits accrue in an equitable fashion to native fishermen.

Any new program, especially one with the complex goal of community development, should be expected to have a start-up period marked by some problems. During this early phase, special attention should be given to working out clear goals, defining eligible participants and intended benefits, setting appropriate duration, and establishing rules for participation. There should be real efforts to communicate the nature and scope of the program to the residents of any participating communities, and to bring state and national managers to the villages to facilitate a two-way flow of information. In addition to these operational concerns, those involved—the residents and their representatives—must develop a long-term vision and coherent sense of purpose to guide their activities.

NOTE: The Committee to Review Individual Fishing Quotas did not have an opportunity to discuss the findings and recommendations of the CDQ report.

Community Fishing Quotas

Building on the concept of community development quotas, community fishing quotas (CFQs) could also be used to direct the flow of economic and social benefits from a fishery to coastal communities. "Community" can be defined at different scales, leading to community fishing quotas specified at a community, regional, or state level (Chapter 1). CFQs may have a variety of objectives and a range of designs beyond the development of fishery infrastructure.

New Zealand's individual transferable quota (ITQ) program contains two examples of community quotas. One example is the quota owned by the Local Authority Trading Enterprise (LATE) at the Chatham Islands, an isolated group of islands about 400 miles east of New Zealand. The LATE owns, on behalf of the geographic community, about 1,200 metric tons of quota for inshore species that it leases only to residents of the Chatham Islands. The other example is one of a quota held by a community of interest and cultural identity. The Treaty of Waitangi (Fisheries Claims) Settlement Act 1992 effectively transferred ownership of almost 40% of the New Zealand ITQ to the Maori people. A large proportion of this quota is held by the Treaty of Waitangi Fisheries Commission. Pending resolution of permanent allocation issues among the tribes, the commission leases ITQs to local *iwi* (tribes) on an annual basis. The iwi may fish the quota themselves, lease it, or get fishermen to use it on their behalf.

Community quotas have also appeared recently in the Scotia-Fundy region of Atlantic Canada, in the context of resistance to adoption of ITQs on the part of small-scale fishermen and coastal communities. The first case, in 1995, involved an agreement to allocate part of the TAC for a particular area to a geographic community (Sambro, Nova Scotia), which could decide how to allocate it, rather than to require ITQs, which are otherwise used widely in the area (Apostle et al., 1998). Subsequent grassroots efforts and a series of demonstrations and occupations of government offices expanded the principle of community-based management to the "fixed-gear" sector of the commercial industry in the Bay of Fundy region (Kearney et al., 1998). Two "community management boards" were formed in 1996. The Canadian Department of Fisheries and Oceans allocates quotas to these boards for three groundfish species, based on their collective catch history. In consultation with their members, each board develops a management plan through a process that is designed to involve all license holders and to be based on consensus and, in one case, consultation with an advisory committee representing community, environmental, and professional interests. The boards have no formal legislative capacity to enforce their management plans; instead, they use contract law, requiring fishermen to sign a contract that they will follow the management plan and accept designated penalties for violation. Fishermen who decline may participate in a government-run management regime. These boards have also become the basis for fishermen's participation in fishery research and an overarching council for the bay as a whole (Kearney et al., 1998).

Another way to implement CFQs would be to modify existing legislation and practice to allow communities and other organizations, such as cooperatives and community development associations, to enter into the markets for IFQs where these have been established. At present, this is not possible in the North Pacific region because of a strong distinction between IFQs and CDQs. This distinction is maintained in many rules, including a rule of the North Pacific Fishery Management Council requiring that the holders of IFQs be on board the vessels, and a congressional restriction of CDQs to the Bering Sea. Also, the halibut and sablefish IFQ programs require quota share purchasers to be fishermen.

Voluntary Cooperatives

An effort to allocate fishing quotas voluntarily among a group of catcher-processor vessels in the Pacific Northwest has shown preliminary signs of success (Box 4.4). In addition, measures recently adopted under Senate Bill 1221 include statutory authority for catcher-processors, shore-based processors, and motherships to form similar cooperatives in the Bering Sea-Aleutian Island pollock fishery. Moreover, the bill included financial inducements that are only available if such cooperatives are formed. This type of cooperative arrangement based on private contract negotiations to sub-allocate quota shares within a group of fishermen with reliance on civil litigation to enforce the agreement is similar to California's adjudicated groundwater basins, as described in Chapter 2.

FEES AND TAXES

Fees and taxes have been used extensively in some common-pool resource settings to control the production of disamenities (e.g., visual and chemical pollution) or to slow the depletion and utilization of natural resources. Economic theory suggests that appropriately designed fees and taxes can lead to socially optimal levels of resource utilization. However, the level of knowledge required to design optimal taxes or fees is difficult to achieve. Consequently, attempts to control resource use through taxation have met with limited success. With few exceptions, the application of fees and taxes in fisheries has been primarily intended as a source of revenue to offset administrative and enforcement costs and to fund product marketing activities.

Fees

Fees are used in many IFQ fisheries to support fishery management, for example, in the North Pacific halibut and sablefish fisheries (being developed) and in the New Zealand and Canadian IFQ fisheries. Fees also have been applied

---

BOX 4.4
The Pacific Whiting Cooperative

Pacific whiting is used primarily for the production of *surimi*, a protein paste used for various products in Japan and for artificial crabmeat sold in the United States. Surimi from Pacific whiting accounts for about 4% of worldwide surimi production.

Thirty-four percent of the 1997-2001 Pacific whiting quota has been allocated to catcher-processor vessels. In April 1997, the four companies holding limited entry permits in the catcher-processor sector agreed to allocate among themselves the portion of the harvest designated for their sector. The principal objectives of their agreement were to eliminate a fishing derby within the sector and to reduce bycatch of other species. They formed a cooperative for this purpose and requested that the U.S. Department of Justice approve their proposal in order to avoid possible antitrust prosecution.

The Department of Justice approved the agreement, noting that it did not appear to have any "incremental anticompetitive effort in the regulated output setting" and could have a procompetitive effect to the extent that it allowed for more efficient processing that increases the output of processed Pacific whiting or reduces the inadvertent catching of other fish whose preservation is a matter of regulatory concern.

The cooperative subsequently announced that implementation of the agreement in the remaining portion of the 1997 fishery had apparently resulted in nearly 20% improvement in yield and significant reductions in bycatch.

The whiting cooperative has at least two characteristics that may have been essential to its success:

• It allocated a known quota that had been specified to the cooperative group (at-sea processing vessels).
• It consisted of a small number of homogeneous participants.

Note, however, that this agreement was not harmless. There is anecdotal evidence that freed-up capacity has been diverted from the whiting fishery and spilled over into other fully capitalized fisheries (e.g., the yellowfin sole and rock sole fisheries in the Bering Sea), exacerbating existing excess capacity and reducing the catch for historic participants in these fisheries. Of course, this negative side effect could be caused by any limited entry system.

Note: This information is based on a letter from Joel I. Klein (Acting Assistant Attorney General, Antitrust Division, U.S. Department of Justice) to Joseph M. Sullivan, dated May 20, 1997.

---

to non-IFQ fisheries. The level of fees can be based on a variety of criteria, including vessel and gear configuration, catch, or past participation in the fishery. Fees are commonly used in fishery management in conjunction with fishing licenses. Fees are generally used in support of the management infrastructure, not as a means of controlling exploitation or increasing efficiency. Targeted fees are sometimes self-assessed by industry to support research or capacity reduction programs, as in the recent proposal by the West Coast trawl sector to finance a permit buyback system (PFMC, 1998). However, in order for a general fishery fee system to limit capacity or control exploitation, fees would have to be sufficiently high to make fishery operations unprofitable for some classes of fishermen or vessels, a politically unrealistic strategy.

Taxes

Taxes, much like fees, can be used to recover the costs of managing a fishery or to generate income for a region, state, or nation. Taxes can be assessed on inputs, raw product, or value-added product. In addition to generating revenues, each of these forms of taxation can be expected to affect the choice of inputs and the demand for products. The extent to which the tax burden is borne by harvesters or processors, or passed on to product purchasers depends on the elasticities[3] of supply and demand in relation to price. Most fishery taxes are currently assessed on the landed value of the fish. They are seldom applied directly as a fishery management tool, although they have the potential to be used in this way because they have the effect of increasing the cost of landing fish (Rettig, 1991). Taxes generally are not effective instruments for controlling the amount of fishing itself, unless they are used in conjunction with other management measures, or are sufficiently high to be an economic barrier to participation. Taxes also tend to lack general political support, but they are likely to be more politically acceptable when the funds are directed toward improving fishery conditions (Rettig, 1991). Taxes might be a particularly useful tool for encouraging bycatch avoidance. Different types of taxes will be discussed in the next chapter.

Fluctuations of fish stocks, fish prices, and TACs are major obstacles to relying solely on taxes and fees as direct management measures because of uncertain short-term reactions of fishermen to changes in taxes. Taxes and fees do, however, offer potential for capturing some fishery rents.

TECHNICAL MEASURES

Technical management measures are those that affect how inputs to the fishery—gear, vessels, or effort—relate to outputs (OECD, 1997). These include limits on fish size and sex and limits on areas fished.

---

[3] Elasticity is a measure of the degree to which supply or demand are sensitive to changes in price. When small changes in price lead to large changes in the quantity demanded, demand is said to be elastic. Conversely, when large changes in price lead to small changes in the quantity demanded, demand is said to be inelastic.

Fish Size and Sex Limitations

Management measures based on fish size or sex attempt to maintain stocks by enhancing their reproductive and growth potential. This type of regulation protects individual fish if they have not yet matured to spawning size or if they are important to reproduction, and it allows fish to be caught at a larger size. Sex and size restrictions are used extensively in crustacean fisheries, such as the West Coast Dungeness and Alaskan snow, Tanner, and king crab fisheries, in which only males with a minimum carapace length can be retained, or the New England lobster fishery, in which only males and non-egg-bearing females above a minimum carapace length may be retained. Minimum fish sizes are frequently used in conjunction with gear restrictions, for example, the minimum size of sablefish combined with a minimum trawl mesh size in the Pacific groundfish trawl fishery. However, minimum fish size rules apply to the retention, not the capture, of undersized fish. Some mortality of undersized fish will result from the process of catching and discarding.

Means of protecting small fish vary from country to country; some make landings of undersized fish illegal so the fish must be thrown overboard at sea, whereas in other countries, all fish—including undersized ones—must be landed (called *full retention*). If captured fish can survive upon release, which is primarily true for crustaceans, mollusks, finfish without closed swim bladders caught in pot or trap fisheries, and some hook-and-line fisheries, requiring immediate release is the most common practice. Release of undersized fish has no conservation benefits in gillnet or trawl fisheries and may have limited benefit in other fisheries because the undersized fish are likely to be dead when returned to the sea. Enforcement of size and sex regulations may be costly. In addition, regulating for size and sex of fish, while appropriate for biological productivity goals, does nothing to alleviate the race for fish (OECD, 1997).

Area Restrictions

Area restrictions limit the geographic region within which fishing is permitted. Area closures are usually temporary—expressed as time-area closures—and are focused on specific types of gear or vessels to prevent harvest during spawning, provide nursery areas for juveniles, or protect species during other vulnerable life history stages. Area restrictions are often applied to geographic regions that have particular conservation needs related to spawning, feeding, or preservation of other ecological services. Area restrictions are also used to allow juveniles to grow to a full, more valuable, size.

For example, fishing in the Shelikof Strait area for Gulf of Alaska pollock is prohibited during the spawning season, and trawl gear is kept out of crab or lobster grounds during molting season. In fact, Bristol Bay and large areas above the Alaska Peninsula are permanently closed to trawling to reduce bycatch mortality of crabs. Area closures have also been used to avoid interactions between fishing operations and marine mammals (e.g., 10- and 20-nautical-mile buffers around Steller sea lion rookeries and haulouts in the Gulf of Alaska and the Bering Sea). In New England, closed areas have been used to protect critical habitat of the northern right whale. Area restrictions also can be used to reduce conflicts between interest groups by setting aside some areas for single gear types or by dividing fishing areas between commercial and recreational fisheries. Area closures or regional segmentation of TACs can have distributional consequences (e.g., Criddle, 1996).

Area restrictions are complementary to IFQs and can be used in IFQ-managed fisheries. For example, residents of Sitka, Alaska, hold more than 1.7 million shares of halibut quota and have proposed a "local area management plan" for halibut fishing. The plan, adopted by the North Pacific Fishery Management Council (NPFMC) in February 1998, would close most of Sitka Sound to larger commercial vessels and charter boats during the months of June, July, and August. Area closures are frequently used in conjunction with gear and vessel restrictions and are relatively easy to enforce. Enforcement of closed areas tends to be more cost-effective when areas are closed to all fisheries than when they allow some fishing (Rettig, 1991).

Some areas are permanently closed to fishing, such as marine reserves and harvest refuges.[4] Set-asides are developed when areas have ecological importance for fish spawning and feeding, as biological communities, or for the preservation of marine biological diversity. Marine reserves may be designed to allow restricted fishing in certain portions of the reserve and banning fishing in other sections. Parts of Georges Bank off the New England coast are closed permanently year-round to protect areas thought to be critical to the spawning and feeding of several groundfish and shellfish species. Australia has established 11 marine protected areas, including the Great Australian Bight in 1997, which won the support of tuna vessel owners and the South Australia Fishing Industry Council. New Zealand has created a number of "no-take" zones and other protected areas under its Marine Reserves Act.

Another function of marine reserves is to serve as a source of replenishment for the stock outside the reserve. The extent to which reserves can be effective in providing these conservation services depends on the size and productivity of the reserve, migration habits and life history patterns of fish and shellfish, and ocean circulation patterns around the reserve (Roberts, 1997b). It also depends on how the reserve is designed, whether it displaces fishing activities from the reserve into greater concentrations outside the reserve, and whether other limits on fishing are in place. Without other restrictions on effort, if marine reserves displace

---

[4] The NRC has another study underway to assess our scientific knowledge of marine protected areas as tools for fisheries management and protecting marine biological diversity.

fishing or increase the profitability of fishing outside the reserve, the resulting intensity of fishing could increase to the point of nullifying biological or economic gains from the reserve. Although there is evidence that closed areas contribute to general conservation and the protection of biodiversity, they may not be sufficient to meet fishery conservation goals when used alone (OECD, 1997). There are many unanswered questions about the design, implementation, and effectiveness of marine reserves in achieving broad-scale conservation goals.

Use Rights in Fishing

Territorial use rights in fishing (TURFs) assign exclusive use rights over a fishery area to an individual or group (Christy, 1982). They are a special case of area restrictions and are analogous to grazing rights. In many cases, traditional territorial use rights are applied in less industrialized and smaller-scale coastal fisheries where management has been based on restricting participation to a localized population in a limited geographical area. The use of TURFs is most suitable for species that are relatively immobile or predictable in location, such as mollusks and crustaceans. Because participants in Bering Sea king and Tanner crab superexclusive[5] area registration fisheries are precluded from participating in other (more lucrative) crab fisheries, few large vessel operators choose to participate, effectively reserving the superexclusive registration fisheries for local, small vessel fleets. Consequently, superexclusive area registration amounts to a form of common property TURF (Hermann et al., 1998).

TURFs may be used in conjunction with fishing gear such as fish-aggregating devices, pound nets, or other entrapment and enclosure devices (Christy, 1996). The establishment of regional lobster zones in the State of Maine shares elements of a TURF, with specific management zones established in geographical regions based on the distribution and historic participation of fishermen (Acheson and Steneck, 1997; Wilson, 1997). These mechanisms can be used to provide continued access for traditional uses of a specific area. Stock-use rights in fishing (SURFs) are similar in concept, establishing exclusive use rights to a fish stock or combination of stocks (Townsend, 1995).

TURFs or SURFs may be held by individuals, cooperatives, corporations, communities, or other organizations (Christy, 1996). The conservation effect of this type of management depends critically on the migration rate of the fish, the number of people with a right of access to an area, and their beliefs about what

---

[5] Superexclusive area registration is a management tool used by the Alaska Department of Fish and Game for some (small) Bering Sea crab fisheries. Participation is open to any vessel, provided that the vessel agrees not to participate in any other crab fishery. Because this tool is applied to crab fisheries with low guideline harvest levels and because participation in the fishery precludes participation in any other Bering Sea crab fisheries, few vessels choose to participate. Most of the vessels that do choose to participate are small and local to the fishery.

must be done to sustain fisheries. For most fish stocks, the TURF would have to cover an extensive area to provide an effective control of an entire stock, or fishing for a particular stock would have to be prohibited outside the territory. TURFs could be useful, however, to minimize spatial conflict among fishermen, for example, those who use different kinds of fishing gear. Fishing areas requiring or prohibiting certain gear types are used in some regions. Exclusive harvest rights to fisheries in geographically distinct regions have been used in many areas to eliminate the open-access problem.

ALTERNATIVE MANAGEMENT PROCESSES

Co-management processes and their more specific forms are a useful element of the discussion of IFQs and their alternatives because they represent collective approaches to decisionmaking that bring together users, regulators, and other stakeholders (Wilen, 1985). As such, they offer the potential to address social and equity goals in the context of IFQ management.

Co-management

Co-management is joint management between resource users and government. It is characterized by two important properties: the sharing of decisionmaking power and a focus on the management process. It is a process, rather than a tool, of management and thus can be used with a variety of management tools. Co-management encompasses different degrees of power sharing between stakeholders and government, from formal power sharing (Jentoft, 1989; Pinkerton, 1989) to "active consultation" (Hanna, 1995; Jentoft and McCay, 1995). The co-management process defines stakeholders and incorporates them, through various forms of representation, into the fishery management process (Costanza et al., 1998). In this sense, the council process under the MSFCMA is a limited form of co-management in which resource users participate in allocation decisions and are appraised of resource conservation and stewardship actions pursuant to the public trust in the fishery resources. Fisheries managed through traditional techniques such as TACs and licenses could involve participants in co-management, as could fisheries managed using IFQs (Box 4.5). When co-management succeeds, it confers a legitimacy on regulations that may increase compliance and reduce monitoring and enforcement costs (Jentoft, 1989; Sen and Nielsen, 1996). However, co-management is often a costly process that requires good relational skills among participants (Hanna, 1997) and is vulnerable to stakeholder fragmentation and incomplete representation. Also, like other management processes, it is vulnerable to sabotage by special interests (Hanna, 1994). Co-management processes have potential for administration of an IFQ program after it is in place (Box 4.6).

BOX 4.5
ITQs in the Scotia-Fundy Small Dragger Fleet of Canada

The ITQ management regime for the Scotia-Fundy small dragger fishery is a case in which ITQs might not seem feasible but have turned out to be. This case also highlights the importance of a user-supported, major investment in monitoring and of cooperation between government and industry in decisionmaking (McCay et al., 1995; Apostle et al., 1997).

The fishery extends along the coast of Nova Scotia, Canada, from the Cabot Strait off Cape Breton to the Bay of Fundy, extending to the border with U.S. waters in the south at a line that bisects Georges Bank. Historically, cod (*Gadus morhua*), haddock (*Melanogrammus aeglefinus*), pollock (*Pollachius virens*), and flounder (e.g., *Pleuronectes ferrugineus*) have been the most valued species. The Canadian Department of Fisheries and Oceans (DFO) uses total allowable catches for individual species, which are allocated according to gear type, vessel size, and management area. Vessels range from 30-foot gillnetters to 100-foot catcher-processor vessels. DFO created individual quotas (IQs) in 1991 for the small draggers, most between 40 and 65 feet and using otter trawls.

The story is a familiar one of overcapacity creating management difficulties. After Canada established its 200-nautical-mile exclusive economic zone in 1977, this inshore small dragger fleet grew dramatically in fishing power, stimulated by government loans and grants and rising prices for fish. By 1989, its capacity was four times that required to harvest the resource at a conservative level ($F_{0.1}$; Burke et al., 1994). Many measures were used to control capacity but none was successful, and the annual management plans became increasingly complex and contentious as this fleet demanded more fish. When negotiations over quota allocations broke down in 1989 and the fishery was closed in midyear, a task force recommended using ITQs as a way to reduce capacity, although some features of the small dragger fleet would not be considered appropriate for an ITQ program. There were 455 licensed vessels, which landed at more than 75 ports and had access to hundreds of others. It was a multispecies and multistock fishery, and other fleets also had shares of the TACs in these stocks as well. The small dragger fleet was known to be uncooperative, independent, and not interested in ITQs.

Nonetheless, within a year and a half, the ITQ program was in place. Among the factors making this possible were determination on the part of DFO, backed by political commitment; increased awareness of how serious the overcapacity problem was when the fishery had to be closed down midyear (Burke et al., 1994); and DFO's use of an industry-government committee for planning and co-managing the system (McCay et al., 1995; Apostle et al., 1997). Also important to its political acceptability were design features intended to preserve the owner-operator, community-oriented nature of the fishery—for example, requirements that holders of ITQs be bona fide, active fishermen; limits on how much of the overall quota could be held by any one person; and limits on transferability.

Because Canada had earlier negative experiences with ITQ programs in this region, unusually careful attention was given at the start to monitoring and enforcement issues, as well as the administrative penalty system. Consequently, when it

(Continues)

BOX 4.5
Continued

began in 1991, with the 327 license holders who elected to participate, a new third-party dockside monitoring program (DMP) was in place, with local weighmasters and centralized operation centers and with the understanding that this would become self-funding over time. The fishing industry was directly involved in setting up and evaluating the DMP. The goal was 100% monitoring of landings. By 1993, the DMP was fully funded by industry, and there was consensus that landings data collection had improved, although problems with at-sea discards, highgrading, and misreporting remained (Angel et al., 1994).

Evaluations of the small dragger ITQ management regime are complicated by drastic declines in TACs because of badly depleted groundfish stocks. Both ITQs and resource declines have resulted in concentration of landings to fewer vessels, ports, and fish buyers (Burke et al., 1994; O'Boyle et al., 1994). Price changes suggest improved fish quality in the ITQ fleet. There is some evidence of reduced administrative costs (Burke et al., 1994). Community studies show greater social stratification due to these changes (McCay et al., 1998). Despite rules against processor ownership of quota and vessels, major processors quickly gained indirect control by financing quota and vessel acquisition. Rules against permanent transfer were quickly removed, enabling a faster pace of consolidation than originally intended. ITQ landings also have shifted within the region, benefiting some communities at the expense of others. The design features intended to help maintain the owner-operated, community-oriented nature of the fishery have not been able to prevail against market-based incentives for accumulation and concentration (McCay et al., 1998).

BOX 4.6
Fisheries Co-management in The Netherlands

In 1993, a co-management scheme was introduced in The Netherlands under which groups of fishermen manage their ITQs. The purpose of establishing co-management was to improve cooperation with the industry and lower the cost of management by making the industry more responsible for its actions. Membership in the management groups is voluntary, but both incentives and disincentives were applied when constructing them. Group members were promised a greater freedom of transferability, and all fishermen were threatened collectively with the loss of a certain portion of the licenses every year if the group formation did not succeed, so that those who wanted to continue their operations as before would have to buy back a part of their license. The critical success rate was set at 75% of the vessels joining a group, which was attained. Each group has the responsibility of managing the ITQs of its members, by setting a limit on the number of days at sea to ensure that the catch stays within the total of the group members' quotas. It is also the responsibility of the manager of the group to ensure that its members do not exceed their total quota allocation. The ITQs are still owned individually, however.

A specific form of co-management with potential application for certain types of fisheries is "contractual co-management." The Alaska CDQ program has been viewed as having the potential of evolving into "contractual shared governance" (Townsend and Pooley, 1995). Contractual co-management, also termed "corporate management" (Townsend, 1997), is based on shared ownership between government and a fishing community or group. A set of rights and obligations is delegated by the government for a specified period to a local fisheries management organization. The government plays a major role in determining the terms of the contract, but during the contractual period the participating organization acts as a "sole owner," taking responsibility for management. Under this arrangement the pool of potential shareholders is larger than those fishing, which distributes the benefits of fishery participation to a wider group of stakeholders (Rieser, 1997b). Used in conjunction with an IFQ program, this arrangement has the potential to address some of the distributional consequences of initial allocations. Alternatively, it may be designed to keep fishing quotas in the ownership of groups rather than individuals.

5 | Considerations for a National Policy on Individual Fishing Quotas

This chapter addresses the specific issues raised by Congress in the Sustainable Fisheries Act of 1996 (Sec. 108[f]). These important issues are considered within the context of *first principles*, which are used to guide the analysis of the specific questions about individual fishing quota (IFQ) design and implementation posed by Congress and serve as the basis for the committee's recommendations. This chapter presents a variety of viewpoints, from which the committee has drawn the findings and recommendations discussed in the following chapter. From this analysis, a national policy on IFQs may be derived.

FIRST PRINCIPLES OF FISHERIES MANAGEMENT AS A POLICY FRAMEWORK

A brief look at IFQ programs that have been implemented around the world reveals that these programs were designed for a spectrum of purposes because nations do not all share the same policy goals for their fishery resources. In New Zealand, for example, the move to institute IFQs on a broad scale was motivated by the desire of the national government to foster rapid development of a domestic fishing industry that could extract maximum benefits from the national resource. In Iceland, an IFQ program was set up to serve a variety of goals: biologic, economic, social, and administrative.

What principles are at work in the United States that can help answer whether the IFQ is an appropriate policy instrument in the U.S. economic setting and help shape the direction of U.S. IFQ programs? A set of foundational principles for U.S. fisheries management is embodied in the Magnuson-Stevens Act; therefore, all discussions of future management logically begins with this act, supplemented by relevant international principles and agreements. The act's dominant purposes provide the reference points by which most, if not all, of the issues concerning IFQs—if, when, where, and how they should be used—can be answered. Since the act will undoubtedly be amended repeatedly in the future, however, recommendations about management should not be limited to its present content.

Over the course of the Magnuson-Stevens Act's history, its dominant purpose and policy goals have changed. The history of these changes is recounted briefly in Chapter 1. Also, the legal responsibilities of the United States as a coastal nation with respect to the biological resources within its exclusive economic zone (EEZ) have been clarified recently through the articulation of international norms and management criteria (Rieser, 1997a). The dominant purposes of U.S. fisheries law and policy may be gleaned from these obligations and from the far-reaching changes to the Magnuson-Stevens Act brought about by the 1996 amendments. The salient features of the 1996 amendments are the following:

1. The express duty to end overfishing and to rebuild overfished stocks (Secs. 303[a][10], 304[e]);
2. A new ecological imperative, with its emphasis on protecting essential fish habitat and reducing bycatch (Secs. 301[a][7], 305[b]);
3. The mandate to consider fishing communities (Secs. 301[a][8], 303[a][9]);
4. A concern with the fair and equitable allocation and distribution of fishing quotas (Secs. 303[d][5][C]); and
5. The cautious approach to market mechanisms evidenced by the moratorium on development and implementation of new IFQ programs (Sec. 303[d]).

In these provisions, the act supports the notion that fish stocks are public resources to be conserved for future generations and for their ecological importance, and managed carefully for the benefit of local communities and the diverse array of citizens who derive benefit from them. From these features, the first principles of U.S. fisheries law appear to be as follows:

1. Conservation and sustainability of biological resources have a high priority.
2. Management programs must take careful account of the social context of fisheries, especially the role of communities and the importance of fishing as both a tradition and a profession.
3. When harvests take place, they should maximize the net benefits (benefits minus cost) that society receives from their use.
4. Management programs must consider equity, fairness, and the distribution of the benefits derived from marine resources.

Other principles stem from the public trust nature of fishery resources. For example, the public trust doctrine prohibits the permanent alienation of use rights to fisheries; access must be balanced with conservation and the public should be compensated for any private, exclusive use of public trust resources.

APPLYING THE FRAMEWORK TO QUESTIONS ABOUT INDIVIDUAL FISHING QUOTAS

With these principles in mind, it is possible now to turn to those questions that Congress posed as forming the basis for a national policy on the use of IFQs. This chapter presents the background information relating to specific issues referred to the committee by Congress and for which recommendations are presented in the following chapter.

Criteria for Determining Appropriate Fisheries for Application of Individual Fishing Quotas

Congress asked the National Academy of Sciences (NAS) to identify, if possible, "threshold criteria for determining whether a fishery may be considered for IFQ management, including criteria related to the geographic range, population dynamics and condition of a fish stock, the socioeconomic characteristics of a fishery (including participants' involvement in multiple fisheries in the region), and participation by commercial, charter, and recreational fishing sectors in the fishery" (Sec. 108[f][1][G]). This query implies the question of whether IFQ programs should be limited to fisheries in which overcapitalization exists or should be made available for application to all fisheries.

Two approaches to the development of threshold criteria are possible. The first is to be proactive and say that all fisheries are potential IFQ fisheries if there is some reasonable expectation that a total allowable catch (TAC) could be specified upon which IFQs could be based (perhaps eliminating fisheries with no TAC, such as some shrimp and crab fisheries[1]). The "threshold criteria" would then evolve more into conditions governing the rational development of IFQ programs (e.g., adequate data, clear objectives, full participation).

The second (reactive) approach, is that only fisheries that meet certain, presumably more restrictive, criteria are appropriate for IFQs. For example, these criteria could include the following:

---

[1] In some fisheries, for example Pacific salmon and some species of shrimp and crab, annual harvests are sustained almost entirely by animals from single year classes and there is substantial variability from year to year in year-class strength that cannot be predicted reliably on the basis of stock size. Because future recruitment is largely unaffected by current harvests, these stocks are not usually managed using TACs. However, fisheries with no TAC could still be amenable to a market-based system with transferable units of effort, rather than catch.

- A TAC can be specified, upon which IFQs could be based.
- An excess of capital, participation, or effort with respect to the available TAC is present or likely to develop.
- Administrative or enforcement difficulties are present that could be remedied by an IFQ program.
- The fishery is in, or appears to be headed into, a "derby" situation.
- Bycatch either is insignificant or can be managed.
- Managers and stakeholders agree that market mechanisms should be relied on to allocate use privileges.
- The goals of improving economic efficiency and reducing the number of firms, vessels, and people in the fishery have higher priority than other goals.

The committee favors the first approach, that all fisheries that can be managed using a TAC are potential candidates for IFQs.

Process and Criteria for the Initial Allocation and Qualifications for Holding Individual Fishing Quotas

Deciding who should receive quota and how much quota they should receive is a difficult, highly political process as participants in a fishery attempt to ensure their continued participation. The controversy about initial allocations results from at least three factors: (1) the "windfall profit" to initial recipients, (2) the choice of criteria for allocation, and (3) the amount of quota received.

1. *Windfall profit*—The initial value of quota shares depends on rational expectations about potential future profits; quota shares increase in value to the extent that an IFQ program improves the profitability of the industry. If the fishery is overexploited or if there is great pessimism about future prospects for the fishery at the time of program implementation, quota share value could be zero. Any value provides the initial recipients with capital they may be able to leverage for additional purchases of quota shares. The recipients of initial allocations of quota shares reap a windfall profit when they sell their shares, which is not available to subsequent holders of the quota shares who must purchase them. The committee knows of no cases in which initial recipients of IFQs have been charged for their quotas. An auction-based allocation could be expected to capture any expected windfall gain.

2. *Criteria used to determine quota allocations*—Dozens of different criteria can be used, each one more or less appropriate and fair, depending on the goals of the IFQ program. The choice of criteria differentially affects fishermen. For example, if the harvest of the past two years is heavily weighted, newer entrants to the fishery benefit, but if the average harvest of the past ten years if used, long-term fishermen, or those who fished in the early part of the time period and then left the fishery, benefit. The particular years used to determine histori-

cal participation and eligibility for IFQs can have profound social and distributional effects, advantaging some groups and disadvantaging others.

3. *Actual amount of quota received by each fisherman*—IFQ programs may reduce the TAC initially, and in an attempt to include a broad class of stakeholders, quota is typically distributed to more individuals than are currently active participants in the fishery and recipients are allowed to use their best catch years as a basis for the initial allocation. This makes the total quota share pool greater than the actual harvest in the target years. Due to these factors, some fishermen who once were able to harvest sufficient fish to remain viable may receive too little quota to continue. Other fishermen may receive no quota at all because their participation did not coincide with the qualifying years. Still others may receive quota even though they are no longer active in the fishery.

The allocation *process* includes the procedures or mechanisms used to allocate IFQs, such as a program that distributes IFQs, an auction, or a lottery. Allocation *criteria* are those conditions or characteristics that individuals must meet in order to participate in the process used to allocate IFQs and to be eligible to purchase quota shares subsequently. There are numerous ways of allocating quota because there are many different combinations of procedures and criteria.

Evaluating Categories of Allocation Processes

The process of initial allocation is probably the most contentious issue in IFQ management. There are basically four different ways to allocate scarce resources:

1. The open-access approach;
2. A rule of equal opportunity—through a lottery, a first-come-first-served principle, or a same-for-everyone allocation (Edney, 1981; Fiske, 1991);
3. The political approach (Edney, 1981) or priority ranking (Fiske, 1991), similar to the triage approach in allocating scarce medical care; and
4. The market device—the scarce resource is distributed to those who are willing to pay the most for it.

All four categories of mechanisms have been used in fisheries to allocate valued things, whether licenses, quota, or prime fishing spots. Each type of mechanism should be evaluated in terms of its ability to provide economic returns to the public and its compatibility with the distribution of IFQs and other limited access permits.

Competitive or Market Mechanisms. This mode is politically attractive because once it is chosen, subsequent choices are decentralized and seem politically neutral (Edney, 1981). Responsibility for negative outcomes resides in an impersonal force, the market, as well as in the actions of individual actors in the market. This mode involves bargaining and negotiated contracts (Fiske, 1991).

A common competitive mechanism for making an initial allocation of a good is the auction. Auctions match buyers and sellers of a good and can be structured in a wide variety of ways. A standard auction involves a seller receiving a number of bids from several buyers. A double auction involves the public posting of bids and offers from multiple buyers and sellers, similar to a securities market. In a first-price auction, which the U.S. Forest Service uses to sell timber, the person making the highest sealed bid purchases the timber for a price equal to his or her bid.

Auctions promote the economically efficient use of resources by allocating goods to their highest-valued uses. To realize the full value of the good, the person who "won" the auction must deploy his or her resources in the most profitable manner possible. For instance, a fisherman who was the highest bidder for a quantity of quota must organize his harvesting activities in the least-cost, or most efficient, manner possible, in order to recover the cost of the quota and earn a profit.

Although auctions may promote the efficient use of resources, they also raise a number of fairness issues. Only those individuals with adequate finances can participate, potentially excluding many people who have historically had access to fishery resources. Furthermore, depending on who is allowed to participate, an auction need not recognize the importance of fishing as both a tradition and profession. It may be possible, however, to address both of these fairness issues while still gaining the efficiency of an auction. People can be given the financial ability to participate in auctions through public loans or other financing mechanisms. Also, eligibility criteria for participating in an auction can be established that allow only those who have historically participated in a fishery to purchase quota in the fishery. Furthermore, auctions can be used as a mechanism by which to decrease windfall profits to initial recipients, allowing the public to be compensated for the private use of a public resource (see discussion of economic returns to the public later in this chapter).

Auctions may also promote resource conservation and biological sustainability in at least two ways. First, by requiring individuals to invest in fisheries by earning or purchasing quota, individuals may be more likely to care for the resource so as to protect their investment. This effect will occur only if the quota is auctioned off only once or at least for a long term. Second, some of the revenue collected by the government through an auction can be invested in the resource to support and enhance its productivity and to mitigate problems caused by human use.

Finally, auctions need not be used only to allocate initial quota shares. If IFQs are transferable, periodic auctions can be used to establish prices for quota, even if they are not used to raise revenue (Box 5.1). Auctions have not been used to allocate quota or to establish quota prices in any IFQ fishery in the United States.

BOX 5.1
The Zero-Revenue Auction

Although auctions have many desirable allocation characteristics, their use has been limited by the reluctance of resource users to pay for the resources they use. One mechanism that has successfully circumvented this barrier is the zero-revenue auction. It is currently used in the Clean Air Act's Acid Rain Program to control sulfur emissions.

After an initial allocation of IFQ shares based on historic catch or some other criterion, under a zero-revenue auction the government would take back some proportion of the allocation each year (approximately 3% in the sulfur allowance program) for sale in an auction. Quota holders are allowed to buy back the quota they put up to bid, but they will succeed only if they are the highest bidder. Revenue is returned to the holders of the auctioned quota shares. In principle, the auction could involve either quota shares (e.g., 0.5% of TAC) or annual quota (e.g., pounds of fish in 1999). Significantly, all components of auction transactions (e.g., price, identification of buyers, quantities transacted) are public information. Privately arranged transfers could also take place any time among eligible participants as long as the control authority was notified and the transfer was approved.

A zero-revenue auction could improve an IFQ program in several ways. First, it provides excellent information about prices, which is helpful not only to fishermen in planning their investments, but also to bankers as they seek to value this uncommon form of collateral. Public information about prices also serves to facilitate private trades outside the auction. Second, the zero-revenue auction guarantees the steady flow of IFQs in the market, ensuring that potential entrants are not precluded from fishing.

Random or Equal Opportunity Mechanisms. A good example of such mechanisms is provided by the people of Bikini, in the Marshall Islands, when they were removed to Rongerik after the U.S. hydrogen bomb tests. Food resources proved to be inadequate on Rongerik. Under their chief, the people pooled their harvests, and the chief divided the pool into equal shares per person.

A common random mechanism that may be used for the initial allocation of quota is the lottery. Lotteries may be structured in a variety of ways, but their essential feature is randomness. A person is randomly selected from a pool of participants to receive a good. Lotteries may be particularly useful when the demand for a good is greater than the supply and it is not desirable to allocate the scarce good on the basis of price.

Lotteries, unlike auctions, do not allocate quota to their highest-valued uses. Rather, lotteries promote equality (but not necessarily fairness) by treating all participants alike. Each participant is equally likely to receive valued goods. Lotteries may be an appropriate mechanism for allocating IFQs to new entrants. If there are more entrants than quota available, and if removing price barriers to entry is an important consideration, lotteries can be a fair allocation mechanism.

Although randomly allocating quota among a pool of eligible participants may at first appear to be unusual, random mechanisms can be used to achieve a variety of ecological and social goals in an equitable manner. For instance, in order to maintain the viability and health of fish stocks, the time and place of harvest may be important. A means of limiting the concentration of fishing effort to particular places and instead spreading it across a fish stock is to denominate quota as a proportion of a TAC and by area of harvest, as is done in the Alaskan halibut fishery. If some areas of harvest are less desirable than others, area quotas may be allocated by lottery, ensuring that fishermen equitably share both the desirable and less desirable areas.

Lotteries could be combined with other allocation rules. For example, considerable concern was expressed that the initial allocation rules for Alaskan halibut and sablefish IFQs were so liberal in recognizing past participation that ensuing quota shares were often too small to fish profitably. An alternative approach would have been to use the eligibility criteria to qualify individuals to participate in the lottery and then randomly award "fishable" quota shares to a subset of those deemed eligible.

Lotteries may be used to promote equality and to address different social and ecological issues, but lotteries do not promote economic efficiency. Other mechanisms, such as transferability of quota, would have to be combined with lotteries to promote efficiency. Lotteries have been used to award limited licenses in developing fisheries in Canada, such as the Newfoundland snow crab fishery. Such cases show the capacity for an initially fair allocation to result in later perceptions of inequity, unless the licenses are transferable or subject to periodic redistribution through additional lotteries or auctions (McCay, 1999).

Procedural or Priority-Ranking Mechanisms. Procedural mechanisms typically involve the development of a set of well-defined allocation criteria and subsequent allocation of valued goods based on these criteria. For instance, priority systems that allocate organs for transplant, or that allocate scarce public resources such as housing, are most familiar (Young, 1994). Allocation of fishing opportunities by the regional councils through various management systems provide other examples of procedural allocation mechanisms. A decision is made to allocate access to resources based on a social criterion. In the Polynesian community of Tikopia (Firth, 1959), people were ranked against one another on the basis of birth order among siblings and within ancestral sibling sets. When a devastating typhoon hit Tikopia, this system of ranking was used to allocate responsibility to take inventory of resources, determine the critical number who could be supported with those resources, and determine that those ranking below this number would go into permanent exile in their canoes.

Given that a group of people is devising a set of criteria that will be used to allocate valued goods, fairness issues, in terms of appropriate representation and participation in the group, are immediately obvious. The interests and values of the individuals participating will strongly influence the set of criteria adopted.

Neglecting particular interests or overrepresenting other interests may weaken the decisionmaking authority of the group, as the neglected individuals oppose the group's decisions or the legitimacy of the group's decisions is questioned if it is dominated by particular interests.

These fairness concerns may be addressed in a number of different ways. The initial design and structure of the group can require representation of all interests directly affected by the decisions of the group. In addition, oversight mechanisms can be adopted that allow for the review of group decisions and perhaps for over-turning decisions that appear to be egregiously unfair or exploitative.

Another issue that often arises is the ability of the group to make decisions in a timely manner in the presence of conflict among its members. This issue may become particularly acute if the decisions that members are asked to make directly affect them and how they achieve their livelihoods, as is the case with some members of the regional fishery management councils. In settings in which members of a group are negotiating solutions to shared problems among themselves and the preferred form of decisionmaking is consensus, win-lose situations are particularly difficult. This is especially likely to arise in designing quota systems. Allocating quota among different gear types, or among different participants such as vessel owners, skippers, and crew members, sets up win-lose situations. Quota allocated to one gear type is not available to another, quota allocated to skippers is not available to vessel owners, and so on. Conflict can become intense, and reaching a decision may be very difficult (Hanna, 1995). Oftentimes, decisions, if they occur, are based on exhaustion and not on reasoned debate.

These decisionmaking issues may be addressed in a number of different ways. For example, fisheries may be disaggregated into more homogeneous subunits that are allowed to develop their own initial allocation formulas (and other program specifications), as in the case of the Pacific Whiting Conservation Cooperative (see Box 4.4). This would mean, for example, allocating portions of the TAC to relatively homogeneous areas, fishing sectors, or communities. This, of course, raises the question of who would allocate portions of the TAC. Another mechanism, used by Congress and state legislatures when faced with particularly difficult and contentious win-lose decisions, is the independent commission. A commission makes recommendations to the decisionmaking body, which then either rejects or accepts the commission's recommendations. This mechanism was used by Congress to close military bases and by the Arizona State legislature to make an initial allocation of groundwater rights. Relying on an independent, external commission to devise alternative allocation schemes removes much of the conflict from within the decisionmaking body and among its members, while still allowing it to make the final decision.

Actual systems of allocation are usually mixtures of the various modes, or movements from one to another, as is very clear for IFQs. How could other modes of allocation apply to an IFQ program? The rule of equal opportunity would suggest giving the initial allocation of a TAC in equal shares to eligible parties. The problem is that with scarce and declining resources, the shares may be insufficient, leading to pressures to buy and sell them, which may result in sharp inequality of access over time unless other resources are available. The third mode of allocation, priority ranking, is found in decisions to restrict initial allocation of IFQs to certain classes of people (e.g., vessel owners; those who fished during a certain period of time and/or caught a certain amount of fish), as well as in rules of transfer and accumulation that favor some groups over others. Reliance on this allocation mechanism has led to the concern that those who are well connected to the council process may be unduly advantaged. After the initial allocation (which may be made on other grounds), the market is intended to provide the incentives and signals to direct economically appropriate individual behavior. Allocating quota shares on the basis of catch history and allowing individuals to buy and sell quota shares make it possible for more viable firms to compensate less viable firms before the latter leave the industry.

To date, federal IFQ programs have been limited in the allocation mechanisms used. Only the council process, a procedural mechanism, has been used. The council process is the central and necessary mechanism because, at the very least, it defines eligibility criteria, but it need not be the only allocation mechanism used. It can be combined with either auctions or lotteries to allocate and reallocate quota. Indeed, a combination of all three mechanisms might best allow IFQ programs to meet the various goals that the Congress (through the Magnuson-Stevens Act) and councils deem important to fisheries management.

In relation to IFQ programs, each of the allocation mechanisms—competitive, random, and procedural—can meet, to a greater or lesser extent, each of the first principles of fisheries management, discussed at the beginning of this chapter, depending on the design of the IFQ program and the allocation criteria used. These mechanisms, however, do not specify who will receive how much and why. Criteria have to be established to determine the eligibility of individuals to participate in one of the allocation mechanisms. Furthermore, criteria will be needed to determine how much of the valued goods should be allocated to each eligible recipient.

These criteria depend, in part, on the goals of the IFQ program and, in part, on the first principles. All federal IFQ programs have used two criteria for allocating quota: (1) historic participation in the fishery (catch history) and (2) vessel ownership. These criteria not only address the importance of fishing as a tradition and a profession, but also recognize the magnitude of investments already made in the fishery. For the other principles to be addressed, IFQ programs and the initial allocation of quota must take into account additional criteria. For instance, objectives related to the conservation and sustainability of biological resources could be reflected in a number of different criteria:

- Individuals who have a history of low bycatch (if this can be documented) could be eligible to purchase additional quota in an auction, allocated more chances in a lottery, or granted additional quota through a procedural allocation process.
- Individuals may be required to use relatively selective gear in order to participate in any one of the allocation mechanisms.
- Individuals need not intend to harvest their quota in order to be eligible to receive it (e.g., conservation groups could be eligible to purchase quota).

Such conservation-oriented criteria would have to avoid penalizing individuals unfairly, based on their use of legal gear and methods.

Objectives related to equity and the social distribution of the benefits derived from marine resources can be met by expanding the criteria for individuals who are eligible to participate in initial allocation mechanisms, rather than only vessel owners (see discussion of allocation to specific participants in the following section). Thus far, federal IFQ programs have been relatively limited both in the allocation mechanisms and in the criteria used to allocate quota initially. New quota programs should consider a more varied mixture of criteria and allocation mechanisms.

Framework for Devising An Allocation System

Given the many diverse fisheries that exist in the United States, it would be impossible to devise a single allocation system appropriate for all settings. Rather, the following questions should be considered when devising an allocation system (see Young, 1994, pp. 164-167).

- What are the eligibility criteria? Who is entitled to receive a share? These are most appropriately determined by the diverse set of stakeholders involved in the fishery. Furthermore, they must be related to the first principles and national standards of the Magnuson-Stevens Act. Eligibility criteria may be divided into several general categories, for example, residency, mode of production (which includes gear and fishing practices), time period of involvement in the fishery, and investment in the fishery (which includes catch levels; ownership of vessels, gear, or processing; employment in the fishery or a fishing-related industry).
- What counts in the distribution? Two people can be eligible to receive a right or bear a responsibility, yet differ in the amount of the right or responsibility they deserve. What is the basis for distribution? What are the relevant principles for making the distribution among all eligible participants? Such principles include (1) proportionality, (2) need, (3) ability, and (4) other factors.

- What are the relevant precedents? The appropriateness of a distributive principle depends on how customary its use is in a particular setting.
- How should competing principles and criteria be reconciled?

An individual transferable quota (ITQ) program does not have to preserve the same distribution of quota shares achieved by the initial allocation if such a distribution later becomes inconsistent with the long-run objectives of the fishery. Modifications can be made over time in such a way as to reward exemplary behavior and punish destructive behavior. The Australian "drop-through" system (see Box 5.2 and Figure 5.1) offers possibilities for subsequent reallocations of permits by taking into account such historical information as bycatch experience, compliance history, and willingness to adopt selective gear.

BOX 5.2
The Australian Drop-Through System

One approach that attempts to address the inevitable tension between the need for administrative flexibility in managing an IFQ program and the need to provide sufficient security to IFQ holders so they can make efficient investments in the fishery involves a cascade of fixed-term entitlements. One variation of this approach has been proposed for the New South Wales fishery (Young, 1995). Under this scheme, initial entitlements of quota share (call them Series A entitlements) would be defined for a finite period, but one long enough to encourage investments (e.g., 30 years; see Figure 5.1). Periodically (e.g., every 10 years), a comprehensive review would be undertaken that would result in a new set of entitlements (Series B), each of which would also have a 30-year duration. These entitlements would confer a similar, but not necessarily identical, set of rights and obligations. Fishermen holding Series A entitlements could have the option to switch to the new set of entitlements at any time earlier than the expiration of their Series A entitlements. Once they switched, they would be able to hold Series B entitlements for the remaining life of the entitlement. This process would continue until it appeared that no more modifications were necessary.

Although many changes in an IFQ program could be directly implemented by changing the governing regulations, in some cases it might enhance political feasibility to introduce changes of a more fundamental nature over a transition period. The definition of each new series offers the control authority the opportunity to phase in new requirements while attempting to provide adequate, if not complete, security for the holder's quota share. It also offers the possibility of combining new obligations with new privileges in such a way as to enhance the political feasibility of the changes. Finally, it provides an opportunity for offering positive incentives to individuals who engage in exemplary fishing (e.g., unusually low bycatch rates) or negative incentives to those who have compliance rates that are low, but not so low as to trigger quota seizure provisions.

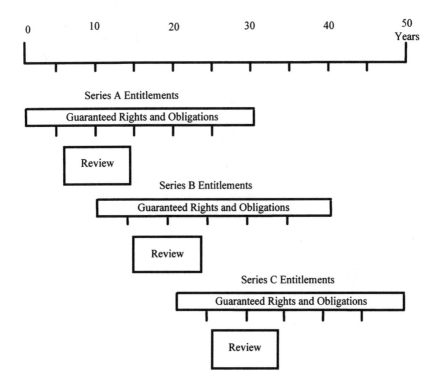

FIGURE 5.1   Australian "drop-through" permit design. SOURCE: Based on Young and McCay (1995), Figure 7-1. Used with permission from the World Bank.

Allocation To Specific Participants

In the short history of IFQs, vessel owners are almost always awarded the initial allocation. This seems to reflect a bias toward capital ownership. Although the practice has been justified in the United States by reference to the lack of adequate fishing records for nonowning captains and crew members, the practice is also found in Iceland where crew participation is fully documented.

Crew Members and Skippers. The failure to include fishing vessel skippers and crew members in the initial allocation of quota shares in the Alaskan sablefish and halibut IFQ programs is perhaps their single most controversial element (see *Alliance Against IFQs v. Brown*[2]). The committee was presented written and oral testimony, particularly related to North Pacific fisheries, indicating that nonowning skippers and crew members believe they should have been allocated

---

[2] 84 F.3d 343.

quota. The co-venturer status of most U.S. fishing crews suggests a strong rationale for considering hired skippers and crew members in the initial allocation. If this is not feasible because of inadequate records,[3] special measures can be designed to compensate hired skippers and crew members for being excluded or to help them enter the market for IFQs if they desire (e.g., the North Pacific loan program).[4] However, if crew members had been included, the initial allocation would have been even more diffuse and subject to increased criticism from vessel owners for allocating overly small quota shares.

Processors. The issue of allocating quota to processors arose in Alaska in the context of long-standing conflict between inshore and offshore processors and in response to the provisions of the Alaskan IFQ programs that make it impossible for processors to hold quota and change their relative bargaining power in negotiating exvessel prices of the landed fish. Some other programs (e.g., for surf clams/ocean quahogs) do not contain such provisions and processors can buy and hold quota.

There is a history under the Magnuson-Stevens Act of government policy favoring U.S. processors over foreign processing vessels. The act defines "U.S. processors" as "facilities located within the United States for, and vessels of the United States used or equipped for, the processing of fish for commercial use or consumption" (Sec. [3][41]). The term is used in the provisions by which U.S. processors are given a preference in access to U.S.-harvested fish over foreign processing vessels (Sec. 204[b][6][B]). Whether processors (1) should be accorded an individual processing quota, (2) should be among those receiving an initial allocation of harvesting quota, or (3) should be eligible to purchase harvesting quota raises different sets of policy considerations.

It can be argued that fish processors are an integral part of the fishing industry. Without processing, much of the fish being brought ashore would never make it to market and thus be useless to catch, but does this entitle processors to be considered for receiving or holding harvesting or processing quotas? The role of harvesting quotas is to ration the use of a limited natural resource, the fish stocks, so as to avoid economic waste. On the processing side there is no such limitation. Unlike fish stocks, fish processing facilities can be replicated to any extent needed, just as in any other manufacturing industry. Some places or areas may have a geographical advantage or disadvantage for fish processing, but this is not likely to be a seriously limiting factor. Some fish processing is most

---

[3] Much of the concern about "inadequate records" was due to an inability to connect landings volume to individual crew members and skippers. A similar problem in the Gulf of Mexico red snapper fishery led to the proposal that the initial allocation be based, in part, on equal allocations to all licensed participants.

[4] Loan programs that focus on providing funding for the purchase of quota shares could make it easier for new individuals to enter a fishery. However, such programs could also inflate the price of quota shares by driving up the demand for them.

effectively done on board factory vessels at sea and is thus an example of a "footloose" industry that can be located anywhere.

Nevertheless, there are arguments for allowing processors either to acquire harvesting quotas, or to be considered for initial allocation of harvesting or processing quotas. Each of these options is considered in turn.

Arguments for Allowing Processors to Hold Quotas. The arguments for allowing processors to acquire and hold harvesting quotas derive from (1) the possible advantages of vertical integration in the fishing industry and (2) the changes in bargaining power resulting from harvesting-only quotas. It may be advantageous to have firms that integrate the entire process from catching the fish to distributing the fish products to retailers integrated in a firm (the industry itself could nevertheless consist of many such firms, to avoid problems of monopoly). A vertically integrated firm might be better able to plan its operations and be more economically efficient than a processing firm having to negotiate with independent vessel owners about prices and delivery conditions. In fact, the committee was presented anecdotal evidence that IFQs may particularly disadvantage processing firms that are not vertically integrated. If the ownership of harvesting quotas is restricted to vessel owners or bona fide fishermen, vertical integration will not be possible, unless processing firms are recognized as vessel owners in their own right. In Iceland, the ownership of IFQs is restricted to vessel owners, but vertically integrated firms are recognized as vessel owners and such firms have been increasing their quota holdings in recent years.

It is possible that allowing processing firms to own IFQs would separate the ownership of quotas from vessel owners and bona fide fishermen, if a quota holder is not required to own a vessel. Under this scenario, vessel owners could become contractors to fish quotas held by processing firms; the firms would contract for fishing their quotas at times and places that would suit them best. This has already happened in the surf clam/ocean quahog (SCOQ) industry and there are tendencies of a similar kind in Iceland (see Appendix G).

Arguments for Initial Allocation of Harvesting or Processing Quota to Processors. The arguments for allocating harvesting or processing quota to processors derive from a desire to compensate those who will have to leave the industry because of excess processing capacity. By allocating harvesting quotas to vessel owners and crew members and allowing quotas to be bought and sold, some vessel owners and crew members can be bought out of the industry by those who are willing to pay a high enough price for their quota shares. This amounts to an industry-financed compensation to owners and crew members associated with redundant fishing vessels (in place of government-sponsored buybacks). However, the harvesting side of the industry is not the only one that may be characterized by redundant production capital; this may also happen on the processing side. This is particularly likely in fisheries characterized by short seasons. The processing sector may be structured to handle large amounts of fish in short periods of time. Not only is the harvesting process restricted to a shorter time period than necessary, the same is true of the processing sector. If an IFQ-managed fishery results in a longer fishing season, some processing capacity will become redundant. Over time, the redundant processing capital will be forced out of the processing industry through price competition among processors. Some processors may be able to sell their businesses, while others may not.

If the decision is made to compensate processors for the introduction of IFQs, processors could be allocated some of the harvesting quotas, so that those who choose to leave the industry will be able to sell out. Another method is to implement processing quotas analogous to harvesting quotas and allow these to be bought and sold. Fish harvesters would then be required to sell their fish to those who hold a similar amount of processing quotas. Processors that choose to leave the industry could then sell their processing quotas to other processors, much as some vessel owners and crew members would sell their harvesting quotas to other vessel owners and crew members; in both cases, excess capacity is reduced. Matulich et al. (1996) suggest consideration of allocations that split harvest quotas among harvesters and processors, create separate harvesting and processing quotas, or pair harvesting and processing rights. Matulich and Sever (1999) suggest that pareto-safe one-pie and two-pie solutions exist in perfectly competitive markets (but are policy-infeasible) while only the two-pie allocation is pareto-safe in the case of a bilateral monopoly.

Arguments Against Allocation of Quota to Processors. There are some arguments, however, against processing quotas (in contrast to harvesting quotas) as a mechanism for compensating processors who leave the fishery following the introduction of IFQ management:

• Processing quotas would, presumably, be a permanent or a long-term arrangement, whereas processors' losses occur only once. It must be carefully considered whether it is desirable to put in place a permanent mechanism of this kind to fix a one-time problem, rather than compensating exiting processors through other means. This argument does not apply to harvesting quota because they have ongoing objectives beyond reducing capacity.

• Processing quotas are likely to strengthen the market power of processors in relation to fishermen.

• The two-pie system is likely to work less smoothly than allocating only harvesting quotas.

• In some fisheries, for example, the Alaska pollock fishery, vertically integrated and nonintegrated firms exist side by side. At some point in the past, firms decided to integrate or not to integrate, knowing that an IFQ program might be put into place. A decision not to integrate could be taken as a decision to forgo the benefits of quota allocation, since such allocations have so far been made only to harvesters.

To establish whether or not there is a case for compensating processors through an initial allocation of harvesting quotas or by setting up processing quotas, the following are among the issues that should be considered:

- Has the processing capacity already been written off as depreciation, or has the government already provided compensation in the form of tax benefits to a processor equivalent to the value of the physical plant?
- What are the relations between processors and fishermen? Are they to some extent integrated through partial ownership of vessels by processors, long-term financial agreements, or other financial arrangements?
- Would processing quotas shift bargaining power too far in the favor of processors?
- What is the degree of foreign ownership in the processing sector?

The committee heard considerable testimony from processors in the North Pacific region that they would be economically disadvantaged and perhaps bankrupted by losing control over their ability to negotiate prices and control the timing of product flow through their plants, because harvesters in IFQ fisheries, as opposed to short derby fisheries, would have much greater opportunity to seek the highest price among processors. Processors also complained that awarding all the windfall gain to harvesters was not fair because processors and harvesters were partners in developing the fishery and the government had encouraged the processors to build processing plants. The committee was not convinced, however, that the solution to the perceived problems lies in the allocation of either harvesting or processing quota to processors.

Foreign Ownership

Congress asked if there are mechanisms available to prevent foreign control of the harvest of U.S. fisheries under IFQ programs (Sec. 108[f][1][B]). In particular, Congress asked whether there are mechanisms to prohibit persons who are not eligible to be deemed citizens of the United States for the purposes of operating a vessel in the coastwise trade under U.S. maritime statutes from holding IFQs (see Secs. 2[a] and 2[c] of the Shipping Act of 1916 in Appendix C). Implicit in this issue is the question of whether it is desirable policy to attempt to control foreign ownership of IFQs.

The history of public resources in the United States is replete with cases of perceived threats of foreign ownership. In the nineteenth century, large areas of rangeland were held by British and other companies who subsequently sold out during a period of severe losses. Fear of foreign ownership of real estate surfaced in the 1970s and 1980s when famous pieces of property such as Rockefeller Center and the Pebble Beach Golf Course were purchased by Japanese entrepreneurs.

Similarly, foreign ownership in the maritime industry in general and the fishing industry in particular has received considerable attention from Congress. If there is a consistent congressional policy, it can be characterized as resistance to foreign ownership of fishing vessels and foreign exploitation of fish resources within the U.S. EEZ (e.g., 16 U.S.C. 1821[a], 1824[b][6]). The concerns giving rise to exclusion of foreign interests fall within several categories:

- Fear of foreign economic domination of the maritime industry and fisheries;
- Difficulties in regulating foreign-owned businesses;
- Threats to the social values of U.S. fishing communities; and
- Loss of potential economic benefits.

Experience with the U.S. IFQ programs and concerns voiced to the commit-tee have dealt more with the *concentration* of quota share ownership rather than whether the holders are foreign or domestic. The issue of concentration, as well as foreign ownership, appears explicitly in the Magnuson-Stevens Act (16 U.S.C. 1851[4]).

Whether the issue is concentration of ownership or acquisition by foreign entities, economic incentives to maximize the value of quota shares may have led to many financial arrangements, for example, quota holders with different names, but the same address, and domestic holding companies for foreign owners. In addition, existing regulations that impose restrictions on the transfer of quota shares for halibut and sablefish have caused involuntary transfers pursuant to court order, by operation of law, or as part of a security agreement (see 50 CFR 679.2, 679.41[f]). Transfer to a foreign entity could result from such involuntary transfers, although the committee is not aware of any evidence of such transfers.

Explicit foreign ownership and control could be prohibited by defining the individuals and entities capable of owning quotas to include only U.S. citizens or entities owned and controlled by them. This, however, is only the beginning of an inquiry that must consider (1) the economic pressures to maximize the effi-ciency of quotas by making them freely transferable and available as collateral for loans; (2) the existing ownership structure; and (3) whether foreign ownership per se or concentration of economic power regardless of its origin is the primary concern; and (4) the cost of detecting and enforcing restrictions on foreign invest-ment in U.S. fisheries.

It is important to acknowledge the presence of significant foreign participa-tion in several U.S. fisheries, most notably the Bering Sea groundfish fishery, a potential candidate for IFQ-based management. Some U.S. fishing and process-ing companies have significant ownership by, or other financial connections to, foreign investors, and foreign companies have equity investments and other ties to these companies (Huppert, 1991). Yet, the notion that foreign companies could obtain controlling interests in the ownership of U.S. fishing rights raises concerns in some quarters. Moreover, even if it were determined that the benefits of freely tradable permits outweighed concerns about foreign ownership, if does not follow that foreign entities should qualify for the initial allocation of quota shares. Admittedly, the creation of tradable fishing privileges could create another avenue for the participation of foreign citizens and companies in certain U.S. fisheries, which may be an attractive prospect for foreign investors given the lucrative overseas markets for certain fish products. It is useful to refer to the four possible idealized scenarios:

	Benefits Returned to the Community	
	(+)	(-)
Quota Holder		
Domestic	1	2
Foreign	3	4

Ranked in order of benefit to U.S. society at large and U.S. communities specifically, the results of these scenarios would be favored as follows: $1 > 3 > 2 > 4$. Thus, foreign holders who provide positive returns to the community (scenario 3), either voluntarily or by law or regulation, would be preferable to domestic holders who provide no return (scenario 2).

If the IFQ or other fishing privilege (e.g., limited access license) is tied to vessel ownership through a standard requiring (for instance) that an eligible quota shareholder be an owner of a U.S. flag vessel that is licensed for commercial fishing, foreign ownership is automatically limited through the operation of the Merchant Marine Act of 1920, commonly known as the Jones Act. Also, the 1987 Commercial Fishing Industry Vessel Anti-Reflagging Act prohibited vessels built or rebuilt in foreign shipyards from operating in U.S. fisheries. This act also requires that owners of all U.S. fishing vessels be U.S. citizens and that the vessels obtain federal licenses. It limits foreign ownership by corporations owning U.S.-flag vessels by requiring that the controlling interest, as measured by a majority of voting shares in the corporation, be owned by U.S. citizens. Recent legislative activity has focused on closing loopholes in this act.

Thus, the issue for IFQ policy is whether there should be additional direct controls or limits on foreign ownership. Many countries seem predisposed to apply much stronger rules about foreign ownership to fishing vessels and fishing quotas than to other industries. For example, New Zealand requires quota shareholders to be either New Zealand residents or companies that have less than 40% foreign ownership. This is in marked contrast to the trend toward globalization in trade and ownership of the means of production.

The North Pacific Fishery Management Council (NPFMC) designed the sablefish and halibut IFQ programs to prevent foreign control of the fisheries via ownership of quota shares (Pautzke and Oliver, 1997). The program's design limits initial issuance and subsequent receipt of quota shares to individuals who are either U.S. citizens or U.S. companies. The council reports that although the level of foreign investment in fishing vessels and fishing companies is not monitored directly by itself or the National Marine Fisheries Service (NMFS), no reports have been received of IFQs being purchased by an individual who is not a U.S. citizen or a company that was not registered or incorporated as a U.S. company. If Congress desires to limit foreign ownership of IFQs in all U.S. fisheries, it could model amendments to the Magnuson-Stevens Act after the provisions of the Alaskan halibut and sablefish IFQ programs.

New Entrants

The charge from Congress to the National Academy of Sciences includes the consideration of mechanisms to facilitate new entrants (Sustainable Fisheries Act, Sec. 108 [f][1][H]). The purpose of IFQs and other limited entry measures is to prevent excessive entry, so new entry must be balanced by the exit of existing quota shareholders. What mechanisms are available to facilitate new entry under IFQ programs? The Magnuson-Stevens Act currently requires that the regional councils and the Secretary of Commerce, in submitting and approving any new IFQ program after the expiration of the moratorium, address the issue of new entry. Specifically, they are required to have considered allocating a portion of the annual harvest in the fishery for entry-level fishermen, small-vessel owners, and crew members who do not hold or qualify for IFQs (Sec. 303[d][5]). The issue of new entrants is related to the issue of transferability, because market prices for quota shares can be significant barriers to new entrants, and without transferability, new entry can be difficult except by lottery or auction. A related issue is the availability of loans for the purchase of quota. The North Pacific loan program was created to make loans more available for quota purchases. The value of a registry of limited access permits, also mandated in the Magnuson-Stevens Act but not yet implemented by NMFS, is that it reduces the risk to lenders and may make loans more available for new entrants.

The committee received the suggestion that new entry could be facilitated by setting aside a certain part of the TAC each year for new entrants. An auction could then be held, with bidders limited to those with certain qualifications that ensure they are truly new entrants.

New entrants after the initial allocation could also be encouraged through the transfer of IFQs through direct sales, lotteries, a first-come-first-served arrangement, or other methods not based on historic use (Huppert, 1991). The zero-revenue auction (see Box 5.1) is another means to promote new entry. Whatever the mechanism, it should not (1) expand the number of quota shares or (2) artificially inflate the price of quota shares.

Diversity

The transferability restrictions adopted by the NPFMC were designed largely to maintain diversity in the fisheries and to protect the involvement of Alaskan coastal communities in these fisheries (Pautzke and Oliver, 1997). To the extent that diversity is valued over economic efficiency, similar limitations on transferability would have to be incorporated into new IFQ programs.

Some governments address diversity in fisheries in a different manner by focusing on the distinction between individual and corporate ownership (although many single-owner vessels are legally organized as corporations). For example, corporations are not allowed to own Alaska salmon limited entry licenses or many Canadian IFQs. This restriction may not be desirable, however, for fisheries in which a large capital investment is necessary and where that investment can be attracted only with the limited liability of a corporate entity.

Recreational Sector

Recreational fisheries have received very little attention in IFQ programs. The allocation of quota to recreational anglers may serve as a way to let the market help solve the often contentious conflicts between the recreational and commercial sectors of a fishery (Squires et al., 1995). Initial allocation methods and increased enforcement needs undoubtedly would be major issues during implementation of IFQs for recreational fisheries.

Recreational fisheries are as diverse as their commercial counterparts in the types of gear involved and their levels of investment, ranging from shore-based anglers to for-hire operators. Cumulatively, recreational fisheries represent a large and growing potential to harvest fish, particularly in near-coastal waters, and there is a tendency for fisheries to evolve from commercial into recreational as coastal populations grow (Smith, 1986). Specification of a harvest quota in the form of a TAC allows fish to be taken by noncommercial interests, including recreational fishermen, but often does not specify how the allowance is to be made. In the United States, the proportion of TAC that goes to the recreational sector is left to the discretion of the regional councils but usually is based on historic use patterns within the fishery. Recreational allocations can also change with growth in the sector, but only through reductions in the commercial share. In some fisheries, the allocation of TAC to the recreational sector already is substantial (e.g., about 70% of the king mackerel TAC in the South Atlantic and Gulf of Mexico regions is allocated to the recreational fishery).

Inherent difficulties are associated with monitoring and enforcement of recreational fisheries because of their wide geographic range, multiple landing locations, and large numbers of fishermen. Consequently, recreational fishery-dependent data generally are of poor quality, especially with respect to the magnitude of recreational catch, effort, and the value of recreational fisheries to regional economies. Data problems are compounded by the commercial sale of fish caught by anglers and by individual fishermen from the for-hire sectors that fish commercially in recreational vessels when not operating for hire.

Recreational fisheries traditionally have been managed on the basis of fishing seasons, gear restrictions, and size and bag limits, and there is widespread resistance by recreational anglers to limited access or licensing. Clear differences between the recreational and commercial sectors can often be observed in the preferred sizes of fish, with recreational fishermen often preferring larger "trophy" fish. Consequently, the optimal stock size for recreational fisheries may be larger due to preference for higher catch rates and larger fish.

Currently, there is little precedent (in the United States or elsewhere) for integration of a recreational fishery into IFQ or other quota management systems (e.g., Arnason, 1996). In some cases (e.g., New Zealand), recreational fisheries are virtually unregulated in harvest, with the estimated recreational catch subtracted from the TAC before the remainder is allotted to IFQ shareholders. However, unrestricted harvest by many noncommercial interests, while fisheries are managed for holders of IFQs, presents major management problems that potentially undermine the integrity of any IFQ program (Ackroyd et al., 1990), particularly when the recreational sector is growing in size. In New Zealand, where the preservation of a satisfactory recreational fishery is an objective of the IFQ program for commercial fisheries, several studies have addressed the problem of recreational fishery management. Ackroyd et al. (1990) identify significant problems presented by recreational fisheries and recommend that the recreational sector be placed under a quota, with trusts established to hold and manage the quota (e.g., similar to the "hunting club" or Ducks Unlimited approach).

Pearse (1991) recommends allocating the recreational sector an explicit quota to be held on behalf of recreational fishermen by local government or by organizations modeled after the regional councils. The New Zealand Fisheries Task Force (1992) also recommends that recreational fishermen be allocated a share of TACs, with establishment of organizations to hold and manage the quota. These studies suggest that IFQ programs for only the commercial sector may benefit and strengthen commercial claims on fishery resources, leaving the recreational sector with no grounds to protect its rights. Conversely, the opposite may be true. One of the greatest challenges to commercial fishing is the growing interest in recreational fisheries worldwide. By sheer numbers alone the recreational fishing community is powerful, and the political clout of recreational anglers is growing (De Alessi, 1998). Consequently, commercial fishermen are concerned that the wealth and power that reside in the recreational sector ultimately will result in its majority ownership of many fisheries if no limits on quota ownership and transferability are in place to protect commercial interests.

In the discussion of IFQs for the recreational sector, a distinction should be made between individual recreational anglers (for whom IFQs are probably not practical; see below) and the for-hire sector that concentrates units of individual anglers and may be practical for inclusion in IFQ programs. Individual quotas for recreational fisheries could be analogous to IFQs in the commercial sector. If feasible, recreational quotas could achieve at least partial integration of recreational fishing into a quota system.

Are quotas for individual anglers feasible? Public testimony indicates that the establishment and implementation of IFQs for recreational fishermen face a formidable problem with respect to equitable initial allocation of quotas among users because catch histories do not exist for most individual recreational fishermen. Thus, the most common basis for initial allocation in commercial fisheries cannot be used in recreational fisheries. Other initial allocation mechanisms, such as lotteries, auctions, charging some predetermined fee, basing quota share on the magnitude of the investment in recreational fishing (vessel, gear), or equal shares for all, also are problematic. Lotteries have been used to allocate big game and waterfowl hunting privileges and could be acceptable for some recreational fisheries. Recreational fishermen generally are great in number, cross many economic classes, and thus vary greatly with respect to economic investment in fishing. They also tend to be spread over a wide geographic area and land their catches at a variety of locations, potentially making quota monitoring a formidable problem. Recreational fishermen have fought strenuously against saltwater fishing licenses in many states. It is likely that recreational IFQs would face similar opposition.

Economic Returns to the Public

Fishing, whether under open access, IFQs, or other limited access programs, provides benefits and creates costs to the nation. In particular, the question arises whether the nation should share directly in the benefits that fishermen derive from their use of the public resource, and particularly of the benefits that IFQs and other limited entry permit recipients obtain from the initial allocation above and beyond capital gains taxes on the sale of the initial allocation. Mechanisms for capturing benefits for the nation include auctions, annual fees, transfer fees, and taxes. Some people are opposed to IFQ programs because they view them as awarding a large financial windfall to quota share recipients, and these windfalls could encourage unproductive behavior (e.g., expenditure of funds to influence the outcome of the implementation process) on the part of the fishery participants. Such windfall gains can be reduced by taxing the rents generated by IFQ programs. To the extent that IFQ programs are subject to taxes that are not used in other limited entry programs, however, support of the industry for IFQ programs will be diminished.

To what extent is the public entitled to a share in fishing rents? Opinions differ about the answer to this question, but it seems fair to some that the public receive some return for the use of public resources. The public trust nature of fishery resources lends greater weight to such a conclusion. In practice, however, schemes that extract a large percentage of the rent can undermine the degree to which the industry will support changes in the management regime; the support of the industry is likely to be related to the prospect of receiving some or all of the resource rent. Alternatively, public support for quota programs may be under-mined when only a small percentage of the rent is extracted for public purposes.

In all but the most extraordinary circumstances (Iceland, for example), taxes on fishing rents would be a nearly insignificant source of revenue for the national government, and even for state or provincial governments (except perhaps Alaska). However, such tax revenues are a significant source of income for local governments in Alaska and elsewhere.

Any fishery can be managed to produce some resource rent (see Chapter 1 for an explanation of rent), although most fisheries in the United States are not managed with this objective. The size of this rent depends on a number of factors, some of which are not influenced by fishing firms or the industry, such as the price of fish (in most cases determined in competitive markets), technology, and the cost of labor and other inputs. Other factors that influence rents, such as modes of organization internal to the firm and cost-cutting measures, are con-trolled by the firm.

The existence of rent is a consequence of an efficiently managed fishery and a naturally limited resource; instead of using too many vessels, employing too many people, or using too much gear and fuel, the redundant factors of produc-tion have been diverted to other purposes, where they create additional value in the production of other goods and services.

Three Rationales for Reclaiming a Public Share of Fishery Rent

Three rather different rationales are usually suggested for public sharing in the rent from a resource. These principles are not necessarily mutually exclusive; they can be applied simultaneously and to different degrees.

1. *The public resources principle*—Under this principle, the public is en-titled to a share of the rent because the resources being exploited are owned by the public. This principle suggests returning to the public some of the value that is rightfully theirs.

2. *The cost recovery principle*—Under this principle, the government is en-titled to reclaim the costs of creating and administering fishery management programs because the beneficiaries of government programs should bear the associated costs. Cost recovery programs can seek to cover some or all of the (1) administrative costs, (2) monitoring and enforcement costs, and (3) research and stock assessment costs.

In practice, cost recovery is becoming increasingly common. This principle is allowed to a limited extent by the Magnuson-Stevens Act in the form of fees levied to cover management and enforcement costs (Sec. 304[d][2]). It is also currently used in air pollution control to fund the administrative cost associated with the permit system (Title V of the Clean Air Act). New Zealand and Canada currently apply the cost recovery principle in their fisheries.

Costs are sometimes placed in two separate categories: attributable and avoidable costs. Attributable costs are those that can be directly charged to a specific activity and are essentially a transaction charge. Examples of attributable costs to the fishing industry in an IFQ program are the costs of allocating IFQs, costs of registering transfer of quota between IFQ holders, the costs of monitoring catch against quota, and costs of dockside monitoring. Avoidable costs for the commercial fishing industry are additional costs that exist because of the presence of a commercial fishing industry and are not transaction costs that can be charged to a specific activity. The presence of a commercial fishery will require fisheries management activities (e.g., research, enforcement, administration) and associated costs that are not assignable to specific fisheries, but are necessary to meet the government's goals.

In New Zealand, the commercial fishing industry has been required to reimburse both the attributable and the avoidable costs of fisheries management (Box 5.3). This has been controversial, with the industry agreeing to pay the attributable costs, but arguing against having to pay the full amount of the avoidable costs, maintaining that these costs are largely for the maintenance of a public good. Cost recovery is also practiced in some Canadian fisheries (see Boxes 4.2 and 4.5).

BOX 5.3
Cost Recovery in New Zealand

In New Zealand, the costs of fisheries management are allocated to associated fisheries in such a way as to provide the correct economic signals and incentives to each fishery. For example, the costs of the research program on hake are charged only to hake ITQ shareholders. There is no research program on arrow squid, so squid ITQ shareholders pay only the general research and stock assessment levy covering general costs that cannot be allocated to specific fisheries. Similar charging arrangements are made for administration and enforcement costs. This system of allocating costs to associated fisheries has resulted in large differences in the cost recovery levies as a percentage of the landed value, from less than 1% in some fisheries to greater than 10% in other fisheries, as well as large year-to-year fluctuations in the levies for a specific fishery. The large year-to-year fluctuations in costs for some specific fisheries have made it difficult for fishing businesses to plan their operations. As a result, the structure of the existing cost recovery system is being reviewed.

The total cost of fisheries management (administration, enforcement, research) is about $NZ45 million, with $NZ37 million being recovered from the commercial fishing industry. With a total landed value for the fisheries resource of about $NZ700 million, this means that the industry is paying about 5% of the landed value of the catch in cost recovery charges (J.H. Annala, unpublished data, 1998).

Cost recovery charges or levies can be calculated in a number of different ways: based on the tonnage of quota held, the tonnage of fish landed, the value of quota held, the value of quota traded, the value of landings, or some combination of these (Huppert, 1991). In New Zealand, cost recovery levies are based on a combination of the tonnage of quota held and the value of the landings.

3. *The compensation principle*—According to this principle, a proportion of the rent should be reclaimed in order to compensate those who may be injured by the process of establishing a fisheries management program. Proponents of this principle may be motivated either by a concern that all parties be treated fairly or by the recognition that failing to compensate victims can undermine the political will necessary to implement the program. Victims who are seen as entitled to compensation under this principle may include both individuals and communities.

Existing examples of the application of this principle include the Canadian and New Zealand "buyback" programs with which the government reduced quota by buying it back from those leaving the industry and then retiring it so it could not be used by anyone else. Another example is the reduction of Alaskan IFQs to be reallocated as part of the community development (CDQ) program.

The compensation principle motivates the wide initial allocation of quota. In a broad allocation of transferable IFQs, compensation occurs indirectly as those with less-than-desired quotas sell or lease out to others. This is sometimes called "exiting with money in your pocket." A compensation system does not simply compensate individuals after some action has been taken; it can also create an incentive for individuals to overinvest in the factor that will be compensated. For example, implementing a buyback program (or even the expectation that one might be implemented) may create an incentive for fishermen to keep capital in a fishery longer than they might otherwise, in order to receive greater compensation for leaving later.

Mechanisms for Reclaiming Rent

The government has a number of different means available to extract rent from IFQ-managed fisheries.

Auctions. Periodic auctions can be held to sell quota shares (percentage of the TAC) or current quota (tons harvested during this fishing season). The government may keep a percentage of the proceeds. Auctions can be combined with a system that allocates quota on a historic basis by requiring a certain percentage of the quota to be placed in the auction every year (see Box 5.1).

Auctions are an ex ante activity, that is, they raise revenue prior to harvesting. In principle, the revenue from a competitive auction of all permits should equal the total of expected rent (Grafton, 1995). In practice, revenues from the auctions will equal actual rent only if the expectations about future TACs, fish prices, and fish costs are accurate and fishermen are not risk averse. Risk aversion will be significant if the quotas are auctioned for a long period, perhaps once and for all, so a one-time auction or very infrequent auctions will not be very effective in extracting the rent, unless one asks for bids in terms of share of profit or in some terms that would alleviate the price and quantity risk.

With auctions, the bidder bears the price and harvesting risk associated with unrealized expectations. Since the payouts for auctions are before the fact, unusually poor harvests or prices mean that auction revenue could exceed the rent. The high cost of auctioned permits can prevent entry for those with inadequate access to capital markets. (This deficiency can be overcome by ensuring that all participants have adequate access to capital markets.)

Quota Shares Reserved for Government Use. Part of the quota can be set aside for use by the government or for compensation to individuals or communities. The revenue received by the public using this mechanism is volatile since it depends on market conditions such as prices and harvest levels. It also depends on how efficiently the shares are utilized. If retained by high-cost harvesters, the resulting value will be small.

The public may bear large administrative costs to secure the value of the quota shares. Shares have to be turned into revenue either by harvesting fish, which are then sold, or by transferring the quota to others. In either case, inexperience can translate into forgone rent.

Quota Attenuation. Under quota attenuation, a share of the quota would automatically revert to the control authority every year. This quota share could be sold by the government to the highest bidder. This works very much like an ad valorem tax on quota holdings, but it would not be necessary to monitor the value of quotas or of fish. A variant of this would be to apply this "tax" only on transactions (leasing and/or selling of quota shares). This has the advantage of easy monitoring and a market-based valuation.

In terms of efficiency of these mechanisms for reclaiming rent, quota shares set aside as compensation may not be used efficiently if the recipients choose not to do so, whereas attenuated quotas work like a tax on quota holders and efficiency is ensured if the shares obtained by attenuation are sold by auction or some other market mechanism.

Fees or Taxes. The government can reclaim rent through the use of fees or taxes levied on several different types of tax bases: landed harvest, quota shares, annual entitlements to harvest, income or profit, or capital gains (Box 5.4). The fees can be lump sum (dollars per fisherman or vessel), specific (a per-unit fee on each unit of the tax base), or ad valorem (a percentage levied on the value of the tax base). Consideration of using taxes to extract rent should be tempered by the fact that if the United States imposes export taxes on the products of foreign-controlled processors or catcher-processors, it could run afoul of the World Trade Organization.

Lump Sum Fees. For this particular mechanism, the revenue collected does not vary with the value of the rent received by any particular harvester. It also does not automatically reflect changes in the rent of the fishery. Compared to

---

BOX 5.4
The "Two-Fee" System

If it is deemed desirable to extract rents from a fishery beyond those needed to cover monitoring, enforcement, and administrative costs, this could be accomplished with a two-fee system. This approach recognizes that the revenue-raising objectives are sufficiently different as to motivate separate fees.

1. Monitoring, enforcement, and administrative costs would be covered by a per-unit fee levied on quota share. The size of the fee would be determined by the magnitude of the costs it is designed to cover. This fee would generally change from year to year based on the enforcement experience in each fishery. Fishermen could collectively lower costs by facilitating programs that are easier and less costly to enforce.

2. Capturing rents beyond these costs could be funded by a per-unit fee on actual catch. This fee would be relatively stable over time, although it could be indexed to some measure of inflation to ensure that its real value did not decline over time.

---

other mechanisms with a similar amount of total rent captured, lump sum fees place a larger burden on small-scale fishermen. Unlike auctions or the allocation of quota share, with lump sum fees the recipient knows with a high degree of certainty how much rent will be extracted. There are at least two other major arguments for using lump sum fees. First, except for entry and exit decisions, lump sum fees do not penalize operational decisions and therefore do not create some of the distortions associated with other taxes. Second, because actions do not need to be monitored, transaction costs may be low. The latter may be the major reason some less developed countries use lump sum access fees for foreign fishing fleets operating in their EEZs.

Fees or Royalties on Quota or Harvest-Specific (Per-Unit) Fees. Fees can be imposed on quota or harvest. A fee on harvest would be paid only if fish are landed. A fee on quota would be paid whether or not the quota is being fished and would thus provide an incentive to use the quota. Fees, even ad valorem fees, may make the industry unprofitable if they are set too high. This would not occur in an auction process unless the bidders are mistaken about their prospects. Harvest fees have less "up-front" risk for harvesters than an auction since they are paid only if and when the harvest is landed. Both quota and harvest fees would lower the price of harvest quota because the quota would generate less revenue for the owner after the fees were paid. As long as the fees are not too high, specific fees on harvest or quota are generally consistent with efficiency incentives. Rent collected for harvest would be less for harvest quotas when some proportion of the quota remains unfished.

Ad Valorem (percentage of value) Fees. Ad valorem fees normally produce a greater variability of rent than specific fees or lump sum payments when prices vary over time. (Prices help determine the degree of rent capture for this mechanism, but not for specific fees or lump sum payments.) Because this is an ex post method of rent capture, it poses less up-front risk for the quota holder than an auction. The payments would depend on the value of the catch.

Transferability and Accumulation

Most IFQ programs used worldwide allow transferability (e.g., Box 5.5). Transferability is one of the most contentious issues in IFQ management. It can be expected that in fisheries that allow easier transferability, consolidation of quota will occur. Transfer of quota shares can lead to a concentration in the ownership of quota, which may have undesirable side effects. Transferability can create unemployment in isolated communities where there are limited economic alternatives to replace the loss of employment caused by a reduction in harvesting and processing capacity. The mechanisms used to dictate the nature of transfers within a fishery and the degree of transferability can significantly alter the nature of the fishery.

BOX 5.5
Transferability of Quotas in The Netherlands

Transferability of Dutch IFQs was allowed officially in 1985, with certain restrictions. Quotas can be held only by those who have a fishing license, although banks and shipyards can hold quotas temporarily, presumably because quotas can be put up as collateral against debts. Quotas can be leased freely, but parts of an ITQ cannot be sold—the allocation must be sold as a whole. Fishermen circumvent this by having their producer organization or management group buy the quota and sell it in parts to individual members. Within a management group, quotas are freely transferable, but between groups, leasing is not allowed after the end of November. Individuals not belonging to any group cannot lease or rent out a quota after the end of February. Unused quotas cannot be transferred between calendar years.

The issues of transferability and concentration limits must be considered in the context of balancing two opposing goals: economic efficiency and social equity. Economic efficiency is maximized when the following occur:

• Quota shares are freely transferable, in the long and the short term.
• Quota shareholders are allowed either to sell their quota shares permanently or to rent them out for any period of time.

• Quota shares are as divisible as practically possible; that is, a quota share holder is able to sell or rent out any portion of his or her quota share.

• The tenure of quota shares is either long term or permanent, in order to minimize uncertainty in the fishing business and encourage long-term planning and stewardship among quota holders.

Each of these factors has social and/or legal implications. A number of negative side effects of free transferability must be considered. To the extent possible, these effects should be reduced with as little infringement on transferability as possible, to minimize the economic losses involved. These side effects will differ from one fishery to another, and they should be analyzed in the context of each fishery to design the most appropriate program.

Economic Aspects of Transferability

Transferability of IFQs has two main, and related, economic purposes:

1. Achieving rationalization of the industry by allowing some participants to leave the industry with a compensation financed by the industry itself, that is, to be bought out by other industry participants; and

2. Ensuring that IFQs are held by those who are willing to pay the highest price for them. This promotes efficiency in the industry because those who are willing to pay the highest price for quotas would normally be those who expect to utilize them most profitably, either by doing so at a lower cost than others or by transforming the fish into a more valuable product.

At this point, a short remark on item (2) is appropriate. It is sometimes alleged that those who are willing to pay the highest price for quotas are the ones with the easiest access to capital. In efficiency terms, this is not a negative factor; there is often a strong relationship between having access to capital and being able to utilize quotas efficiently. The value of quotas as investment objects derives from the ability to use them for generating net profits from fishing. To achieve a return on investment in a high-priced quota, the quota holder will have to use it efficiently himself or to lease it to someone who can do so. It is not likely, however, that persons or financial institutions will invest their money in quotas unless they can be assured of a reasonable return on their investment, which again would contribute to greater economic efficiency in the fishing industry.

In addition to achieving greater economic efficiency, transferability is also intended to mitigate imbalances that may occur in the initial allocation. For example, although crew members did not receive initial allocations of IFQs in the Alaskan IFQ fisheries, they now own 11.2% of the halibut quota share and 4.6% of the sablefish quota share. This would not have been possible without transferability.

It is useful to distinguish between the ways in which transferable quotas may promote economic efficiency in the short and the long run. In the short run, transferability leads to lower operating costs and a higher production value in fisheries plagued by harvesting overcapacity. Those who can fish at the lowest cost or produce the most valuable product are able to buy or lease fishing quotas at a price that is acceptable to both buyer and seller. In the long run, transferability of quotas can be expected to produce optimally sized fishing fleets. A person or firm with a given quota will have no economic incentive to invest in more or larger fishing vessels than needed to take this quota. Alternatively, if there are economies of scale in fishing for the target species, those who wish to invest in vessels of an optimal size but have insufficient quota to utilize the vessels fully will be able to buy additional quota for this purpose.

Quota transactions may produce inequitable results, in the fishing industry as in other industries. Various methods can be employed to avoid or reduce inequitable results that otherwise would emerge from the marketplace, for example, restricting the types of vessels or the areas in which quota can be traded, establishing accumulation limits, and requiring owners to be on board vessels. However, since unfettered market transactions normally lead to enhanced efficiency, some trade-off between efficiency and equity is likely.

Other rules may be necessary to govern the subsequent transfer of initially allocated shares. It may be desirable to allow individuals who did not receive initial allocations to buy and hold IFQs. There must also be consideration of whether the distribution of quota shares among classes of holders and regions should be allowed to vary over time.

In designing the sablefish and halibut IFQ programs, the NPFMC considered a prohibition on transferability to avoid consolidation of ownership, divestiture of coastal Alaskan residents from the fishery, and creation of windfall profits from transfers (Pautzke and Oliver, 1997). Ultimately, the council decided to allow transfers, albeit restricted, to permit new entry into the fisheries and maintain significant Alaskan ownership. The amount of the total quota share pool that can be owned or controlled by individuals and companies is restricted, and quota transfer provisions and ownership limits are specified to prevent overconsolidation of quota share in the fleet. The council also created vessel size and operational quota share categories within which transfers are limited. Other controls include the requirement that certain categories of quota share may be purchased only by individual fishermen who must be on board the vessel and fish the quota share.

Social Aspects of Transferability

Free transferability of quota shares is likely to have a range of social implications, as judged from both theoretical predictions (Copes, 1986; see Figure 3.1) and the empirical evidence available (McCay et al., 1995; Pálsson and Pétursdóttir, 1997). These effects occur both among and within communities.

Impacts Among Communities. Freely transferable quota shares may concentrate over time in some communities while other communities lose part or all of their quota (Eythorsson, 1996). It is difficult to predict the pattern and overall movement of quota in advance, since these will depend on a host of contextual factors and the design features of the program in question; generally, however, one may expect communities with a large share of quota to gain more because of more infrastructure and better access to capital. Some smaller communities dependent on fisheries and without alternative means of support are likely to suffer severe unemployment and related social and economic problems. McCay et al. (1995) demonstrate a clear geographical shift in quota holdings for the SCOQ IFQ program, where quotas are freely transferable. The same applies, they believe, for the Canadian program they studied; here more constraints were placed on transfers as well as on accumulation and ownership by nonfishermen, but the constraints were generally ineffective against strong economic incentives to consolidate holdings. In contrast, the Alaskan IFQ programs, which include some area restrictions, have maintained (to date) similar participation by Alaskan and non-Alaskan fishermen both before and after the program was implemented. In Iceland, the main accumulators of quota are companies in the larger towns of the northern part of the country. Small communities, with less than 500 inhabitants, have lost a much greater share of their quotas than larger communities.

To some extent, regional concentration of quota shares is unavoidable, a healthy sign of increased economic efficiency. The social costs, however, may outweigh the gains in economic efficiency. As was the case when agriculture became increasingly intensive and took advantage of gains to scale, negatively affecting traditional farming communities, some fishing communities will undoubtedly thrive, whereas others' valued life-styles and traditions will be threatened.

One way of dealing with undesirable flows of quota is to limit the transferability of quotas from one community or region to another (see Box 3.3). It has to be kept in mind, however, that changes in fishing technology, fish processing, and transportation may make the location of the fishing industry in certain communities or regions obsolete and economically inefficient. Rather than preventing the realization of economic benefits from such changes by limiting the transferability of quota shares, it may be preferable to allow transfers, but also to compensate disadvantaged communities with a fair share in the gains of the overall fishery through payments or buyouts. An alternative approach is to design an IFQ program to allow municipalities, regional organizations, or other entities representing the needs of local communities to purchase IFQs and create local rules about their allocation and transferability.

It may be argued that in overcapitalized fisheries, transferable IFQs could lead to downsizing of fishing technology, smaller vessels, and greater efficiency. For example, testimony from some Alaskan fishermen stated that IFQs have helped the small-vessel fisheries because they are able to operate in better weather with IFQs. Generally, however, the concentration of IFQs will depend on the economies of scale available for a specific fishery.

Impacts Within Communities. Transferability may have far-reaching repercussions on the internal dynamics of fishing communities. The social distribution of quota shares is one variable to consider. It is difficult to predict general effects, due to the existence of confounding factors. In some cases, for instance in the Mid-Atlantic SCOQ fishery (McCay et al., 1995), some of the large firms have broken up since the implementation of IFQs, countering the otherwise strong tendency for concentrated ownership in this industry. However, even in that case quota shares tended to concentrate in the hands of those with the largest shares at the initial allocation (McCay and Creed, 1994; Weisman, 1997). In the Icelandic case, those individuals or firms that own more than 1% of the total quota have increased their share from a quarter to about a half of the total in just over a decade, and in the SCOQ case, those with the largest allocations in 1990 had significantly increased their share of the quota by 1994 (see Appendix G).

Such concentration may make the issues of equity and social distribution pressing concerns, important features in the moral landscape of the affected fishing communities. In some extreme cases, resistance to conservation measures in the fishery may reduce or invalidate potential economic and ecological benefits of an IFQ program, resulting in fishermen's strikes and increased highgrading and bycatch.

Not only can transferability increase conflicts among quota holders, it also can alter relations of power between vessel owner and crew, with the latter increasingly losing power to quota-owning vessel owners. Iceland and Alaska provide some examples of this process. Again, however, the social impact of transferability will depend partly on the design features of the specific program, particularly the ways in which shares are initially allocated and the degree to which rents are returned to fishing communities.

One relevant community concern relates to the ways in which IFQs affect the prospects of marginal participants in fisheries, including "native" groups and women (regarding gender, see, for instance, Macinko, 1993, on Alaska and Skaptadóttir, 1996, on Iceland). As quotas tend to be concentrated and rights to the resource are removed from the communal frameworks to which fishing has been subjected, they tend to freeze or exaggerate existing patterns of occupational participation, making it more difficult for marginal participants to advance. In the New Zealand case, tribal claims were not anticipated and, when exercised, resulted in costly changes to the system (Cheater and Hopa, 1997).

Leasing of Quota Shares

Some of the opposition to IFQs centers on leasing. The reactions to leasing vary from one context to another. In the Icelandic case, fishermen have gone on strike three times in four years to protest against what they see as unfair "quota

profiteering" by absentee owners. The economic efficiency gained by the introduction of IFQs may be lost due to such strikes. It is important to differentiate
between leasing to absentee owners and other fishermen. Although both have the
same general economic effects, leasing to absentee owners (those "sitting on the
beach" in fishermen's jargon) is much less acceptable in the fishing community
than leasing to other active fishermen.

Absentee ownership can develop when the transfer of quotas is unrestricted.
Rather than selling their quotas, quota shareholders may choose to lease them and
gain unearned income on their quota wealth. This may tear at the social fabric in
fishing communities, where absentee ownership is often seen as unfair. Fisheries
that traditionally have been owner operated may be altered by the ability to lease
quota shares, and relationships in a community can become more sharply divided
between the "owners" of the resource and the "tenant" fishermen. In some cases,
the crew members aboard leased vessels pay for the cost of the lease, effectively
reducing both their average share and their overall income. One way to deal with
absentee ownership is to require active participation of the quota holders in
fishing—for example through "owner-on-board" provisions—or to impose geographical restrictions on transferability. However, this is not an economically
efficient instrument in fisheries that are subject to economies of scale and scope;
in these industries, economic efficiency is increased when large, vertically integrated firms, rather than individuals, hold the fishing quotas. Moreover, such
provisions can work against attempts to meet social goals through community-
based control of IFQs (or CDQs).

Leasing is often seen as a way for quota shareholders to fine-tune their
operation to meet short-term needs arising from fluctuations in local, regional,
and national markets and to deal with bycatch problems. Additionally, leasing
can allow individuals to learn how an IFQ program and market works before they
buy into or sell out of the program (perhaps prematurely). For example, the
British Columbia halibut IVQ program allowed leasing, but not sales, of quota
shares during the first two years of the program, to provide such learning time.
Thus, in the beginning, the lessors are likely to be firms or individuals actively
engaged in the industry, using their own IFQs for fishing. In time, however, IFQ
holders may come to discern that profits might be made through leasing IFQs on
a larger scale.

With increasing concentration of quotas, new and more formalized modes of
leasing may emerge. In such transactions, the supplier of the IFQs is likely to be
a large vertically integrated firm and the recipient a small-scale harvester. In
time, this situation may create a new kind of social structure with permanent
divisions among those who live from leasing quota shares and those who rent
them and do the fishing. A similar social arrangement characterized the salmon,
herring, and halibut fisheries of the Pacific Northwest, British Columbia, and
Alaska from the mid-1800s through the early 1900s (Bay-Hansen, 1991; Newell,
1993). In the case of Alaska, the desire to eliminate this feudal structure was a

---

BOX 5.6
Concentration of Quota in The Netherlands' IFQ Program

Since 1988, quotas have gravitated to the largest and the smallest vessels in The Netherlands, as shown in Table 5.1. ITQs are not concentrated in the hands of a few large operations; the largest holder owns no more than 3% of the total Dutch plaice and sole TAC. The leasing of ITQs by people who do not own fishing vessels (e.g., retired fishermen) is becoming an issue in the industry, with active fishermen resenting this practice.

---

primary motivation for statehood. A residual of this sentiment is at the root of the owner-on-board provisions of the Alaskan halibut and sablefish IFQ programs.

There are several ways of avoiding the problems of the tenant fisherman; the applicability of various approaches will depend, again, on the design of the IFQ program in question, the fishery, and the cultural context. The proportion of quota that is leasable during any single year can be limited; this ceiling can be kept low, if desired, and subject to restrictions on the frequency of leasing permissible for individual quota shareholders. In addition, parts of the income from leasing can be distributed to the larger community through taxation. These measures are powerful tools for ensuring economic returns to the community and social equity.

Accumulation and Concentration

An ITQ program will almost inevitably lead to some accumulation of quota shares as excess capacity leaves the fishery (e.g., Box 5.6 and Table 5.1). If an overcapitalized fishery is put under an ITQ regime, some vessels will leave the

TABLE 5.1   Changes in Quota Holdings for Sole by Vessel Size in The Netherlands Between 1988 and 1996

Horsepower Group	Mid-1988		January 1996	
	No. of Vessels	% Sole Quota	No. of Vessels	% Sole Quota
< 260	141	1.0	109	2.2
261-300	125	4.7	134	8.5
301-1,500	201	38.5	43	6.9
> 1,500	139	55.8	171	82.4
Total	606	100	457	100

SOURCE: Salz (1996).

fishery sooner or later, and vessel owners will sell their quotas to others who can improve the utilization of their vessels. This is, indeed, part of the purpose of ITQs. However, even in a fishery that is not overcapitalized, some accumulation of quota will result if there are economies of scale in the industry, because larger and more efficient firms will be able to buy out smaller and less efficient firms.

Concentration of quota among a small number of quota-holding firms or individuals may unduly strengthen the market power of quota shareholders and adversely affect wages and working conditions of labor in the fishing industry. This could be a particular problem in rural coastal areas in which the alternatives for employment are limited. The ability of quota owners to dominate labor markets will depend on a number of factors, most importantly the supply of crew members and the general labor market. For example, the West Coast groundfish industry, in a region of positive economic growth and healthy economies, is currently experiencing a serious shortage in supply of experienced crew members, and accumulation of quota share would not be likely to have as great an effect on labor markets there as in other areas. The market power created by concentration can also marginalize smaller fishing firms, hurting their position vis-à-vis larger harvesting firms in competing for fish buyers in the markets.

One way of dealing with the problem of unreasonable power in the labor market is to set an upper limit on how large a share of the total quota pool can be held by any one firm or individual (concentration or accumulation limits). This method is applied in New Zealand, Icelandic, Canadian, and Alaskan IFQ fisheries; however, the limits vary from 0.5% in some Alaskan IFQs to 35% in some New Zealand IFQs. It is not possible to provide a general rule regarding an optimal percentage concentration; this will undoubtedly vary by fishery and the goals of fishery managers in the region. It seems necessary, however, that concentration limits of some degree be included in all new IFQ programs, because National Standard 4 of the Magnuson-Stevens Act prohibits the holding of excessive share (not defined by the act) by any individuals or entities and antitrust laws have not been effective in controlling concentration.

In regard to the conflict between concentration and efficiency, it is important to assess whether this conflict is serious and what trade-offs might exist. The usual antitrust arguments do not seem very relevant to management of IFQs. The fishing industry is not in a position to dominate its market to any appreciable extent. Fish from a particular IFQ-managed fishery compete with fish from other sources. In addition, fish markets are global and there are many possible substitutes for most fishery products from any given region; moreover, fish is but one particular food item and other food items compete substantially with and substitute for fish. Even if the IFQs of one particular fishery were owned by a single company, the company's influence on the market price of its fish is likely to be negligible. In contrast, such a firm might exert a strong influence in local factor markets (e.g., labor).

Effective Monitoring and Enforcement

Regardless of how well any fishery management plan is designed, noncompliance can prevent the attainment of its economic, social, and biologic objectives. Plans containing IFQs are no exception. Noncompliance not only makes it more difficult to reach stated goals, it also makes it more difficult to know whether the goals are being met, due to data fouling. Much of our understanding about the health of a fishery is derived from an analysis of its commercial catch. Therefore, if the landed catch is unrepresentative of what actually is harvested (as would be the case with highgrading or high rates of bycatch discards), incorrect inferences would be drawn from the landed catch. Not only would true mortality rates be much higher than apparent mortality rates, but the age and size distribution of landed catch would be different from the size distribution of the initial harvest (prior to discards).

Consequences of Implementing an IFQ Regime for Monitoring and Enforcement

Although it is true that any management regime raises monitoring and enforcement issues, regimes based on IFQs raise some special issues. One of the most desirable aspects of IFQs, their ability to raise income levels for fishermen, is a two-edged sword because it also raises incentives for quota busting and *poaching* (catching fish for which no quota is held). In the absence of an effective enforcement system, higher profitability could promote illegal fishing. Insufficient monitoring and enforcement could also result in failure to keep a fishery within its TAC (Box 5.7).

Do monitoring and enforcement costs rise under IFQ programs? The answer depends both on the level of required enforcement activity (greater levels of enforcement effort obviously cost more) and on the degree to which existing enforcement resources are used more or less efficiently. As has been argued above, there are some good reasons to expect that the degree of enforcement activity will increase with the implementation of an IFQ regime. On the other hand, IFQs also seem to introduce the opportunity to use existing resources more efficiently. Because IFQ fisheries have longer seasons, monitoring activity may be spread over more days of the year, decreasing the likelihood that the monitoring capacity would be overwhelmed, but possibly increasing the number of days that monitoring would be necessary. In fisheries that currently require onboard observers, the incremental cost of monitoring and enforcing an IFQ program may be inconsequential.

What has been the actual experience? In practice, the outcomes have varied. According to a survey by the Organization for Economic Cooperation and Development (OECD, 1997): "Higher enforcement costs and/or greater enforcement problems occurred in 18 fisheries compared to five that experience[d] improve-

---

BOX 5.7
Effects of Inadequate Monitoring and Enforcement

Prior to 1988, the expected positive effects of ITQs did not materialize in the Dutch cutter fisheries. Fleet capacity increased further, the race for fish continued, and the quotas had to be supplemented by input controls such as a limit on days at sea. The reason for this appears to have been inadequate monitoring and enforcement; the race for fish continued because some fishermen overfished their quotas with impunity, and the fishery would be closed when the TAC had been taken, leaving other fishermen with a part of their quota unfished (this type of circumstance motivated the committee's recommendation that IFQ holders be provided legal recourse to pursue civil actions against quota busters). Fishermen who were unable to catch their quota sometimes sued the Dutch government, but without success. The continued race for fish is likely to have provided incentives to maintain or even increase fishing capacity. An additional incentive to invest was provided by a subsidy of up to 12% of the value of the vessel, a scheme that remained in effect until 1986 (these subsidies were not unique for the fishing industry but applied to other sectors of the economy as well).

In 1988, enforcement of the Dutch quota program was tightened substantially. At the peak, there was one controller for every five vessels (or twenty fishermen). The cost of this system was quite high, about 2% of the value of landings or 5-6% of the value added. Since 1993, groups of fishermen have been given some autonomy to manage the activity of their members in a co-management system with self-enforcement.

---

ments" (p. 84). Appendix H demonstrates that enforcement costs have increased for the Alaskan halibut and sablefish fisheries with IFQs.

Higher enforcement costs are not, by themselves, particularly troubling because they can be financed from the enhanced profitability of the fishery. Eliminating the race for fish provides an additional source of revenue to finance enhanced monitoring and enforcement efforts, through reimbursement of costs by the industry. Not only has the recovery of monitoring and enforcement costs become standard practice in some IFQ fisheries (New Zealand, for example), but funding at least some monitoring and enforcement activity out of rents generated by the fishery has already been included as a provision in the most recent amendments to the Magnuson-Stevens Act. (Note, however, that there are few, if any, instances in which a significant share of monitoring and enforcement costs is recovered from open-access or limited-access fisheries.)

Monitoring

In addition to the obvious potential for quota busting or poaching, unreported highgrading and bycatch discards may either increase or decrease with the introduction of an IFQ regime. Gilroy et al. (1996) demonstrated that highgrading did not appear to change in either the Alaskan halibut or sablefish fisheries. It was estimated that bycatch mortality should decrease because fishermen can own quota shares for both species, so that regulatory discards are reduced.

Whether these problems are intensified or diminished by the implementation of an IFQ program depends (in part) on the economic incentives confronting fishermen. The incentives for highgrading, for example, depend on the magnitude of price differentials for various types and sizes of targeted species. As the price premium for fish of a particular size and type increases, the incentive to use quota for especially valuable fish increases along with the incentive to discard less valuable fish.

Incentives for bycatch can vary considerably as well. The more leisurely pace of fishing afforded by IFQs allows fishermen to avoid geographic areas or times when bycatch is more likely. At the same time, the more leisurely pace reduces the opportunity cost of hold space and, consequently, may also provide fishermen with new opportunities to retain a greater proportion of the bycatch as joint products. For example, although the halibut fishery encounters significant bycatches of rockfish (*Sebastes* spp. and *Sebastalobus* spp.) and although most rockfish and thornyheads command high exvessel prices, most of this bycatch was discarded during the derby fishery because halibut were even more valuable. A greater portion of this bycatch is now being retained. On the other hand, implementing an IFQ regime may favor some technologies over others. If the favored technologies typically involve more bycatch, bycatch rates can rise in the absence of enforcement.

Ultimately, therefore, whether highgrading, bycatch, and bycatch discard increase or decrease under an IFQ regime depends both on local circumstances, whether highgrading and bycatch discards are legal (or even required[5]), and on the enforcement response. One way to assess the likelihood of one outcome in relation to another involves comparing fisheries before and after they have implemented some form of IFQs. According to a survey conducted by the OECD (1997):

- "[B]ycatch was reduced in a few IQ [individual quota] and ITQ fisheries and increased in nearly as many" (p. 83); and
- "Highgrading is a concern in many IQ and ITQ fisheries" (p. 83).

---

[5] Bycatch discards are required for undersized fish, those of the wrong gender, or when the allowable biological catch has been reached.

Every monitoring system must identify both the information that is needed to monitor the operation of the IFQ program and the management component that will gather, interpret, and act on this information. Data should also be collected on quota share transactions so that monitoring and analysis of the quota share market can take place. Effective monitoring systems are composed of data, data management, and verification components.

The Data Component. In general, the smooth implementation of IFQ programs requires two different kinds of data. First, periodic data on the condition of the fish stock are needed to evaluate the effectiveness of the program over time. These data are used as the basis for adjusting TACs as conditions warrant. Second, regional councils need sufficient data to monitor compliance with the various limitations imposed by the regulatory system.

Monitoring compliance with an IFQ program requires data on the identity of quota holders, amount of quota owned by each holder, quota use (harvest levels or cage tags), and quota transfers. Where programs have additional restrictions on quota use (such as type of equipment allowed or geographic areas fished) or allow quota transfers only to "eligible" buyers, the data must be complete enough to contain this information and to identify noncomplying behavior in a timely manner.

The precise data needed to implement any specific program depend on the nature of the program. Although it would be impractical to deal with all possibilities in this report, it is also unnecessary since the large number of operating programs now provide a ready supply of models for initiating new monitoring systems.

The Data Management Component. One key to a smoothly implemented IFQ program is ensuring that all data are input to an integrated computer system that is accessible by eligible users on a real-time basis. Such a system would provide up-to-date information on quota use to both users and enforcement agencies. It would ideally also allow short-notice transfers, such as when a vessel heading for shore has a larger than expected bycatch and needs to acquire additional quota for the bycatch species before landing. Facilitating this kind of flexibility would reduce the enforcement burden considerably by giving quota holders a legal alternative to illegal discarding without jeopardizing the objectives of the program. Such an approach is used in New Zealand.

The computer system should also provide easy data entry. Card swipe systems, such as used in the Alaska halibut and sablefish IFQ fisheries, automatically input all the necessary identification data so that only landings (and hence quota use) need to be recorded. It is also possible to have the harvest level recorded directly from the scales (with appropriate adjustments for "ice and slime" or the degree to which the fish are already processed). Entry terminals that are connected to the master computer system should be available at all authorized landing sites.

Data management systems should also facilitate periodic reviews of the effectiveness of an IFQ program. It is noteworthy that the committee's analysis of existing IFQ programs was hindered by the unavailability of the kinds of data needed to examine the costs and benefits of existing IFQ programs or to contrast social and economic conditions before and after program implementation (see Appendix H).

The Verification Component. To ensure the accuracy of reported data, it is necessary to build a number of safeguards into the program. The first of these involves a notification component.

In many fisheries, landings at particular locations are sufficiently infrequent that it is not cost-effective to have an enforcement representative on station at all times. To provide adequate oversight of the recording of landings in these circumstances, it is necessary to ensure that enforcement agencies have some advance notice of the intention to land fish. In the Alaskan halibut and sablefish fisheries, for example, at least six hours' prior notification is required. Not only does this provide an opportunity for enforcement personnel to be present at the landing if desired, it also provides a means of identifying those who are intercepted heading for a landing site to discharge fish illegally without notifying authorities.

Proper control procedures include both onshore and at-sea components. An onshore system of checks would normally include a requirement that sales only be made to registered buyers and that both buyers and quota shareholders cosign the landings entries. These measures would create an audit trail that could be electronically monitored for instances in which a comparison of processed product weight and recorded purchases suggests suspiciously high product recovery rates. The at-sea component would include both onboard observers, where the fishery is profitable enough to bear the cost, and random checks at sea by the U.S. Coast Guard (or perhaps by video monitoring). Some believe that onboard observers are generally needed in IFQ fisheries in which bycatch and highgrading are expected to be problems.

Enforcement

Successful enforcement processes involve three essential components: (1) effectively coordinating onshore and at-sea enforcement activities, (2) devoting adequate resources to the enforcement process, and (3) targeting resources at the most important noncompliance problems.

In the United States, coordination of enforcement responsibilities is more difficult for fisheries than for other natural resources because multiple agencies are involved and fishing activities take place over a broad geographic area. Since many fisheries involve landings in multiple states and foreign nations, onshore enforcement may need to be coordinated among several agencies.[6] Furthermore,

---

[6] The need for coordination can be illustrated by an example from the Canadian halibut IVQ fishery. Canadian authorities require that all commercially harvested halibut and sablefish be permanently marked to distinguish them from sport harvested fish. Canadian fishery managers have recently asked the IPHC to encourage the NPFMC to require that all commercially harvested Alaskan

at-sea enforcement is the responsibility of the Coast Guard, a federal agency with an entirely separate chain of command from states or NMFS.

This multiplicity of enforcement agencies not only makes coordination of activities difficult, but also makes the coordination of funding difficult. Different agencies have different priorities (the Coast Guard also has drug interdiction and search and rescue responsibilities, for example), and this may lead to inadequate funding of fisheries enforcement. (At the committee's hearing in Alaska, the view that Coast Guard and NMFS enforcement funding is inadequate was expressed repeatedly by state officials and by a letter from the NPFMC to NMFS.)

A successful enforcement program also requires a carefully constructed set of sanctions for noncompliance. Penalties should be commensurate with the danger posed by noncompliance. Penalties that are unrealistically high may be counterproductive if authorities are reluctant to impose them and fishermen are aware of this reluctance. Unrealistically high penalties are also likely to consume excessive enforcement resources as those served with penalties seek redress through the appeals process. In many cases, predetermined administrative fines can be imposed by the enforcing agency itself for "routine" noncompliance. For example, the Alaskan IFQ programs allow overages of up to 10% above the fisherman's remaining IFQ balance to be deducted from the next year's IFQ permit amount. Overages greater than 10% are considered a violation and are handled by enforcement personnel. In an ideal system, more serious noncompliance in terms of either the magnitude of the offense or the number of offenses could trigger civil penalties (fines and possible seizure of catch, equipment, and quota). Criminal penalties should be reserved for falsification of official reports and the most serious violations.

Income levels from fishing are generally bolstered by the implementation of an effective IFQ program. An effective program presumes effective enforcement. Honest fishermen should be willing to contribute some of their increased rent to ensure the continued existence of an effective IFQ management regime.

Duration of Individual Fishing Quota Programs

Some arguments have been made for limiting duration of IFQ programs, and in fact, the House of Representatives version of the Sustainable Fisheries Act of 1996 contained a "sunset" provision for existing IFQ programs. Such approaches are based primarily on equity considerations, to prevent quota from being assigned in perpetuity to the original recipients. Although the committee does not

---

halibut and sablefish be similarly (but uniquely) marked before sale. The motivation for their request arises from a concern that Canadian fishermen may have an opportunity to conceal black-market sales by neglecting to mark their fish and claiming that the unmarked fish were obtained legally from Alaskan IFQ harvesters for subsequent resale.

favor sunset provisions, it provides suggestions in the next chapter for how a limited duration IFQ program might be developed.

Impact on Fishing Communities and Other Fisheries

In social science literature, a community is often considered to be a relatively small, usually residential, spatially bounded unit, whose members deal with one another on a daily basis. For example, it has been defined as "the maximal group of persons who normally reside together in face-to-face association" (Murdock et al., 1945, p. 79). This type of definition has problems, in that such spatially bounded entities vary a great deal in the extent to which the people residing in them exhibit what is often referred to as a "sense of community." On the other hand, people who do not reside in the same locality, but who regularly work together in the same place or are fellow members of the same religious congregation, may have a strong sense of community. What makes for a community is not necessarily proximity but awareness of shared interests and concerns (Dyer and McGoodwin, 1994). Those who live and work together in close daily association are likely to share a number of interests. The greater the number of interests they share and the more intensely they feel about them, the greater is their sense of community likely to be. Small, relatively isolated, residential units that depend for their livelihood on a single industry are likely to have a strong sense of community in regard to this industry as a focus of intensely felt, common interest. Such communities of interest can be found in urban neighborhoods, and also in occupational enclaves within large urban settings and in farming and fishing communities. The effect of change on a community is larger or smaller in proportion to how the common interests that give rise to a sense of community are affected.

Culture, Community, and Individual Fishing Quotas

Many fisheries are based in coastal communities that receive significant economic inputs from the fishing industry. Coastal communities in turn, with assistance from state and federal governments, often provide the shoreside infrastructure (harbor jetties and docking facilities) that support the fisheries. In addition, they may supply important services to fishermen, such as food, fuel, and maintenance facilities. Many coastal communities are made up of multiple generations of families engaged in fishing. For others, there is significant movement of individuals into and out of the fishery over time so that even though there is a relatively constant presence of fishermen in the community, different people and families are represented over time. Coastal communities sometimes offer only limited alternative employment opportunities for displaced fishermen and fishing industry workers.

The social, economic, and cultural characteristics of U.S. fisheries are as diverse as the characteristics of the fish stocks and fish habitats. Some are small-scale, artisanal fisheries conducted in estuarine areas or close to shore in small vessels with low levels of technology. Others are large-scale, industrialized fisheries conducted coast- if not worldwide in large, highly capitalized vessels with sophisticated technology (McGoodwin, 1990). Some fisheries are traditionally conducted by members of communities with specific national or ethnic characteristics, whereas others are composed of large proportions of fishermen relatively new to the occupation and with no common social, economic, or cultural background. Some fisheries are primarily recreational (Spanish mackerel, striped bass); others are primarily commercial. Subsistence and ceremonial fishing are components of many U.S. fisheries.[7]

Significant regional and geographical variation can be observed in the social, cultural, and economic characteristics of fisheries. In Alaska, the Western Pacific region, Caribbean Sea, and many rural areas of the contiguous United States, fishing communities are small and relatively isolated, with few occupational alternatives and high levels of community cohesion. On the other end of the spectrum, participants in other fisheries are embedded within large metropolitan areas or may be composed of a significant number of fishermen with higher levels of education or training who have gravitated to fishing as a life-style choice or have relatively marketable skills in other occupations. All of these differences will have implications for the degree to which various policy objectives may be achieved with particular management options.

Communities can serve as important participants in fisheries management in situations in which institutional arrangements are developed by resource users and others to manage the resources (McCay and Acheson, 1987; Ostrom, 1990; Rieser, 1997b). The public debate about access rights in fisheries management has focused primarily on exclusive individual harvest privileges such as IFQs, neglecting alternative harvest access regimes, including those that build upon human communities and involve contractual "co-management" relationships between fishing communities (whether communities of place or of interest) and government (Rieser, 1997b). The concepts of "human ecology" (Pálsson, 1991) and "embeddedness" (Granovetter, 1985; McCay and Jentoft, 1998) emphasize unavoidable interactions among culture, ecosystems, and the economy, and the methodological flaws of the theories representing economics and public choice as merely the summation of individual actions. In the context of questions about

---

[7] The role of processors and fish buyers and the extent of harvester dependence on them may vary among communities. Sometimes processors provide jetties, docking facilities, and other infrastructure for fisheries, which are one dimension of the dependence of fishermen on processors. Processors may also lend fishermen money to buy quota shares or even buy quota shares on behalf of fishermen who promise to deliver their catch to the sponsoring processor, although some U.S. IFQ programs prohibit nonfishermen from purchasing quota shares.

proper governance or management of "the commons," these authors argue for more attention to the search for alternatives and complements to (1) top-down "command-and-control" approaches and (2) private property, market-based approaches (e.g., IFQs) to environmental problems. In particular, they support current efforts to (1) improve the degree and nature of public participation in natural resource management; (2) develop and build on systems of community-based resource management; and (3) experiment with public-private partnerships in resource management, or co-management (Felt et al., 1997). IFQs and other fishery management tools have profound effects on human communities and can be designed, through participatory processes, to meet the needs and concerns of these communities (McCay et al., 1998). That is, they can be designed to reinforce community structures and to formalize common property regimes. The tendency for IFQs programs is to create a new community of interest, the holders of IFQs, whose interests and goals can diverge sharply from the rest of the community (McCay et al., 1998). Consequently, attempts to link IFQs with larger community values and interests must take great care to design IFQ programs that will in fact reinforce desired community structures and formalize common property regimes.

In many cases, common-pool resources have been rationally managed for centuries (Hanna, 1990). The likelihood of an effective agreement on resource use will depend on the resource in question, the chances of effective monitoring and enforcement, the ability to exclude outsiders, and the sense of community among resource users. IFQs are most likely to be successful when they reinforce communities by creating legal protection for informal mechanisms of common property management. This will be true only if the legal protections build on, rather than break apart, community capacities for common property management.

Community-Based Governance

The implicit assumption deriving from the influential arguments of Gordon, Hardin, Christy, and Anderson, is that fishermen will not take into account the effects of their own actions on each other and on the fish stocks they target. Even if they do attempt to consider these impacts, including expending ever greater effort to maintain harvest levels in the presence of declining populations or increasing competition for fish, they continue to face incentives to renege on any voluntary arrangements to limit harvesting effort, leading to the collapse of such arrangements. Consequently, under these assumptions one of two approaches must be taken: (1) a government agency must be given sufficient authority to impose and enforce rules and regulations that would induce fishermen to change their actions and limit the adverse impact of their uses of fisheries, or (2) fishermen must be given well-defined individual privileges to a portion of the harvestable fish stock. Individual harvest privileges presumably focus fishermen's attention on effectively harvesting their own share of the quota. If fishermen are granted secure interests in a proportion of a TAC, they may attempt to harvest it as efficiently as possible, thereby reducing the undesirable effects on other fishermen and on the fish stock experienced under an open-access situation.

However, these two approaches to addressing the inefficient harvesting of fish ignore the hundreds of examples of fishing communities that organized themselves and have effectively managed their access to and use of fish stocks on which they were heavily dependent (e.g., NRC, 1986; McCay and Acheson, 1987; Berkes, 1989; Martin, 1989; Bromley, 1992). These community-based governing arrangements are not historic anomalies. Rather, they represent a viable alternative to central government and market-based approaches to addressing biologic, economic, and social problems related to fishing.

Communities of fishing people that have developed long-standing, successful (legal and extralegal) arrangements for governing their use of fish stocks and fishing grounds share several general characteristics:

• They are capable of effectively excluding outsiders from their fishing grounds. Effective exclusion may be based on physical, economic, or cultural isolation, legal authority, or other mechanisms.
• They have existed for long periods of time. Most likely, their families have fished in the areas for decades, and they want their children and grandchildren to have the opportunity to fish there in the future.
• They have extensive experience with their fishing grounds and the stocks they fish. They possess good information concerning the structure and functioning of their fishing grounds and variations over time.
• They share norms of trust and reciprocity.
• Community forums, whether the local bar or a social club, provide opportunities for community members to discuss and resolve shared problems.
• Community members have access to trusted conflict resolution mechanisms that allow them to settle their differences relatively peacefully (Ostrom, 1990; Schlager and Ostrom, 1992; Schlager, 1994).
• The communities are relatively small and largely homogeneous.
• The harvested stocks are largely independent of stocks outside the community's sphere of influence (largely nonmigratory), or the community is able to coordinate its actions with those of other communities.

The governing arrangements that guide and constrain communities' uses of fishing grounds also share several general characteristics:

• The rules are carefully matched to the situation or problem. The problems are those that community members have some control over and that can be resolved or reduced by changing actions and strategies.
• The rules are easily monitored, typically while fishermen are on the water

harvesting fish. It is oftentimes in the self-interest of individual fishermen to monitor the rules; otherwise, they may be prevented from fishing.

• Rules are typically enforced first by threats and modest social sanctions. Only after repeated rule breaking are stronger sanctions imposed.

• The rules reflect communities' notions of fairness, particularly in the distribution of costs and benefits.

• Communities' norms and values interact with the rules governing access and use of fisheries, with each supporting the other. Rules governing fisheries resolve problems and reduce conflict in ways that support and help maintain the community of fishermen. Conversely, the community of fishermen in designing, implementing, and enforcing the rules supports their maintenance (Schlager, 1990, 1994). Depending on how the rules related to initial allocation and transferability are set up, IFQs could increase or decrease the security of a community's stake in a fishery.

Community-based governing structures possess several advantages over central government management[8] and market-based management. First, community-based governing structures are based on local norms, values, and information and are matched to the situation. Government officials rarely, if ever, have access to the type of information that would allow them to design appropriate governance structures, unless they sponsor research that provides the necessary information. Second, community-based governing structures maintain the community and its norms of fairness. The interests of central government and the values of market-based approaches do not routinely give a high priority to the value of maintaining a community as such, nor are they likely to reflect a community's interests and values (Goodenough, 1963), although regulated market-based systems such as IFQs can be designed to do so (see Box 4.5). Third, monitoring and enforcement may be less troublesome and costly with a community-based system. Individuals who devise rules by which they will be governed are more likely to follow them and monitor others for compliance (Ostrom, 1990; Tang, 1992; Schlager, 1994). It is reasonable to conclude that IFQ programs are more likely to be successful if representatives of the relevant fishing communities have been active participants in devising the program and/or if such communities are themselves recipients of IFQ shares and are left to devise their own procedures for allocating these shares and monitoring their use.

Community-based governance is not necessarily incompatible with central government or market-based designs for managing fisheries. In cooperation with them, community-based management can deal with issues it cannot easily manage alone. Cooperation with central governments may make it possible to deal

---

[8] The regional council system is an approach to co-management and is decentralized in many of its components. It generally is not, however, community-based management.

effectively with larger, regional issues that extend beyond any single community in ways that meet the concerns of government (including the public trust), economic efficiency, and the maintenance of community.

Certain cases of fisheries management, such as those involving transnational migratory fishing fleets, highly migratory or anadromous fishery resources, or fisheries with a broad range of geographically dispersed constituencies pose particular challenges to the development of efficient, effective, equitable systems of management that allow some degree of responsibility or authority on the part of fishery constituents while conforming to overall public trust principles. The development of fishery policy and management with the maximum involvement of all constituents and sensitivity to both cultural traditions and broader public trust principles is clearly an appropriate goal.

The impact of IFQs on fishing communities is likely to be more or less stressful depending on how fishing and fishing-related activities are organized, the isolation of communities, and their ability to switch among fish species as stocks and exvessel values fluctuate. The members of some fishing communities derive significant income from work in processing plants located in their communities. The extent to which the impact of IFQ programs on processor-dependent communities can be mitigated is a matter to be determined for each fishery and each community and merits serious discussion when IFQ programs are contemplated.

Less isolated fishing communities, such as those found in New Bedford and Gloucester, Massachusetts, pose somewhat different problems. These and other communities like them in the northeastern United States have been intensively engaged in commercial fishing for at least several generations. Fishing is built into the established way of life and social values, and is a main contributor to people's sense of worth as respectable citizens. These communities have resisted the imposition of TACs and any kind of fishery management program that would effectively reduce overcapitalization by significantly limiting the number of people and vessels engaged in fishing. In doing so, they have contributed to developing crises in their fisheries that earlier, prudent management programs may have mitigated.

Some fishing communities may suffer a tension between the demands of fishing and other aspects of life. In New Zealand (Levine and Levine, 1987), IFQs shifted the competition among fishermen in the community from competition for fishing the ocean to competition in the marketplace for quota shares. At the same time, in other respects, the community's members valued maintaining a sense of community in spite of the divisiveness of fishing. The way management programs are set up can serve either to exacerbate the tension between competition and community or to reduce it. The Canadian Pacific coast fisheries managed on the basis of individual vessel quotas (IVQs) have proposed significant measures to encourage community development and proper treatment of crew (Box 5.8).

BOX 5.8
Groundfish Development Authority in Canada

In the Canadian West Coast groundfish trawl fishery, the Pacific Region of the Department of Fisheries and Oceans has established procedures whereby 20% of the annual groundfish trawl TAC is set aside and managed by the Groundfish Development Authority, which is made up of representatives from coastal communities and fishermen's unions. The Groundfish Development Authority (GDA) is composed of seven voting members. The GDA was established in 1997 as the result of agreement between the Department of Fisheries and Oceans and various fishing industry and coastal community participants. The GDA reviews joint proposals from processors and shareholders who must commit IVQ shares to match the quota shares granted by the Authority. The purpose of the GDA is "to ensure fair crew treatment, to aid in regional development, to promote the attainment of stable market and employment conditions and to encourage sustainable fishing practices" (GDA, 1998). The proposals are rated on the basis of things such as processing and catch history, and fair treatment of crew members, and with respect to development objectives such as stabilization of employment, sustainable fishing practices, and training opportunities for new entrants.

The 20% of the groundfish trawl TAC controlled by the GDA is authorized to be used in two separate programs: (1) 10% of the total TAC can be allocated to a Groundfish Development Quota to aid in regional development in coastal communities, and (2) 10% of the total TAC can be allocated to a Code of Conduct Quota for the purposes of protecting the interests of crew members under the new IVQ management plan. The GDA receives proposals for both Groundfish Development Quotas and Code of Conduct Quotas. Based on the criteria established for the programs and the ranking given the proposal by the Minister of Fisheries, an area- and species-specific quota will be allocated to those submitting proposals.

Groundfish Development Quota proposals can be submitted by properly licensed vessel owners and processors. The GDA will evaluate how well the proposal meets qualifying criteria. These criteria include contributing to market stabilization, maintaining existing processing capability, stabilizing employment in the groundfish industry, contributing to economic development in coastal communities, providing economic benefits, increasing the value of groundfish production, providing training opportunities, and maintaining sustainable fishing practices. The proposals are ranked by the GDA, and quota is allocated for the Groundfish Development Quotas.

Code of Conduct Quotas are allocated to vessel owners unless there have been valid complaints received from crew members of a vessel indicating that they have been asked to contribute to the cost of the vessel's original IVQ allocation; if they are coerced into contributing to the leasing of additional IVQ or any other costs not traditionally associated with the operation of the vessel; or if there are indications that crew safety is being compromised. If the complaints are verified, the GDA can recommend that the Minister of Fisheries withhold the Code of Conduct Quota from the vessel owner.

Participants in Decisionmaking Processes

Who should participate in decisionmaking processes engenders considerable debate. A consensus is emerging among scholars and practitioners that those most directly affected by natural resource management should participate directly in devising management rules and regulations (Ostrom, 1990; Hanna, 1995, 1997). Users possess critical time and place information about natural resource stocks and about community norms and values. Furthermore, resource users bear the benefits and burdens of management regulations. In many cases, adopting new rules profoundly changes their lives.

Although it may be accepted that resource users should actively engage in defining, implementing, monitoring, and enforcing any management system, there is less agreement as to who should be considered a resource user and the level of participation that should be allowed.

Many people who do not directly harvest a resource nevertheless have made substantial investments in it. For instance, the families of harvesters, businesses that provide supplies and equipment to harvesters, processors of the harvested product, and the communities in which harvesters live, all have direct ties to the resource. Whether, and to what extent, such people and organizations should participate in designing management systems is subject to debate. Each group has different ties to the resource, a different set of interests, and perhaps, a different mixture of values. Likewise, there are nonconsumptive users of fish who may be interested in preserving fish, their predators, or the existing ecosystem structure. Defining the participation boundaries too narrowly increases the likelihood that important individuals and interests will not be represented, affects the quality of decisions, and makes these decisions vulnerable to external intervention. Excluded interests may take advantage of appeals processes or other avenues to nullify decisions (Hanna, 1994, 1995).

On the other hand, defining participation boundaries too broadly increases the likelihood that conflicts among represented interests will increase, making decisions more difficult to achieve and/or compromising their quality (Ostrom et al., 1994). Individuals without a direct stake in the natural resource, who have the authority to participate in decisionmaking, may introduce values and issues that are tangential to natural resource management. Although scholars and practitioners understand the problems of defining the set of participants too narrowly or too broadly, there are no widely accepted procedures for determining who should participate in decisionmaking. One factor that has been important for fishery management in Iceland is the political capacity of various components of fishing industries to respond. This raises the issue of the organization of various dimensions of fisheries and their political influence. In some areas, sportfishermen and environmentalists have considerable influence; in some, processors are well organized and influential; and in some, vessel owners are organized.

Decisionmaking Structure

Not only do the participants involved affect the quality of decisionmaking, so does the structure of decisionmaking. There are several issues involved in the structure of decisionmaking: (1) the role of resource users, (2) the types of problems resource users are asked to address, (3) the transparency of decisionmaking processes, (4) the deliberative aspects of decisionmaking processes, and (5) the information that is part of the decisionmaking process.

In decisionmaking processes, the role of natural resource users varies considerably. At one end of the participation spectrum, users may be asked to comment on management plans devised by managers. Under this scenario, *managers* are entrusted with the authority to define natural resource problems, devise regulations to address these problems, devise implementation structures for administering and monitoring the regulations, and solicit resource users' comments at various points in the development and implementation of the plans. At the other end of the participation spectrum, *resource users* define natural resource problems, devise regulations to address these problems, and devise implementation structures for administering and monitoring the regulations. In this scenario, managers act to support and facilitate users' decisionmaking activities.

How well a natural resource management system performs is determined in part by how well it is accepted by users of the resource. A management system is most likely to be well received if it addresses the problems experienced by resource users in ways that they perceive as legitimate and fair. This implies that resource users' participation in decisionmaking must extend beyond commenting (Pinkerton, 1989; Ostrom, 1990; Hanna, 1995).

Natural resource user participation may be structured in a variety of ways. Many U.S. federal agencies are experimenting with various procedures. For instance, the U.S. Environmental Protection Agency supports the creation of community advisory groups in relation to Superfund sites and national estuaries. Likewise, cattle ranchers and others participate in resource advisory committees sponsored by the Bureau of Land Management. Also, of course, fishermen and others participate in regional fishery management councils. Creating procedures that integrate resource users' participation in decisionmaking, however, does not ensure that they will choose to participate or will participate in meaningful ways. Participation requires a substantial commitment of resources. If resource users believe that their participation will be of no consequence in devising regulations or if they believe that their decisions may be easily overturned by managers, resource users are unlikely to invest in participation.

Norms, values, and expectations shape the outlooks and actions of individuals as they make use of fisheries. By ensuring the active participation of resource users in designing and implementing fishery management systems, users' norms, values, and expectations are more likely to be taken into account. Since management regimes are simply collections of rules, and rules guide and shape people's behavior only to the extent to which they are followed, management approaches must, to a large extent, be consonant with a community's norms if they are to be meaningful. Norms and values concerning fairness, reciprocity, and work effort, among others, constrain and guide the types of rules and regulations most appropriate for a community. A community's sense of fairness concerning who should have access to a resource, how the resource should be used, and how rights of use should be transferred to others or passed to future generations must be accounted for in designing management systems if these systems are to be followed and not fought.

Furthermore, crafting management systems to better fit the characteristics of the fishery and the groups of people who use the resource may reduce the costs of implementing, monitoring, and enforcing a system, as well as enhance the probability that desired outcomes will be achieved. Management systems that closely match the physical and cultural environments in which they operate should reduce monitoring and enforcement costs because rule-following behavior is likely to be enhanced. People are more likely to commit to a set of rules if they believe these rules are fair, if they believe that other members of the community are committed to following the rules, and if noncompliance is observable and subject to meaningful sanctions. Well-designed management systems to which resource users are committed will also elicit significant levels of mutual monitoring and enforcement.

Process for Design and Adoption of Individual Fishing Quota Programs

Are regional councils the best forum for making the decision to adopt an IFQ program for a particular fishery? Testimony to the committee reflected the concern that at any given time, a council may include neither adequate voting representation of all of sectors that would be affected by the design choices in an IFQ program nor the broader public interest. The council appointment process is a political one, carried out by the governors of states in the council region and the Secretary of Commerce, and politics can skew the voting membership in favor of a more powerful sector of the commercial or recreational fishery. To some extent, however, the participatory nature of the council process can moderate unequal voting representation on a specific council.

The Secretary of Commerce is limited by the Magnuson-Stevens Act and by political realities in the degree to which he or she can correct for unbalanced representation and decisions in the design of IFQ programs. Currently, the Secretary may not approve or implement any fishery management plan, plan amendment, or regulation that creates a new IFQ program before October 1, 2000 (Sec. 104[d][1][A]). The Gulf of Mexico Fishery Management Council is further restricted from submitting and approving an IFQ program for the commercial red snapper fishery unless the preparation and submission of the IFQ program to the Secretary are approved in two separate referendums (Sec. 407[c]).

Even if the moratorium on the approval and implementation of IFQ programs were lifted, there are other procedures that the Secretary of Commerce must follow in order to approve any proposed IFQ program. The Secretary may disapprove or partially approve a council-developed IFQ program for inconsistency with the national standards (Sec. 304[a]), other provisions of the Magnuson-Stevens Act, or other applicable law, but the Secretary cannot rewrite a proposed IFQ program to change, for example, the criteria for the initial allocation (Sec. 304[a][1]). The Secretary may not adopt a provision establishing an IFQ program unless such a system is first approved by a majority of the voting members of the council (Sec. 304[c][3]). The act gives the Secretary the power to reject an IFQ program when he or she deems the initial allocation to be skewed in favor of sectors that have more weight on the councils, under National Standard 4, which requires any allocation of fishing privileges to be fair and equitable to all fishermen, to be reasonably calculated to promote conservation, and to give no particular individual, corporation, or other entity an excessive share (Sec. 301[a][4]).

Both the design and the implementation of IFQ programs can be delegated to other units. Examples of such co-managed or delegated authority regimes are found in the Canadian mobile gear groundfishery of Nova Scotia (see Appendix G; McCay et al., 1995, 1998) and the IFQ fishery of The Netherlands.

6 | Findings and Recommendations

GENERAL CONSIDERATIONS ABOUT INDIVIDUAL FISHING QUOTAS

The committee believes that individual fishing quotas (IFQs) can be used to address a variety of social, economic, and biologic issues in fisheries management. Alternative management systems can also achieve some of the objectives that can be achieved with IFQs. There are no general threshold criteria for deciding when IFQs are appropriate; the use of IFQs should be considered on a fishery-by-fishery basis. **IFQs can be used in a preventive manner with stocks that are not overfished or to remedy existing overfishing, overcapitalization,[1] and incentives to fish under dangerous conditions.** In general, the committee believes that IFQ programs will be more successful when the following conditions exist:

- **The total allowable catch (TAC) can be specified with reasonable certainty.** Where TAC-based management is not possible, other types of individual quota systems, such as individual transferable effort quotas, may be more appropriate.
- **The goals of improving economic efficiency and reducing the numbers of firms, vessels, and people in the fishery have a high priority.**

---

[1] Capitalization is the total dollar value invested in a fishery. Overcapitalization is therefore the existence of a greater financial investment in harvesting or processing capacity than is efficient to catch and process the available fish. As a financial concept, it differs from the idea of overcapacity, or excess capacity.

- **Broad stakeholder support and participation is present.** Although consensus is not necessary, active stakeholder involvement throughout the design, implementation, and operation of an IFQ program is crucial.
- **The fishery is amenable to cost-effective monitoring and enforcement.**
- **Adequate data exist.** Because of the long-term impacts and potential irreversibility of IFQ programs, it is important that **sufficient data are available to assess and allow the mitigation of, insofar as possible, the potential social and economic impacts of IFQs on individuals and communities.**
- **The likelihood for spillover[2] of fishing activities into other fisheries is recognized and provision is made to minimize its negative effects.**

IFQ-based management can be particularly useful, but more difficult to develop and administer, when fisheries have evolved to a point of overcapitalization. When IFQs are applied to overcapitalized fisheries, they can be expected to result in a reduction in the number of participants. However, IFQs can be even more valuable as a preventive measure when applied to fisheries that are not already in trouble.

As discussed in greater detail later, decisions to implement IFQs or to use alternative methods of fishery management should be handled by regional councils, rather than at the national level. The committee believes that fishery management—including the development of IFQs and other management programs—should continue to be the responsibility of the regional councils, subject to review by the Secretary of Commerce.

Evidence of the effects of IFQs for the conservation of fish stocks is mixed and there are few generalizable statements of fact that can be made (ICES, 1996, 1997). However, to the extent that IFQs are enforced, they can keep harvests within a TAC; open-access fisheries often exceed their TACs. Both OECD (1997) and the committee's examination of U.S. and foreign IFQ programs indicate that IFQs may increase or decrease bycatch discards and highgrading, depending on the fishery. It can be demonstrated, however, that highgrading is unlikely to be profitable (see Box 3.4). Neither the existence of quota busting nor the lack thereof have been demonstrated as a general feature of IFQs.

Discussed below are a number of recommendations to Congress, the Secretary of Commerce and the National Marine Fisheries Service, the regional fishery management councils, states, and other fishery stakeholders related to a national policy for IFQs. Action on some of the recommendations in this report will require changes to the Magnuson-Stevens Act, regulatory language developed by the Secretary of Commerce, or rules implemented by regional councils.

---

[2] Spillover occurs when one fishery becomes more restrictive in area, time, or number of licenses available and fishermen shift to other fisheries, increasing the capitalization and effort in these fisheries. This shift is possible when fishing skills and equipment are relatively transferable among fisheries with minor adjustments.

POLITICAL STRUCTURE AND JURISDICTIONAL ISSUES

Moratorium

Findings: The individual fishing quota is one of many legitimate tools that fishery managers should be allowed to consider and use. Sufficient experience and analysis of existing programs is available, both nationally and internationally, to suggest that IFQs can address some fishery management problems that are not easily addressed with other measures. Specifically, IFQ programs can have advantages over alternative management measures in addressing problems of overcapacity, efficiency, and utilization, if appropriately designed in relation to other objectives.

Recommendation: Congress should lift the moratorium on the development and implementation of IFQ programs established by the Sustainable Fisheries Act of 1996, provided the other recommendations and suggestions of this report are considered and followed. Furthermore, the existing federally managed IFQ programs (Mid-Atlantic surf clams/ocean quahogs, Southeast Atlantic wreckfish, and North Pacific halibut and sablefish) should be allowed to proceed under the stewardship of their respective councils, again with the committee's recommendations in mind.

A related issue involves the two proposed systems that were stopped by the moratorium, the Pacific coast fixed-gear sablefish fishery and the Gulf of Mexico red snapper fishery. It would be desirable in both fisheries to take advantage of the work expended in developing the plans and to avoid changing control dates that would force the fisheries to start from a new, more intensely capitalized condition than before the moratorium. Both fisheries show evidence of overcapitalization (MRAG, 1997; PFMC, 1997).

Roles of Regional Councils Versus the National Marine Fisheries Service (NMFS) and Congress

Findings: Circumstances in fisheries vary widely and require different mechanisms to address the diverse conditions. Regional management is more likely than a national authority to be able to respond effectively to regional biologic, economic, and social conditions. It is a general principle that in dynamic, complex systems, it is better to design interventions as close as possible to the source of the problem (see later section on the delegation of management to local authorities). The Magnuson-Stevens Act creates a forum for the development of fishery management plans (FMPs) by those directly involved in the fisheries of each region. However, the history of the act's implementation provides many examples of congressional intervention in the regional management process (Shelley et al., 1994; NRC, 1997). The committee received testimony and has found examples nationwide of congressional action to prevent approval of council-submitted management programs or to stall implementation of Secretary-approved plan amendments.

Recommendation: Congress and the Secretary of Commerce should allow the regional fishery management councils flexibility to adjust existing IFQ programs and develop new ones, subject to normal review by the Secretary and consistent with the national standards of the Magnuson-Stevens Act.

State-Federal Interaction and Communication

Findings: Adoption of a federal IFQ program can have significant impacts on state management authority in state waters and vice versa. Fishing effort displaced from newly developed limited entry programs in federal waters can shift to state waters (e.g., in surf clam fisheries in New Jersey and New York), placing additional stress on fisheries therein. A lack of coordination among federal and state management authorities can undermine the effectiveness of fishery management programs such as IFQs. A lack of consistent measures can create loopholes that encourage quota violations, underreporting, and other problems. Coordination and cooperation between state and federal managers can increase the effectiveness of limited enforcement resources and improve the scope and quality of data collection programs. Because the division of management authority between legislative and executive branches of government varies among states, state fishery officials who are members of the regional councils may or may not have the authority to ensure that compatible state measures are adopted to complement a federal IFQ program for transboundary stocks.

Recommendations: Regional councils should—at their earliest opportunity—officially inform affected state fishery agencies that they are considering adoption of an IFQ program for fisheries that occur in both federal and state waters. Councils should seek the assurance of state authorities that complementary measures will be adopted in the state waters, including consistent controls on recreational or other fisheries that are not included in the IFQ program. Consistent controls could include coordination of enforcement activities, common open fishing periods, and cooperative agreements for transboundary stocks and limited access programs. The committee's intention is not to advocate the usurpation of state authority by the councils. Instead, the committee simply seeks to stress that the consequences and effectiveness of management measures adopted in one jurisdiction depend to a large extent on management measures adopted in adjacent jurisdictions. For example, implementation of a restrictive limited access program in federal waters could lead to a flood of effort into adjacent state waters or vice versa. Proposed regulations implementing a federal IFQ program should identify the manner in which relevant state fishery policy and regulations could be made compatible with the federal program and which state regulations will continue to apply to vessels fishing in the federal IFQ program. Conversely, if states in a region have developed coordinated and effective limited entry programs in state waters, including IFQs, the regional councils should complement these programs in fishery management plans, where consistent with the national standards of the Magnuson-Stevens Act.

Delegation to Local Authorities

Findings: The use of IFQs in fisheries management does not preclude, and may in some circumstances benefit from, delegation of some management decisions to subregional or local authorities. For highly localized and relatively discrete fish stocks and/or for geographically bounded fisheries, delegation to the local level can bring biologic, economic, and social benefits. Examples of such functions might include gear restrictions and partnerships between government and industry for purposes such as scientific research, selection and monitoring of closed areas, and decisions about allocation among individuals or groups. Among the expected benefits of delegation are greater input of locally derived knowledge and experience into the decisionmaking process, greater compliance with the rules due to improved participation of stakeholders, and the ability to tailor rules to local conditions. Costs are also associated with such delegation, including the need to coordinate actions taken by such management units within frameworks established by regional councils.

Traditionally, fisheries management under the Magnuson-Stevens Act has not taken full advantage of the use of local communities and authorities in the development and implementation of policy. However, the act allows the regional councils and the Secretary of Commerce to delegate management under certain situations, and the councils were authorized (before the Sustainable Fisheries Act's moratorium) to create IFQ programs that may be received or held by associations or units of local government. Local delegation and the pooling of resources are illustrated by the Dutch sole and plaice beam trawl fishery, Japanese inshore cooperatives, and certain Nova Scotia fisheries.

Recommendations: In considering the range of management options for a specific fishery, the regional councils should not be precluded from considering proposals for delegated management authority that would operate within the framework of the Magnuson-Stevens Act's national standards and NMFS regulatory guidelines.

DESIGNING AN IFQ PROGRAM TO MEET SOCIAL AND ECONOMIC OBJECTIVES

General Issues

Finding: Fishery management systems, including IFQ programs, are sometimes designed with unspecific or conflicting objectives. Uncertainty about the relative importance and measurability of the objectives confuses the design of an IFQ program and makes its implementation less effective. Confusion, conflict, and ambiguity about the relative importance and value of the objectives of an IFQ program can result in contradictions and inconsistencies in its design and implementation, making the program more vulnerable to unintended consequences and less likely to succeed. Goals and objectives are central to IFQ program design. If economic efficiency and rapid downsizing of a fleet are the major objectives, quota shares should be freely transferable, be as divisible as possible, and have long-term tenure. If other major design objectives are paramount or there are conflicting objectives, these central design features may have to be changed. Objectives of fishery management often differ by region and even by fishery within a given region.

Recommendation: The biologic, social, and economic objectives of each fishery management plan (and how a limited entry or access program, including IFQs, will achieve the objectives) should be specified clearly through a process that invites broad participation by stakeholders. Similarly, at the plan development stage, the potential impacts of each alternative management option considered should be specified clearly. This could be accomplished by requiring that limited entry programs proposed in FMPs document the likelihood of the possible outcomes and alternatives to achieve plan objectives (e.g., through a "limited access assessment"). Congress should recognize that the design of an IFQ or other limited entry program in relation to concentration limits, transferability, distribution of quota shares, and other design questions will depend on the objectives of a specific plan, underscoring the importance of providing flexibility to the regional councils.

Path Dependence

Findings: Once an IFQ program is considered, a series of events ensues that may lead to unintended or unexpected consequences that may be difficult to reverse or mitigate. This path of events occurs to some extent with all management measures, but it is particularly true for limited access programs, including IFQs, that involve a fundamental restructuring of social and economic relationships and the creation of expectations about secure privileges of individuals. Depending on the particular fishery and the design of the IFQ program, it may create a new class of stakeholders—those granted IFQs—with potentially different interests and views than existing shareholders, many of whom may not qualify to hold IFQs despite their previous or current involvement in the fishery. Moreover, resulting perceptions of unfairness and inequity may affect the manner in which stakeholders interact with the management process in the future.

In addition, it is widely recognized that although IFQs are limited privileges and may be legally revocable, political pressure from permit and quota shareholders concerned about protecting their investments will resist revocation. This is evidenced in other natural resource sectors, such as mining and ranching, when reduction in privileges of access to public resources are challenged by those who benefit from them.

The extensive literature and testimony received indicate that insufficient attention and resources have been devoted to socioeconomic impact assessments prior to decisions about IFQs, and to monitoring and evaluating the performance and consequences of IFQ programs once in place.

Recommendations: **Councils should give high planning priority to the question of social, economic, and biologic consequences of an IFQ program or alternatives to it.**   This requires projections of likely consequences based as much as possible on rigorous impact assessment and monitoring and evaluating the subsequent development of a limited access management regime. The regional councils and NMFS must allocate more resources and attention to impact assessments, which are now required by law but often are given inadequate attention.

IFQ programs should include a commitment to monitor both short- and long-term impacts and to include in the program the political, financial, and administrative ability to make changes as required to meet original objectives. At a minimum, the regional councils and the Secretary of Commerce should ensure that a preliminary study of the relevant socioeconomic aspects of a fishery being considered for IFQs be done prior to the design of the management program, that alternative limited access management programs be considered, and that a monitoring and evaluation program be part of the initial design (Sec. 303[d][5][A]). These activities could be undertaken relatively quickly and should not provide an excuse for inaction on the part of managers.

Discussion, Data Collection, and Speculative Behavior

Findings: **The committee received considerable evidence that the discussion of programs in council meetings and the initiation of certain kinds of data collection and research, as well as implementation of limited access measures, have led to speculative behavior on the part of participants, encouraging new entry into the fishery and increased harvesting effort before IFQ programs were in place.** Delays and postponements of control dates specified by councils can exacerbate speculative entry and capital stuffing. They can also generate a basis for claims of inequity and unfairness.

If the initial allocation of IFQs is based primarily on catch history, modification of the original control date will reward speculative entrants to the disadvantage of earlier participants. The committee heard testimony that this has been a widespread problem affecting the halibut and sablefish IFQ programs, the proposed Pacific sablefish program, and the surf clam/ocean quahog program. Failing to adhere to strict control dates can encourage capital stuffing and the buildup of excess harvesting capacity, one of the primary problems that IFQ programs are designed to mitigate. Testimony indicated that delays and changes in control dates appear to be due not only to the desirable (and required) public involvement process but also to the unnecessary administrative inertia of NMFS.

Recommendations: To minimize the potential for speculative investments, the regional councils and the Secretary of Commerce should ensure that data collection and studies be undertaken as part of long-term, routine activities, separate from the consideration of specific management alternatives for a fishery. Data collected on a regular basis will facilitate evaluation of the social, economic, and biologic impacts of various allocation actions, including IFQs.

Finding: Early adoption of and adherence to control dates and moratoria on new entry, licenses, and effort will greatly reduce the incentive for speculative entry. If the IFQ program is not implemented, the moratorium can be lifted if new entrants to the fishery are deemed desirable.

Recommendation: Control dates should be set early in the development of an IFQ program and be strictly adhered to throughout the development of the program. The public involvement process prior to implementing IFQs should be speeded up as much as possible without compromising its purpose; administrative delay in implementation should be minimized.

Nature of the Right or Privilege

Definition of the Right

Findings: If one of the goals of an IFQ program is to engender stewardship behaviors in the fishing sector, it is desirable to create a long-term stake in the fishery that can be defended against other private actions that threaten the health of the resource. The Magnuson-Stevens Act currently states that the IFQ should not be construed as creating any right, title, or interest to any fish before the fish is harvested (Sec. 303[d][3]). This language properly reflects the fact that the public trust nature of quotas precludes takings claims for their revocation or modification. However, it also may prevent fishermen who have not yet harvested their quota share from bringing a civil action against fishermen who exceed their quota share or trigger bycatch limits and cause a premature closing of the fishery. Likewise, it may preclude actions by IFQ shareholders against other parties who damage a fish stock through pollution or habitat degradation. Although users in California's adjudicated groundwater basins do not have the right to bring action against the government for restricting total withdrawals from a groundwater basin, they can bring civil action against other private users who damage the resource through pollution or through exceeding agreed-to withdrawals.

Recommendations: The Magnuson-Stevens Act should be amended to make it clear that the nature of the privilege embodied in an IFQ encompasses the right to protect the long-term value of the IFQ through civil action against the private individuals or entities whose unlawful actions might adversely affect the marine resource or the environment. The act should be clear that it does not authorize actions by IFQ shareholders against federal, state, or local governments for actions designed to protect marine resources and the environment through area closures or other modifications or revocation.

Findings: IFQ programs will achieve greater benefits if the interests they create are stable enough to encourage long-term investments, to be useful as loan collateral, and to engender in quota holders a sense of long-term stake in the resource. To the contrary, the moratorium on new IFQ program development and proposals to amend existing IFQ programs (e.g., to allocate increases in the TAC among a new set of stakeholders) could undermine the security of the interest, discourage transfers and purchases of additional quota shares, and destabilize the lending environment. The revocable nature of an IFQ and congressional discussions of uniform sunset provisions may have impaired the security of these interests. Because larger economic entities have access to better means of balancing risk, actions and proposals that undermine the security of the interest can be expected to have a disproportionately large impact on small economic entities.

The goal of a sunset law is to counter the tendency of government programs and bureaucracies to be self-perpetuating and to institutionalize policies that favor one group over another. A sunset law does not signify that policymakers know in advance that a particular policy or program has a useful life of a fixed number of years. It signifies instead that policymakers need to reevaluate the utility and effectiveness of an existing policy or program after a period of time to ensure that the best policies are in place, that they address objectives of the law, and that they are not continued merely due to bureaucratic inertia.

Under the Magnuson-Stevens Act, IFQ programs are not permanent and may be limited or revoked without compensation (Sec. 303[d][3][C]). However, as noted in Pautzke and Oliver (1997), the idea of "sunsetting" an IFQ program is fundamentally inconsistent with the nature of IFQs, given the investment made. They stated also that any IFQ program that contemplates a sunset date should be very specific from the outset about the termination date so that fishermen and their lenders can gauge the value of the quota share or IFQ purchase accurately.

Within the economic literature on IFQs, the consensus appears to be that permanent fishing rights are the preferred option (Huppert, 1991). Huppert makes the distinction between elimination and modification, and reports that the academic consensus recognizes that modifications should be achieved through buyback programs or annual changes in the TAC. Elimination, by contrast, should be only for cause, for example, repeated and egregious violation of catch limits and other regulations.

A limited duration IFQ is likely to reduce the holder's incentive to conserve the fish stocks because its value decreases over time and investments in stewardship of the resource will be reaped by someone else. A sunset provision would largely undermine the purpose of an IFQ program. It could inhibit the downsizing of fleets and probably lead to an investment outburst just prior to the sunset, because everyone would want to be poised for the new allocation mechanism that comes into place after the sunset.

When designing an IFQ program, councils should carefully consider the issues of stable, long-term privileges against the public trust nature of fisheries. The committee particularly urges councils to explore alternative ways to gain the benefits of long-term security while preserving the ability to modify programs. For example, the Australian "drop-through" system (see Box 5.2) or simple 10- to 25-year leases may achieve the same degree of security as an allocation in perpetuity.[3] These are among the alternatives available to councils to discourage behavior that degrades resources and to reward exemplary behavior without disrupting the security of the harvesting privilege.

Recommendation: Regional councils should be authorized to decide on a case-by-case basis whether to limit the duration of IFQ programs through the inclusion of sunset provisions. A blanket national policy of sunset provisions should not be adopted. In deciding on sunset provisions, councils should take into account the effects that limited duration may have on the ability of IFQs to meet their objectives, including the willingness of lending institutions to provide loans and the ability of participants to enter and leave the fishery.

An individual council could design an IFQ program that included a sunset provision, a design element that would cause the program to expire after a speci-

---

[3] However, the experience with 25-year rock lobster IFQs in New Zealand is that they do not seem to achieve the same security as IFQs that are indefinite in time. There are few limited duration IFQ programs, however, so it is difficult to generalize this experience.

fied number of years, with or without a renewal provision. In this case, the question arises whether the Secretary of Commerce should be allowed to use the partial approval authority to disapprove a sunset provision while approving the remainder of an IFQ program.

Central Registry System

Findings: **Individuals who do not receive an initial allocation, or those who received a small quantity of quota, may find it difficult to obtain bank financing to purchase shares because they lack acceptable collateral.** A concern raised by some lenders is that in some cases a lien has been placed on quota shares as a means of collecting delinquent taxes from the quota holder. Such a lien could be passed on to the purchasers of quota shares without their knowledge, undermining the confidence of lenders in the security of the loan. In response to these concerns, the Magnuson-Stevens Act mandated the development of a central registry system for limited access system permits (Sec. 305[h]). NMFS has published an advance notice of proposed rulemaking for a limited access lien and title registry in the *Federal Register* (NMFS, 1997b). None of the four U.S. IFQ programs has a registry in place yet and it appears that development activities have been initiated only for the Alaskan halibut and sablefish programs.

Recommendation: **NMFS should implement the central registry system (as required by the Sustainable Fisheries Act of 1996) as soon as possible for each U.S. IFQ fishery, to increase the confidence of lenders in the security of loans to purchase quota share and provide opportunities for individuals to obtain financing to enter or increase their stake in IFQ-managed fisheries.**

Initial Allocation

General Considerations

Findings: **Initial allocation of quota share is the most controversial aspect of the implementation phase of IFQ programs. Controversy focuses on who should be eligible for initial allocations and the criteria that should be used to allocate shares.** Furthermore, initial allocation of quota can result in windfall gains to the recipients if the quota shares are transferable and measures are not taken to address this issue. The potential for windfall economic gains has created both support for and opposition to IFQs, depending on the particular case and constituency.

The initial allocation has been characterized as enabling initial recipients to obtain loans to buy additional quota, resulting in significant shifts in the power of quota holders versus others in the fishery and changes in the composition of stakeholders involved in managing the fishery. In the existing IFQ programs, initial allocations went to vessel owners even though fisheries include harvesters (vessel owners, hired skippers, and crew), processors, fish buyers, and consumers. Some participants have contributed capital; others also have risked their lives and health to develop successful fisheries. All participants have reacted over time to changing incentives created by the free market and by government regulations and development programs. Owner-on-board provisions are used in some fisheries, but tend to preclude ownership by processors and communities and may not be appropriate in industrial fisheries. Broader initial allocations will lead to more equitable distribution of benefits and compensation of more individuals as shares become concentrated. At the same time, broad distributions are more likely to leave initial recipients with small initial allocations.

Recommendations: The committee recommends that the councils consider a wide range of initial allocation criteria and allocation mechanisms in designing IFQ programs. Councils could avoid some of the allocation difficulties encountered in the past by more broadly considering (1) who should receive initial allocation, including crew, skippers, and other stakeholders (councils should define who are included as stakeholders); (2) how much they should receive; and (3) how much potential recipients should be required to pay for the receipt of initial quota (e.g., auctions, windfall taxes). Moreover, councils may be able to avoid many of the problems associated with the initial allocation process if they consider allocating a cascade of fixed-term entitlements rather than a permanent exclusive privilege. This approach may give the councils a better, more finely tuned, instrument to reward stewardship and other positive behaviors over time. An example from the New South Wales fishery has been described in Box 5.2.

Catch History

Finding: Catch history has been used as the primary factor for determining the initial allocation of quota among participants in U.S. IFQ fisheries. Catch history is perceived by fishermen as a reasonable and fair measure of participation in a fishery. It is typically a quantifiable and verifiable indication of participation. Catch history, however, has focused on the species that is managed through the IFQ program and may disadvantage fishermen who shift back and forth among different species of fish. Catch history also can be distorted or substantially shifted from historical trends by speculative entry into the fishery. Furthermore, catch history can reward fishermen who have increased their catch at the expense of good stewardship (e.g., had high bycatch rates). Other factors used in initial allocations include dividing part of the quota equally among all verified participants (as done for wreckfish) and basing part of the quota share determination on vessel size (e.g., as done for surf clams and ocean quahogs).

Recommendations: **The committee recommends that councils consider a broad range of criteria for determining participation in and allocation of initial quota shares in addition to catch history. The specific criteria may vary from fishery to fishery and from region to region.** Examples of factors that may be taken into account beyond catch history include (1) the extent of dependence and commitment to fishing as a way of life, as in the Alaskan Limited Entry program; (2) evidence for or against good stewardship and acceptance of conservation goals (e.g., bycatch rates, violation histories, types of fishing gear used); (3) whether rule following is the norm in the fishery; and (4) other criteria that councils deem appropriate. These factors reflect the conservation and equity goals of the Magnuson-Stevens Act, as discussed in Chapter 5.

Skippers and Crew Allocations

Findings: **In the existing IFQ programs, non-owning captains, mates, and deckhands have not been allocated quota shares in the initial allocation.** Testimony and documents provided to the committee indicate that this is due partly to alleged difficulties in obtaining information on the historical participation of non-owners in the fisheries and partly to the philosophical position that those who have put their capital at risk are the proper recipients of quota share. Crew members and skippers (whether or not they own the vessel) are an integral part of the harvesting process in many fisheries. In a number of fisheries, crew members and skippers are considered co-venturers who have invested their time and risked their lives, even if they have not risked their capital. Moreover, in many fisheries, skippers and deck crew are paid on a share basis and consequently assume much of the financial risk as well as all of the physical risk associated with fishing.

Measures have been devised to help hired skippers and crew members participate in IFQ holding, including the "block" system and the loan system activated in 1998 for the Alaska halibut and sablefish fisheries. These measures partially redress the inequity created by making the initial allocation only to vessel owners, and in fact, crew members owned 11.2% of halibut quota and 4.6% of sablefish quota as of December 31, 1997. Detailed skipper and crew catch data are not necessary for allocating quota to them, unless the allocation rule requires the allocation to be proportional to landings.

Recommendations: **In order to achieve the stewardship and equity goals of the Magnuson-Stevens Act, regional councils should consider including hired skippers and crew members in the initial allocation of IFQs where appropriate to the fishery and goals of the specific IFQ program.** Detailed skipper and crew catch data are not necessary for allocating quota to them, because quota could be allocated in equal shares. As with other measures, the appropriateness of skipper and crew allocations is expected to vary among fisheries. For example, some fisheries are more industrial than others (e.g., the Alaska pollock fishery) and have not involved crew members as co-venturers in the same sense as other fisheries.

The committee also recommends that councils designing IFQ programs evaluate the block system and loan program in the North Pacific region for possible applicability elsewhere.

Processor Allocations

Findings: In the existing IFQ programs, processors were not awarded shares in the initial allocation unless they also were vessel owners during the control period. In some programs, they are also constrained from purchasing or leasing shares. Some processors and economists argue for allocating part of the quota to processors or for creating separate processor and harvester quotas. Just as the harvesting sector is overcapitalized in some fisheries, so too is the processing sector. Some processors told the committee they have been adversely affected by the introduction of an IFQ program or would be harmed by potential programs. Others benefited or were not greatly affected. Adversely affected processors assert that harvester-only IFQs may result in stranded capital, lower profitability, and significant impacts on isolated rural communities. These consequences would result from the fishery becoming more efficient, shifts in the timing of deliveries, and harvesters gaining bargaining power in relation to processors over exvessel prices. In some fisheries, processors seem to have responded effectively to the changes brought about by IFQs through a variety of contractual methods and vertical integration. If avoiding processor losses is considered an appropriate social goal, this could be accomplished by allocating separate harvester and processor quotas.

Recommendation: On a national basis, the committee found no compelling reason to recommend the inclusion *or* exclusion of processors from eligibility to receive initial quota shares. Nor did the committee find a compelling reason to establish a separate, complementary processor quota system (the "two-pie" system). If the regional councils determine that processors may be unacceptably disadvantaged by an IFQ program because of changes in the policy or management structure, there are means, such as buyouts, for mitigating these impacts without resorting to the allocation of some different type of quota, with a concomitant increase in the complexity of the IFQ program. For example, coupling an IFQ program with an inshore-offshore allocation would preserve the access of shore-based processors to fishery resources. Whatever method is chosen, it should not have the effect of subsidizing excess processing capacity. Depending on regional considerations, some councils may choose to allow processors to acquire quota share either through transfer or through ownership of harvesting vessels that are entitled to an initial allocation of quota.

Allocations to Communities

Findings: **Catch history, as a measure of participation in a fishery, reflects the participation not only of individuals and occupational groups, but also of fishing communities. From this perspective, communities may be entitled to initial quota allocations.** Community development quotas (CDQs) have been implemented in the Western Alaska-Bering Sea region to stimulate development of commercial fishing activities in communities adjacent to the fishing grounds and having few other economic opportunities. A separate National Research Council (NRC) committee has reviewed the Alaskan CDQ program and considered whether a similar program would be feasible and desirable for communities in the Western Pacific region (see NRC, 1999a). The definition of IFQs in the Magnuson-Stevens Act includes quotas held by a community association or local government but does not include community development quotas (Sec. 2[21]).

"Community fishing quotas" could contribute to community sustainability in areas that are heavily dependent on fishing for social, cultural, and economic values and/or are lacking in alternative economic opportunities. They may also be considered as ways of delegating some management responsibility and authority to communities. Shares of a TAC may be awarded to designated communities—whether politically defined or defined as groups of fishing crews working in or from the same area—which then have the opportunity to determine how they will allocate the shares and manage the fishery, whether by open access, trip limits, or individual transferable quotas.

Recommendations: **The committee recommends that councils consider including fishing communities in the initial allocation of IFQs, where appropriate, and that the Secretary of Commerce interpret the language in the Magnuson-Stevens Act pertaining to fishing communities (Sec. 303 [b][6][E] and National Standard 8) to support this approach to limited access management.** Congress should allow, through a change in the Magnuson-Stevens Act if necessary, councils to allocate quota to communities or other groups, as distinct from vessel owners or fishermen. Where an IFQ program already exists, councils should be permitted to authorize communities to purchase, hold, manage, and sell IFQs. These communities could use their quota shares for community development purposes, as a resource for preserving access for local fishermen, or for reallocation to member fishermen by a variety of means, including loans. If the communities chose to allocate the rights to individuals, they could be constrained by covenants or other restrictions to be nontransferable.

Regional fishery management councils should determine the qualifying criteria for a community that is permitted to hold quota. A range of factors, such as proximity to the resource, dependence on the resource, contribution of fishing to the community's economic and social well-being, and historic participation in the fishery, may be among the factors that a council considers when setting criteria for establishing which communities may hold quota. The range of criteria will have to be carefully considered and weighted and the implications of defining these criteria would have to be examined fully.

Returns to the Public and Nation from Initial Allocations

Findings: The typical approach for allocating fishing quotas is for a council to establish a set of criteria identifying who will receive initial allocation and how much each recipient will be awarded. In other settings, a variety of other mechanisms has been used to allocate access to scarce resources. For example, auctions or other forms of competitive bidding have been used to allocate air pollution quotas, Outer Continental Shelf oil exploration rights, and timber rights in federal and state forests. Lotteries have been used to distribute public lands and determine access to recreational opportunities (wildlife viewing, hunting, camping, hunting, sportfishing). Queuing (first-come-first-served) is another allocation mechanism commonly applied to commercial and publicly provided recreation opportunities. U.S. history is replete with examples of queued resource allocations: the Oklahoma land rush; the California gold rush; and the various homestead, reclamation, railroad, and timberlands acts, to name a few. The race for fish is queuing in a commercial fishery setting.

How can the public be compensated when awarding exclusive privileges for use of public resources to private persons? At this time, the privileges to fish are granted for free and the public benefits only from normal taxation measures plus indirect benefits from improved efficiency in the industry and lower costs for fish products. Another question is who should make the decisions about initial allocations, whether an impersonal market (via auctions), fate (via lotteries), or an interested, participatory body such as a regional council.

Recommendations: Regional councils should avoid taking for granted the option of "gifting" quota shares to the present participants in a fishery, just as they should avoid taking for granted that vessel owners should be the only recipients and historical participation the only measure of what each deserves. Councils should consider using auctions, lotteries, or a combination of mechanisms to allocate initial shares of quota.

Transfers

General Guidelines

Both the literature on IFQs and the testimony to the committee show that transferability is one of the most controversial aspects of IFQs. Although transferability promotes economic efficiency in the fishery at large, the structural changes accompanying transferability are often perceived as a threat by some fishermen and other members of fishing communities. The perceived threats are the concentration of quotas in the hands of a few individuals and/or communities, a lopsided distribution of economic gains accompanying IFQs, and a change in social relations among the members of a community.

Recommendations: The decision whether quota shares should be transferable, one of the most critical elements in the design of an IFQ program, should be left up to the regional councils or other regional or local groups because it depends entirely on the specific goals and objectives of the management regime. If economic efficiency and rapid downsizing are the primary goals of an IFQ program, transferability should be as free as possible. However, if other goals are more important, such as protecting an owner-operator mode of organizing production, preventing absentee ownership, or protecting fishery-dependent coastal communities, it may be necessary to restrict transferability— either geographically, between groups of fishermen, between bona fide fishermen and others, with respect to time, or possibly all of these.

Temporary Transfers (within fishing year or within season)

Findings: Some degree of leasing may be important to allow fisheries to adapt to change, address concerns of overages and bycatch of the non-target species, and lower the enforcement burden. For many people, however, the social relations of tenancy that may be established through repeated leasing violate community values. The practice of leasing is likely to alter the relations of vessel owners and crew members. Leasing can benefit some and disadvantage others, and thus violate deeply felt concerns and generate conflicts and moral debates. Opinions on leasing by fishermen testifying to the committee were divided. The committee also heard how New Zealand and other countries reduce bycatch in their fisheries by allowing fishermen to lease quota to cover their bycatch species, rather than discarding them.

Recommendation: Leasing of quota should generally be permitted but with restrictions as needed to avoid undesirable side effects such as absentee ownership. Restrictions on the proportion of total quota that can be leased, the frequency that individuals can lease quota, and the taxation of leased quota to help affected communities are means to reduce the negative effects of leasing.

Permanent Transfers

Findings: Many objectives of IFQ programs require some degree of transferability or flexibility for industry participants (particularly for purposes of economic efficiency). However, unrestricted transferability can lead to socially negative side effects such as an excessive degree of consolidation or

regional shifts in access to a fishery. Increased limitation on transferability can also create additional monitoring and enforcement costs. The use of individual allocations (whether IFQs, other output allocations, or input allocations) makes transferability, in some degree and form, desirable. Even if transfers are prohibited or sharply constrained, illegal transfers are likely to occur to some degree.

Recommendation: Permanent transfers of quota shares should generally be allowed without any restriction among eligible quota holders. If the desire is to promote an owner-operated fishery and prevent absentee ownership, or to conserve geographic or other structural features of the industry, it may be necessary to restrict long-term transfers of quota shares to bona fide fishermen or to prohibit transfers away from certain areas or between different vessel categories.

Accumulation and Monopoly Issues

Findings: The Magnuson-Stevens Act provides that an IFQ management program must prevent "...any person from acquiring an excessive share of the individual fishing quotas issued" (Sec. 303[d][5][C]). **Some IFQ programs define the upper limits on individual holdings, whereas others, notably the surf clam/ocean quahog (SCOQ) and wreckfish programs, rely on federal antitrust law.** Issues such as concentration of quota among firms or communities can be addressed through setting upper limits on accumulation of quota share and instituting measures such as compensating disadvantaged communities. If, on the other hand, important objectives include maintaining owner-operated fisheries and fishery-dependent coastal communities, transferability may have to be constrained and greater attention given to equity considerations in setting upper limits on accumulation, boundaries to transfer of quota share among communities, and other restrictions.

The SCOQ IFQ experience suggests that reliance on antitrust law and procedures will not be sufficient to prevent the excessive share problem referred to in the Magnuson-Stevens Act. A lack of accumulation limits may unduly strengthen the market power of some quota holders and adversely affect wages and working conditions of labor in the fishing industry, particularly in isolated communities with limited employment alternatives.

The desirable speed and level of concentration of quota shares will depend on technology, culture, and other characteristics of a particular fishery. Three general points about concentration limits should be considered in designing an IFQ program:

1. Accumulation limits are one way to promote equity, but their use varies. Recently, the Icelandic Parliament enacted a ceiling of 10% for the most important species (e.g., cod). In the United States, concentration limits vary from 0.5%

of the halibut quota in certain regions to no limit in the SCOQ and wreckfish IFQ programs.

2. Concentration limits may not be very effective if there are ways to circumvent them. Restrictions based on different classes of shares (crew members, vessel owners) may be more effective than concentration limits from a social perspective.

3. Other concerns may be much more important than concentration limits, in particular the issues of initial allocation, transferability, leasing, and the return of rent to the community. These are more fundamental for the structure of fishing societies than simply restricting the size of quota holdings.

Recommendations: Congress should recognize that the design of an IFQ, or other limited entry system, in relation to concentration limits will depend on the objectives of each specific plan, underscoring the importance of providing flexibility for regional councils in designing limited entry programs. Vertical integration, monopolization, and regional aggregation of quota shares can be addressed through setting upper limits on concentration and limits on the transferability of quota share among regions.

Recommendation: Congress should require the creation of limits on the accumulation of quota share by individuals or firms in each new IFQ program. These limits should be fishery specific and may also be area or class specific. Proposals for IFQ programs should specifically define "excessive share" for the fishery. The definition of an excessive share and limits on accumulation in specific areas should be left up to the regional management councils and other groups engaged in fine-tuning IFQ program design. Care should be taken to define excessive share, how to measure it, and what to do when it occurs rather than rely on federal antitrust law.

Mechanisms for New Entrants

Finding: The Magnuson-Stevens Act (Sec. 303 [d][5][6]) requires that new IFQ programs provide opportunities for new individuals to enter IFQ-managed fisheries, and Congress specifically asked the National Academy of Sciences (NAS) to address this issue (Sustainable Fisheries Act, Sec. 108 [f][1][H]). The implementation of IFQ programs may restrict opportunities to enter the fishery. New individuals normally enter an IFQ-managed fishery through transfer of quota shares, and measures to facilitate new entry could defeat the purposes of IFQs if such measures either expand the quota share pool or hinder the consolidation of quota share and associated economic efficiency. In existing IFQ programs, some quota shareholders or potential entrants are disadvantaged with respect to the financial capital market if they lack collateral or credit history or if they did not receive an initial allocation of quota share. In some cases, it was reported to the committee, the price of the quota share has risen to the point that its debt service is greater than its expected revenue stream, so normal rates of return are not possible. Inflated share prices may be created either through speculation or through irrational expectations.

Recommendation: Individual quota programs that contain mechanisms to facilitate the entry of new participants to meet the Magnuson-Stevens Act requirements should do so without expanding the total number of quota shares. The zero-revenue auction (see Box 5.1) is one promising technique. If quota shares are transferable, provisions for ownership qualification, purchasing mechanisms, and limits on share concentration should be contained in program documentation. The price of quota share can be reduced by taxing quota rents, and provision of the central registry of liens on quota could make loans more available for new entrants to buy quota being offered for sale.

Foreign Ownership

Findings: Substantial foreign ownership already exists in the harvesting and processing sectors of some U.S. fisheries; limits on the extent of foreign ownership would have major implications for the potential effects of introducing IFQ programs in these fisheries. The exact level and nature of foreign ownership and the degree to which the income generated by foreign interests is transferred outside the United States are uncertain. Many countries restrict foreign investment in the fishing industry, but this runs counter to the present trend of liberalizing direct foreign investment worldwide.

Assessing the extent to which fishery resource rents are expropriated by foreign nations is beyond the scope of this consideration of IFQ and limited access systems. Similar concerns exist in U.S. fisheries under traditional fishery management measures. At stake are some of the same policy issues that led to the adoption of the Fishery Conservation and Management Act in 1976. This issue should be a topic for separate study by Congress. If Congress were to decide to control foreign ownership, criteria could be established for IFQ and other fisheries. Enforcement would require careful analysis of the financial and corporate records of all processors and harvesters, and the economic conditions of the fishery, as well as improved access to certain types of proprietary data.

Recommendation: The committee recommends that Congress establish a policy related to the eligibility of foreign individuals or companies to receive IFQ shares in an initial allocation. The committee notes that foreign individuals and firms have invested significant amounts of capital in harvesting and processing capacity and that identification of foreign firms would be at best problematic and could discriminate against U.S. co-owners and investors.

Recreational Fisheries

Findings: In many fisheries, recreational participation is significant and should be considered in the development of any IFQ program. The recreational component of IFQ fisheries has at least two important implications. First, there may be significant allocation issues between commercial and recreational sectors. For example, increasing participation and catches in the recreational sector may have an effect on the allocation of a TAC between commercial and recreational sectors, with effects on the functioning of an IFQ program in the commercial sector. Second, IFQs may have some application within the recreational sector itself. In the case of "for-hire" recreational fisheries, for example, IFQ programs may be considered to address social or economic objectives within this sector of the fishery and to integrate the recreational and commercial sectors in an IFQ program.

Recommendation: In any fishery for which an IFQ program is being considered, attention should be given to the implications of recreational participation in the fishery and, where appropriate, to the potential application of the IFQ program to both commercial and recreational sectors. For cases in which monitoring catch proves to be unreasonably burdensome, transferable effort quotas may be used in place of transferable catch quotas.

Individual Bycatch Quotas

Finding: Individual bycatch quotas (IBQs) are expressly permitted under the Magnuson-Stevens Act. Although IBQ programs have not yet been developed by regional fishery management councils, they may be a useful tool for controlling the magnitude of bycatch and through individual accountability, encouraging fishermen to avoid bycatch. Thus, they could serve as complements to an IFQ program or be used in conjunction with other management systems. Effective implementation of IBQs necessitates close monitoring of actual catches and would probably require onboard observers. In fisheries with low bycatch rates, it may be necessary to assign bycatch quotas to vessel pools to achieve confidence intervals on bycatch estimates that are narrow enough to support prosecution of overages.

Recommendation: The councils should be encouraged to explore the use of individual and pooled bycatch quotas to control overall bycatch and to give fishermen the incentive to minimize their bycatch rates. The councils may also wish to consider using fishing histories developed during a period when IBQs are in force in determining the initial allocation of limited entry permits or IFQs.

PUBLIC AND PRIVATE COSTS AND BENEFITS

This section discusses the issue of windfall gains to initial quota share recipients. Such gains (and normal operating profits) can be taxed at two potential levels. First, the cost of IFQ programs can be recovered. The Magnuson-Stevens Act presently permits cost recovery of up to 3% of the exvessel value of landings. Beyond recovery of costs related to IFQ management, a case can be made for implementing means to extract some of the extra value created by IFQ programs, for return to the public.

The Magnuson-Stevens Act currently prohibits taxation beyond 3% of the exvessel value of fish landed under the IFQ (or community development quota) program and only for cost recovery (Sec. 304[d][2][A&B]). Up to 25% of this amount may be used for special programs submitted by regional councils to aid in financing the purchase of IFQs by small-vessel fishermen or new entrants (Sec. 303[d][4][A]); the rest goes into a dedicated fund for administering limited access fisheries (Sec. 305[h][5][B]).

Windfall Gains

Finding: Because it has been the practice in existing IFQ programs to award the initial allocation without charging the recipient for the use of public resources through royalties or taxes, concerns have been raised that this allocation provides a substantial competitive advantage for the initial recipients. Not only do they have privileged access to the fishing quota, but they also may have competitive advantage in raising capital for future investments in the fishery, especially if the quota shares are treated as collateral by lenders. This was an issue raised by a spectrum of participants in the committee's public meetings, including some harvesters, processors, and environmental groups. For many other natural resources, the use of public resources requires specific compensation to the public at the time of transfer (e.g., Arizona water rights, mineral leases, timber contracts).

When quota shares are allocated initially for free, quota shareholders are able to obtain a windfall economic gain when they choose to sell their shares at a later point in time. This represents unearned income, which is regarded by many as unfair, although part of this windfall is recovered at first sale in the form of capital gains taxes. Rather than prevent this gain from occurring by banning transferability, it would be preferable to extract some of this unearned income through a suitable system of fees or taxes. However, it is important that rent extraction not be so large as to eliminate transfers totally and thus counteract the economic efficiency objectives of quotas, and that it be explicitly defined as part of the initial program development rather than added on after fishermen and processors have made investment decisions.

Recommendation: The committee recommends that the Magnuson-Stevens Act be amended to allow the public to capture some of the windfall generated from the initial allocation of quota in new IFQ programs. This could be accomplished by taxing the first transfer of shares, via leasing or sale, to reduce the windfall gains to the initial recipients. Other mechanisms, such as auctions, assessing an annual fee on quota share, or other taxes, would also reduce windfall gains to the initial recipients.

Cost Recovery

Finding: The implementation of an IFQ program introduces exclusive privileges to harvest a portion of a public resource. Establishment of these privileges should be accompanied by the assignment of obligations and responsibilities to quota shareholders. Shareholders should be obligated to pay an appropriate share of the continuing costs of managing an IFQ program, including a share of the costs of fisheries administration, enforcement, and research.

Recommendation: Congress should authorize the collection of fees from the transfer and/or holding of IFQs to provide funding for the attributable costs of research and management associated with establishing and maintaining an IFQ program. Although the Magnuson-Stevens Act is compatible with this recommendation in principle, in practice the limit of 3% may well be too low for some IFQ programs and should be increased (New Zealand currently collects about 5% of the exvessel value of landed fish for cost recovery).

Costs that are attributable to specific transactions (such as the recording of a quota transfer or the certification of a registered buyer) should be recovered through fees on those benefiting from the transaction. Other costs, such as monitoring, enforcement, and stock assessment research for particular species, should be borne by holders of quota for these species. They should be recovered by means of a levy either on quota or on landed harvest.

Rent Extraction Above Cost Recovery

Finding: For some natural resources (e.g., timber, minerals, and oil and gas), the government captures a significant portion of the rent above cost recovery. The extraction of rent depends on having an effective system for capturing and reallocating the rent generated. In so doing, it is important that the rent recaptured not create a system that would encourage overharvesting of the fish stock to maximize the short-run revenue flow to the government. Extracting rent will decrease the value of quota, which has some benefit in terms of making it easier for new individuals to enter a fishery.

Recommendation: The Magnuson-Stevens Act should be amended to authorize the capture of rent in excess of cost recovery. A variety of means could be used by the IFQ programs to recover this rent, including the two-fee system described in Box 5.4. Rent recovery components of IFQ programs should include specification of the use of any extracted rent (see following finding and recommendation).

Dedicated Funds from Rent

Finding: Channeling rent incomes to dedicated funds would make them visible and important. The beneficiaries of the dedicated funds would acquire an interest in maximizing the rent income, and hence the total rent, if the captured rent is a set proportion of the total rent (for example, through auctions or tax as a share of rent). Because the rent reflects the economic efficiency of the fishery, beneficiaries of the dedicated funds would acquire an interest in having the industry managed efficiently and sustainably and could be expected to promote this goal.

What purpose should dedicated funds serve? Most individuals would probably agree that using such funds for the benefit of communities that depend on the fisheries from which the rents are extracted is one legitimate objective. Such dedicated funds could support fisheries research; could finance retraining of fishermen displaced by IFQ programs and support other forms of education, health care, and infrastructure in these communities; or could be allocated by direct cash transfers to inhabitants of affected communities. The exact legal and administrative form of dedicated fishing rent funds, should they be established, must be left for the political process to decide. One example of such a dedicated fund is the 25% of the maximum 3% of landed value of catch that can be devoted to loan programs, according to the Magnuson-Stevens Act. Another example is the dedicated fund associated with the Florida spiny lobster fishery, which receives 90% of revenues from fees and charges in the fishery for the purpose of research, monitoring, enforcement, and education related to the fishery.

Recommendation: If the Magnuson-Stevens Act were amended to allow rent capture beyond what is needed to cover the administrative cost of fisheries management, the option of channeling the captured rents to dedicated funds should be given serious consideration. The amounts of money involved are likely in most cases to be small and to make little difference for federal or even state public revenue (Alaska might be an exception). Priority should be given to dedicating any rent extracted from fisheries beyond administrative costs to improving the fisheries and the fishing communities dependent on them. In general, these rents should not be directed to the general treasury. One suggestion is that a community trust be established and co-managed by representatives of quota shareholders, regional councils, communities, and NMFS. Management of dedicated funds at a local level provides incentives for proactive maintenance and enhancement of marine resources.

Monitoring and Enforcement

Finding: **Regardless of how well any fisheries management plan is designed, noncompliance can prevent the attainment of its economic, social, and biologic objectives.** Noncompliance not only makes it more difficult to reach stated goals, it also makes it more difficult to know whether the goals are being met, if data fouling occurs.

Incentives for quota busting, poaching, and highgrading may increase with the introduction of an IFQ program, due either to the higher profitability of fishing activity or to the perception in the community and the industry that the program is unfair and inequitable. IFQs also improve incentives for quality over quantity, but this in turn may lead to more highgrading and bycatch problems.

Monitoring and enforcement costs have risen in some fisheries with the introduction of an IFQ program, whereas in others, costs have fallen. Testimony to the committee indicated that enforcement has apparently become easier and less costly in the wreckfish fishery, for example, and more costly in the Alaskan halibut and sablefish fisheries (see Appendix H). Furthermore, highgrading or bycatch may either increase or decrease with IFQ regimes, depending on how they are designed and enforced. The increased value of the fishery that results from the elimination of the race for fish provides an additional source of revenue. This revenue could be used to finance enhanced monitoring and enforcement efforts if the revenue is channeled appropriately.

Recommendations: **Councils should design IFQ programs in such a way as to enhance enforcement.** The committee recommends that the regional councils and the Secretary of Commerce consider the following three principles for effective monitoring and enforcement in designing IFQ programs.

1. Agreements are more likely to be perceived as fair and desirable if the fishermen participate in their creation and are also more likely to be enforceable.

To the extent that the participants in a fishery understand the need for regulation and concur with the form the regulation is taking, enforcement will become easier. Enforcement that is considered fair and desirable to participants is most likely to be respected as a norm and to result in higher compliance rates and less necessity for enforcement actions and associated costs.

The rules and regulations governing a fishery are more likely to be supported by fishermen if the administrative process of establishing an IFQ program involves co-management schemes that allow fishermen to participate in their development and implementation. Fishermen are also likely to have the best wisdom about what monitoring and enforcement measures would be most effective.

2. Incentives to cheat should be identified, and IFQ programs should be designed to reduce these incentives.

Program design can have a great effect on the enforcement burden. For example, rather than placing the entire burden for bycatch reduction on enforcement, it is possible to provide alternatives to court-imposed sanctions. Concrete examples are provided by the New Zealand approach, which includes "as-needed" quota transfers and the Alaskan "underage and overage" system and graduated penalties. Sanctions should take the principle of marginal deterrence into account, because it is important to consider the likely effect of a set of penalties on the incentive to commit more serious crimes. In fisheries, one application of this principle would ensure that the penalties for bycatch were not so severe that they would make discarding the catch, rather than landing it, the preferred option. Unobserved discarded catch not only is wasted, but causes data fouling.

In practice, applying the principle of marginal deterrence implies establishing a set of graduated sanctions. Administratively imposed sanctions should be established for minor violations with specified increases in penalties for each additional offense. Criminal penalties (jail sentences and/or seizure of catch, vessel, and equipment and forfeiture of quota) should be reserved for serious offenders and for intentional falsification of reports.

3. Adequate funding should be provided for monitoring and enforcement, and it should be obtained from fees assessed on quota or harvest.

RESEARCH, MONITORING, EVALUATION, AND EVOLUTION

Improve Adequacy of Research

Finding: The data needed to manage a fishery become more extensive as management becomes more complex. Different management approaches require different types of data. IFQs require enough biologic data to set a reasonably accurate TAC and enough socioeconomic data to anticipate some of the effects of proposed systems on individuals and communities.

In general, labor statistics for the fishing industry are not as complete as for other industries. Assessing the effects of proposed limited entry programs requires information on the range of fishing activities, and other activities in which they are embedded, especially for communities in which fishing is important. Such information makes it possible to estimate the probable effects of different proposals for implementing IFQs or other limited access programs, in much the same way that systems analysis enables assessing the effects of proposed changes in manufacturing procedures and the way an industry is organized (Goodenough, 1963; Lieber, 1994).

Recommendations: Funds should be made available through NMFS to strengthen research on the design, performance, and impacts of IFQ programs.

IFQ programs can vary greatly in the details of their implementation and in their effects, intended and unintended, on fish stocks and fishermen. It is important for the nation to learn from its mistakes and successes so that it can develop more effective and efficient management systems. The committee recommends that the Secretary of Commerce promulgate guidelines to the regional councils for all new IFQ programs to monitor their effectiveness. At a minimum, this would include the following:

1. Maintaining a registry of shareholders and all share transactions, including the names of buyers and sellers, and the quota share amounts and value;

2. Assessing the biological status of the stock in a timely fashion, given earlier NRC recommendations (1998a);

3. Measuring the economic performance and characteristics of
  • Commercial fisheries (e.g., number of participants, annual operating income and costs, investment in gear and vessels, investment in shoreside processing and other infrastructure);
  • Recreational fisheries (e.g., angler days, consumer surplus); and
  • Subsistence use patterns;

4. Assessing the performance of the quota share market, including sales price, sale and leasing frequency, and changes in the distribution of quota shares;

5. Collecting data on a routine basis on the system's administrative and enforcement costs; and

6. Monitoring the translocation effects on other fisheries.

The committee further recommends to the Secretary of Commerce that by the year 2005, this monitoring information from all existing and new IFQ programs in the United States be analyzed in a comprehensive manner and reported to the councils and other interested parties. The committee was hindered in its efforts to assess the relative economic costs and benefits of IFQ programs because of a lack of IFQ market data (as shown in Appendix H).

The Secretary of Commerce should require the regional councils to plan research to allow for systematic evaluation of the effects of proposed IFQ programs and alternatives on the way of life of fishing communities and the way fishing activities are conducted.

Complementary to the routine monitoring described above, NMFS should significantly expand its routine collection of social and economic data to allow baseline descriptions of fishery users, monitoring of impacts associated with individual quota and other management programs, and an im-

proved understanding of the human dimensions of fisheries. Economic data on operating unit budgets, including costs of equipment, fuel, and gear, are particularly lacking. The Secretary of Commerce should direct NMFS to incorporate socioeconomic variables in routine and case-specific data collection and to fully implement existing plans to do so, such as the Atlantic Coastal Cooperative Statistics Program (ACCSP). Issues of confidentiality of data and data as property must be resolved in order to obtain and use data on a per vessel, skipper, and processor basis for monitoring and enforcement.

Concerns about the equity of the initial allocation of quota shares is a major obstacle to the implementation of any IFQ program. It is important that the initial allocation process be transparent and perceived to be fair; this requires adequate data. To accomplish this goal, the committee recommends that the Secretary of Commerce encourage regional councils to undertake the following tasks as soon as is practicable for fisheries under their jurisdiction:

1. Review the adequacy of catch history or other records that might be used for the initial allocation of quota shares.
2. Establish registries of crew and skipper participation, including information on fishing activities since 1990 or as soon after as is feasible.
3. Gather data on the effects of management on communities.

Periodic Independent Assessment of the Performance of IFQ Programs

Findings: IFQ programs are still relatively new, and the effects of various program characteristics on different types of fisheries are still being tested. Adaptation of programs to changing conditions and design of new programs depend on evaluation of existing programs. The development of any limited access program, especially one that is designed to allocate harvest rights and access to the resource by individual fishermen or firms, will be a complex and controversial process, and the performance of such programs should be monitored carefully.

The greater the degree to which stakeholders in the fishery (e.g., vessel owners, hired skippers, crew members, processors, managers) can agree on the process for handling allocation, appeals, enforcement concerns, concentration limits, transferability, and broader socioeconomic considerations, the less controversial or contentious are these elements likely to be. It is possible that a well-designed program can prevent some of the unforeseen consequences of the initial allocation process on the socioeconomic equilibrium in the fishery. However, even in a well-designed program with strong consensus from the stakeholders, the initial allocation process and its possible effects on the distribution of capital within the fishery could create significant long- and short-term difficulties for managers and participants in the fishery. A key to maintaining the stability of and support for a limited access program are the processes and institutions used to design, manage, review, and change the program.

In its public meetings, the committee heard concerns from a number of participants that a process and institution external to the regional council-NMFS system should play a significant role in the design, implementation, and management of any limited access program, especially IFQ programs. Several participants expressed the opinion that the existing council process has not responded adequately to their concerns, specifically about initial allocation mechanisms. Others have expressed dissatisfaction with the NMFS appeals procedures, leasing, the setting of limits on the level of quota concentration, enforcement, and other issues related to the management and regulation of IFQ programs.

Although an external institution might provide some stakeholders with greater confidence that their information and recommendations would be evaluated and used objectively, it is just as possible that other stakeholders would feel alienated from a new institution in which they had not participated previously. Both the regional councils and NMFS have well-established structures and mechanisms for gathering and analyzing fishery information, both are familiar with characteristics of the regional fisheries for which they are responsible, and both are familiar to the stakeholders.

However, areas for which an external review process might be helpful are (1) reviewing the information-gathering process and (2) reviewing the fishery management plans containing IFQ provisions before they are submitted to the Secretary for approval. Such an external review process could serve several functions: (1) assist the councils and NMFS in identifying the concerns of stakeholders and help direct efforts to gather information on these issues of concern; (2) ensure that the objectives identified by stakeholders are adequately addressed in the proposed program; and (3) provide recommendations to the proposing council if deficiencies or concerns about the proposed IFQ program are noted.

Such an external review process could be useful to stakeholders only if it avoids unduly burdening or slowing the decisionmaking processes of the existing fishery management system. The value of an external review body would be to provide recommendations to the councils and NMFS on the development of programs, without requiring additional effort from the councils or NMFS, and to provide an independent, objective review before implementation of IFQs or some other form of limited entry. An external review panel could be organized as a group that understands a fishery but has no direct financial interest in it. Membership on such a panel could include members of the council's Scientific and Statistical Committee (for councils in which they are used), fishermen from other regions or from different fisheries who do not have a financial stake in the program being considered but are familiar with issues in the fishery, outside academics, and fishery managers from other regions or countries having familiarity with the issues and processes involved. Additionally, it might be helpful to include individuals from outside fisheries management to provide new perspectives to the councils and NMFS. This review process could help reduce future challenges to the program being established by providing an opportunity for an additional perspective that is not affected by the proposals being developed and considered. Another option would be to develop and manage IFQ programs through a subset of a council, rather than the entire council. This option could have several advantages, including freeing the council to concentrate on broader management issues and allowing the subgroup to work more quickly and with greater focus.

Recommendation: The committee does not believe that creating an institution and process separate from the existing councils and NMFS to design, implement, and/or manage an IFQ program would best address the concerns of stakeholders. The major aspects of designing, implementing, and managing any potential IFQ program should remain within the purview of the regional councils and NMFS. The Secretary of Commerce should ensure that each fishery management plan that incorporates IFQs include enforceable provisions for the regular review and evaluation of the performance of IFQ programs, including a clear timetable, criteria to be used in evaluation, and steps to be taken if the programs do not meet these criteria. Provisions should be made for the collection and evaluation of data required for this assessment. The process could include review by external, independent review bodies.

Finding: During its review, the committee found that as IFQ programs have been developed and implemented throughout the United States, the regional councils and NMFS have learned from the implementation of previous IFQ programs. In examining the evolution of IFQ programs nationally from surf clams/ocean quahogs to the proposed red snapper IFQ program in the Gulf of Mexico region and the sablefish IFQ program in the Pacific region, newer programs have been designed to avoid past difficulties. It appears that design features related to accumulation limits, transferability, leasing, quota shares for crew members, and other aspects have been modified based on observations of the effects of previous programs. However, the exchange of information among various regional councils, NMFS, and other interested stakeholders does not appear to be well coordinated. The committee could not find a centralized source of data describing the effects of existing programs. Although some of the programs have documented important aspects of their effects, such as the accumulation of quota share, the degree of transferability, and the geographical distribution of quota, these data are often difficult to obtain. It appears that in some IFQ programs, this lack of readily available information has contributed to the controversy surrounding the implementation and management of the program. With the exception of the Alaskan halibut and sablefish IFQ programs, there is no regular periodic review of the changes in trends in critical elements in the distribution of quota shares. In the IFQ programs reviewed, information concerning the trends in the price of quota shares appears to be lacking.

Recommendation: Existing and future IFQ programs should provide an annual report describing trends in the fishery and the effects of the IFQ program on important management variables. Where possible, it would be worthwhile to examine how these variables have changed since implementation of the program. Factors in an annual review could include the number of quota shareholders, the distribution of quota shareholders among various sectors or vessel classes, changes in the number of vessels, changes in the number of crew members and their holdings of quota shares, and the trends in the price of quota shares over time. This report should be available to IFQ managers in other regions, as well as participants in the fishery and the general public, through the World Wide Web and other venues.

Changes to Existing Programs

Finding: Holders of quota shares in existing IFQ programs have often made major investments in purchasing IFQs and adjusting their business capital and practices to the IFQ program. For example, the committee received testimony from some Alaskan fishermen who received little or no initial allocation, subsequently invested in quota shares, and are concerned that their investments will be eroded by changes in the program that diminish the value of their quota share (e.g., increasing the quota share pool). Fishermen told the committee that they believe lending institutions will be less willing to make loans for purchases of IFQs if the programs are unpredictable.

Recommendation: Councils should proceed cautiously in changing existing programs, even to conform to the recommendations of this report. In spite of initial windfall gains (or even in the absence of them), many individuals have made subsequent investments in quota shares. Changes should be designed to maintain the positive benefits of IFQs that result from their stability and predictability. One means to accomplish gradual change is through use of an Australian "drop-through" system (described in Box 5.2), with different conditions and requirements for each level.

Every IFQ program should establish at the outset a process for the review and evaluation of the program and a mechanism for timely, nondisruptive, equitable consideration of program changes. The Magnuson-Stevens Act requires such procedures for review and revisions (Sec. 303[d][5][A]). The North Pacific Fishery Management Council, for example, has set up an annual management cycle for considering proposals for adjusting the IFQ programs for sablefish and halibut. Proposals are solicited in the summer, and the council decides in December of each year which proposed changes warrant further consideration. Recommendations for change also come from the council's Industry Implementation Workgroup, which assists the council in reviewing proposals and overseeing implementation. This process has led to changes such as the block "sweep-up" provision, allowing consolidation of smaller blocks of quota share into larger blocks. Evaluation should be focused on whether the IFQ program is meeting the biologic, economic, and social objectives of the program and the Magnuson-Stevens Act.

Some committee members believe that the evolution of an IFQ program to feature broader participation and cooperative management should be one of the key objectives of the program's initial design. This process could be assisted by requiring holders of IFQs to participate in management decisions and to assume responsibility for some of the management functions, such as the observer program and dockside monitoring. The evaluation process could include criteria such as whether the holders of IFQs have acted together to address common concerns and objectives, for example, adopting experimental gear modifications to improve selectivity, or voluntarily closing areas.

The Transition Process

Findings: The transition from open-access conditions to IFQs can be eased by advanced planning and design. It is important to define the process for change in advance, so that fishermen who are considering investment in the quota share understand and can evaluate the terms of their investment.

Recommendations: Several components of program design and implementation influence the effectiveness of the transition to IFQs, and some should be established prior to implementation of an IFQ program:

- *Output controls*—A TAC or some other form of output control is in place.
- *Moratorium*—An effective moratorium on entry is in place (see below).
- *Consultation*—All affected stakeholders are consulted on program design elements (e.g., initial allocation, transfer mechanisms, accumulation limit, cost recovery).
- *Control date*—The control date is set as close as possible to implementation **and adhered to.**
- *Qualifying period*—The qualifying period is of sufficient length to capture the relevant participation in the fishery with respect to the goals of the program. For example, the qualifying period could be set before the control date to reflect historic participation, or afterwards to reward clean fishing.
- *Program development*—The program is developed and implemented as quickly as possible after the control date.
- *Initial allocation*—An appeals process perceived to be fair and equitable is established.

• *Compensation*—A plan is in place to compensate those who are excluded by the program or whose initial allocation is less than their historical level of participation.

• *Research monitoring and evaluation*—Plans for funding and implementation of research, monitoring, and evaluation are embedded in program design.

The committee received testimony indicating that the consideration of an IFQ or other limited access program can cause considerable speculative entry into a fishery. A first step in developing an IFQ program should be to ensure that speculative entry is prevented by limiting new participants. Removing those participants with recent and limited activity in the fishery, and preventing the reentry of latent permits (those permits qualified for use, but not currently being used) would be primary goals of a moratorium. Historical participants with a long history and current participation would continue in the fishery. Once a moratorium is in place and speculative entrants have been removed, the subsequent IFQ program can be established. It might be advantageous to begin using a combination of catch history, stewardship, or other criteria measured after a moratorium has been established rather than prior to it. One advantage of using catch history after the establishment of a license moratorium as a criterion is that all the participants would be using the same qualifying years for the allocation and would have similar conditions under which they are fishing. An additional advantage could be that stewardship criteria may be easier to incorporate after a license limitation and could be used as a complement to catch history in making allocation decisions.

Establishing the criteria for determining initial allocation in the absence of speculative entry and with all fishermen operating under the same conditions could make the process of moving from an open-access regime to an IFQ program more lengthy. However, it is possible that defining all potential participants in the program first by limiting speculative entrants, and then gaining stakeholder support by developing criteria for measuring allocation for an IFQ program after this moratorium, could result in an improved transition. This mechanism would provide an opportunity for participants to improve their catch history, decrease bycatch rates, or make other adjustments in fishing practices to make them more likely to qualify for some, or a greater quantity of, the initial allocation.

References

Acheson, J.M., and R.S. Steneck, 1997. The role of management in the renewal of the Maine lobster industry. Pp. 9-25 in G. Pálsson and G. Pétursdóttir (eds.), *Social Implications of Quota Systems in Fisheries.* Nordic Council of Ministers, Copenhagen.

Ackroyd, P., R.P. Hide, and B.M.H. Sharp. 1990. New Zealand's ITQ System: Prospects for the Evolution of Sole Ownership Corporations. *Report to MAF Fisheries.* Minister of Fisheries, Wellington, New Zealand.

Adelaja, A., B.J. McCay, and J. Menzo. 1998. Market power, industrial organization, and tradable quotas. *Review of Industrial Organization* 13:589-601.

Alverson, D.L., M. Freeberg, S.A. Murawski, and J.G. Pope. 1994. *A Global Assessment of Fisheries Bycatch and Discards.* FAO Fisheries Technical Paper 339. U.N. Food and Agriculture Organization, Rome.

Anderson, L.G. 1976. The relationship between firm and fishery in common property fisheries. *Land Economics* 52:179-191.

Anderson, L.G. 1977. *The Economics of Fisheries Management.* The Johns Hopkins University Press, Baltimore, Maryland.

Anderson, L.G. 1980. Necessary components of economic surplus in fisheries economics. *Canadian Journal of Fisheries and Aquatic Sciences* 37:858-870.

Anderson, L.G. 1994. An economic analysis of highgrading in ITQ fisheries regulation programs. *Marine Resource Economics* 9:209-226.

Anderson, T.L., and P.J. Hill. 1975. The evolution of property rights: A study of the American west. *Journal of Law and Economics* 18:163-179.

Angel, J.R., D.L. Burke, R.N. O'Boyle, F.G. Peacock, M. Sinclair, and K.C.T. Zwanenburg. 1994. *Report of the Workshop on Scotia-Fundy Groundfish Management from 1977 to 1993.* Canadian Technical Report of Fisheries and Aquatic Sciences. Minister of Public Works and Government Services Canada, Catalog No. Fs 97-6/1979E, Ottawa, Canada.

Annala, J.H. 1993. Fishery assessment approaches in New Zealand's ITQ system. Pp. 791-805 in G. Kruse, D.M. Eggers, R.J. Marasco, C. Pautzke, and T.J. Quinn II (eds.), *Proceedings of the International Symposium on Management Strategies for Exploited Fish Populations.* Alaska Sea Grant College Program Report No. 93-02, University of Alaska, Fairbanks.

Annala, J.H. 1994. *Report from the Mid-Year Fishery Assessment Plenary, November 1994: Stock Assessments and Yield Estimates.* MAF Fisheries, Greta Point, Wellington, New Zealand.

Annala, J.H. 1996. New Zealand's ITQ system: Have the first eight years been a success or a failure? *Reviews in Fish Biology and Fisheries* 6:43-62.

Annala, J.H., and K.J. Sullivan. 1997. *Report from the Fishery Assessment Plenary, May 1997: Stock Assessments and Yield Estimates.* Ministry of Fisheries, Wellington, New Zealand.

Annala, J.H., K.J. Sullivan, and A. Hore. 1991. Management of multispecies fisheries in New Zealand by individual transferable quotas. *ICES Marine Science Symposium* 193:321-329.

Anonymous. 1984. *Inshore Finfish Fisheries—Proposed Policy for Future Management.* New Zealand Ministry of Agriculture and Fisheries, Wellington, New Zealand.

Anonymous. 1987. *Economic Review of the New Zealand Fishing Industry 1986-87.* New Zealand Fishing Industry Board, Wellington, New Zealand.

Apostle, R., and G. Barrett. 1992. *Emptying their Nets: Small Capital and Rural Industrialization in the Nova Scotia Fishing Industry.* University of Toronto Press, Toronto.

Apostle, R., G. Barrett, P. Holm, S. Jentoft, L. Mazany, B. McCay, and K. Mikalsen. 1998. *Community, Market and State on the North Atlantic Rim: Challenges to Modernity in the Fisheries.* University of Toronto Press, Toronto, Canada.

Apostle, R., B.J. McCay, and K.H. Mikalsen. 1997. The political construction of an IQ management system: The mobile gear ITQ experiment in the Scotia Fundy region of Canada. Pp. 27-49 in G. Pálsson and G. Pétursdóttir (eds.), *Social Implications of Quota Systems in Fisheries.* Nordic Council of Ministers, Copenhagen.

Archer, J.H., D.C. Connors, K. Laurence, S.C. Columbia, and R. Bowen. 1994. *The Public Trust Doctrine and the Management of America's Coasts.* University of Massachusetts Press, Amherst.

Arnason, R. 1996. On the individual transferable quota fisheries management system in Iceland. *Reviews in Fish Biology and Fisheries* 6(1):63-90.

Bay-Hansen, C.D. 1991. *Fisheries of the Pacific Northwest Coast: Traditional Commercial Fisheries.* Vantage Press, New York.

Beal, K.L. 1992. Surf Clam/Ocean Quahog ITQ Evaluation Based on Interviews with Captains, Owners and Crews. Exhibit 2. MacLeod 1992. Feb. 10-13, 1992. Focus on cage tags and enforcement.

Benson, G., and R. Longman. 1978. The Washington experience with limited entry. Pp. 333-352 in R.B. Rettig and J.J.C. Ginter (eds.), *Limited Entry as a Fishery Management Tool.* University of Washington Press, Seattle.

Berkes, F. (ed.). 1989. *Common Property Resources: Ecology and Community-Based Sustainable Development.* Belhaven, London.

Bevin, G., P. Maloney, and P. Roberts. 1989. *Economic Review of the New Zealand Fishing Industry 1987-88.* New Zealand Fishing Industry Board, Wellington, New Zealand.

Bigford, T.E., and C. Bribitzer. 1985. *Fishery Management: Lessons from Other Resource Management Areas.* NOAA Technical Memorandum. Department of Commerce, Washington, D.C.

Bjorndal. T., and G. Munro. 1998. The economics of fisheries management: A survey. Pp. 153-188 in T. Tietenberg and H. Folmer (eds.), *The International Yearbook of Environmental and Resource Economics: 1998/9.* Edward Elgar, Cheltenham, UK.

Blomquist, W. 1992. *Dividing the Waters: Governing Groundwater in Southern California.* ICS Press, Oakland, California.

Botkin, D.B. 1990. *Discordant Harmonies: A New Ecology for the Twenty-First Century.* Oxford University Press, New York.

Botsford, L.W., J.C. Castilla, and C.H. Peterson. 1997. The management of fisheries and marine ecosystems. *Science* 277:509-515.

Boyd, R.O., and C.M. Dewees. 1992. Putting theory into practice: Individual transferable quotas in New Zealand's fisheries. *Soc. Nat. Res.* 5:179-198.

Brander, K. 1988. Multispecies fisheries of the Irish Sea. Pp. 303-328 in J.A. Gulland (ed.), *Fish Population Dynamics*. Second Edition. John Wiley & Sons, New York.

Brandt, S.J. 1994-1995. Effects of limited access management on substitutable resources: A case study of the surf clam and ocean quahog fishery. *Journal of Environmental Systems* 23(1):21-49.

Bromley, D. (ed.). 1992. *Making the Commons Work: Theory, Practice, and Policy*. Institute for Contemporary Studies, San Francisco.

Brooks, W.K. 1891. *The Oyster: A Popular Summary of a Scientific Study*. The Johns Hopkins University Press, Baltimore, Maryland.

Brownlow, J., and J. Ropes. 1985. *Annotated Bibliography of the Ocean Quahog,* Arctica islandica *(Linnaeus)*. National Marine Fisheries Service, Northeast Fisheries Science Center, Woods Hole, Massachusetts.

Brubaker, E. 1996. The ecological implications of establishing property rights in Atlantic fisheries. Pp. 221-251 in B.L. Crowley (ed.), *Taking Ownership: Property Rights and Fishery Management on the Atlantic Coast*. Atlantic Institute for Market Studies, Halifax, Nova Scotia.

Burke, L., C. Annand, R. Barbara, L. Brander, M.A.-Etter, D. Lieu, R. O'Boyle, and G. Peacock. 1994. The Scotia-Fundy Inshore Dragger Fleet ITQ Program: Background, Implementation and Results to Date. International Council for the Exploration of the Sea. C.M. 1994/T:35.

Caddy, J., and J. Gulland. 1983. Historical patterns of fish stocks. *Marine Policy* 7:267-278.

Carrier, J.G. 1987a. Marine tenure and conservation in Papua New Guinea: Problems in interpretation. Pp. 142-167 in B. McCay and J. Acheson (eds.), *The Question of the Commons*. University of Arizona Press, Tucson.

Carrier, J.G. 1987b. Fishing practices on Ponam Island (Manus Province, Papua, New Guinea). *Anthropos* 77(5-6):904-915.

Carrothers, W.A. 1941. *The British Columbia Fisheries*. University of Toronto Press, Toronto, Canada.

Casey, K., C. Dewees, B. Turris, and J. Wilen. 1995. The effects of individual vessel quotas in the British Columbia halibut fishery. *Marine Resource Economics* 10:211-230.

Cheater, A.P., and N.K. Hopa. 1997. Representing identity. Pp. 208-223 in A. James, A. Dawson, and J. Hockey (eds.), *After Writing Culture: Epistemology and Praxis in Contemporary Anthropology*. Routledge, London.

Chopin, F.S., and T. Arimoto. 1995. The condition of fish escaping from fishing gears—A review. *Fishery Resources* 21:315-327.

Christy, F.T., Jr. 1969. Session summary: Fisheries goals and the rights of property. *Transactions of the American Fisheries Society* 2:369-378.

Christy, F.T., Jr. 1973. Fishermen's quotas: A tentative suggestion for domestic management. University of Rhode Island, Law of the Sea Institute, Occasional Papers 19, Kingston, Rhode Island.

Christy, F.T., Jr. 1977. The Fishery Conservation and Management Act of 1976: Management objectives and the distribution of benefits and costs. *Washington Law Review* 52:657-680.

Christy, F.T. 1982. *Territorial Use Rights in Marine Fisheries: Definitions and Conditions*. FAO Fisheries Technical Paper 227: FIPP/T227. Food and Agriculture Organization, Rome.

Christy, F.T. 1996. The death rattle of open access and the advent of property rights regimes in fisheries. *Marine Resource Economics* 11:287-304.

Christy, F.T., Jr., and A. Scott. 1965. *The Common Wealth in Ocean Fisheries: Some Problems of Growth and Economic Allocation*. The Johns Hopkins University Press, Baltimore, Maryland.

Cicin-Sain, B., J.E. Moore, and A.J. Wyner. 1978. Limiting entry to commercial fisheries: Some worldwide comparisons. *Ocean Management* 4:21-49.

Clark, C.W. 1985a. *Bioeconomic Modelling and Fisheries Management*. John Wiley & Sons, New York.

Clark, C.W. 1985b. The effect of fishermen's quotas on expected catch rates. *Marine Resource Economics* 1:419-427.

Clark, I.N. 1993. Individual transferable quotas: The New Zealand experience. *Marine Policy* 17:340-342.

Clark, I.N., and A.J. Duncan. 1986. New Zealand's Fisheries Management Policies—Past, Present and Future: The implementation of an ITQ–based management system. Pp. 107–140 in N. Mollett (ed.), *Fishery Access Control Programs Worldwide: Proceedings of the Workshop on Management Options for the North Pacific Longline Fisheries.* Alaska Sea Grant College Program Report No. 86-4, University of Alaska, Fairbanks.

Clark, I.N., P.J. Major, and N. Mollett. 1988. Development and implementation of New Zealand's ITQ management system. *Marine Resource Economics* 5:325-349.

Commercial Fisheries Entry Commission (CFEC). 1996a. *Changes Under Alaska's Halibut IFQ Program, 1995.* Alaska Commercial Fisheries Entry Commission, Juneau, Alaska.

Commercial Fisheries Entry Commission (CFEC). 1996b. *Changes Under Alaska's Sablefish IFQ Program, 1995.* Alaska Commercial Fisheries Entry Commission, Juneau, Alaska.

Commercial Fisheries Entry Commission (CFEC). 1997. State of Alaska commercial fisheries. *CFEC 1996 Annual Report.* Commercial Fisheries Entry Commission, Juneau, Alaska.

Commercial Fisheries Entry Commission (CFEC). 1998. *Holdings of Limited Entry Permits, Sablefish and Halibut Quota Shares Through 1997 and Data on Fishery Gross Earnings.* CFEC Report 98-SP. Commercial Fisheries Entry Commission, Juneau, Alaska.

Copes, P. 1979. The economics of marine fisheries management in the era of extended jurisdiction: The Canadian perspective. *American Economic Review* 69:256-260.

Copes, P. 1986. A critical review of individual quotas as a device in fisheries management. *Land Economics* 62:278-291.

Copes, P. 1997. Social impacts of fisheries management regimes based on individual quotas. Pp. 61-90 in G. Pálsson and G. Pétursdóttir (eds.), *Social Implications of Quota Systems in Fisheries,* Nordic Council of Ministers, Copenhagen.

Cordell, J. (ed.). 1989. *A Sea of Small Boats.* Cultural Survival Report 26. Cultural Survival, Inc., Cambridge, Massachusetts.

Costanza, R., F. Andrade, P. Antunes, M. van den Belt, D. Boersma, D.F. Boesch, F. Catarino, S. Hanna, K. Limburg, B. Low, M. Molitar, J.G. Pereira, S. Rayner, R Santos, J. Wilson, and M. Young. 1998. Principles for sustainable governance of the oceans. *Science* 281:198-199.

Creed, C.F., and B.J. McCay. 1996. Property rights, conservation, and institutional authority: Policy implications of the Magnuson Act reauthorization for the Mid-Atlantic region. *Tulane University Environmental Law Journal* 9(2):245-256.

Criddle, K.R. 1994. Economics of resource use: A bioeconomic analysis of the Pacific halibut fishery. Pp. 37-52 in *Proceedings of the Fourth International Symposium of the Conference of Asian and Pan-Pacific University Presidents,* Alaska Sea Grant, Anchorage, Alaska.

Criddle, K. R. 1996. Predicting the consequences of alternative harvest regulations in a sequential fishery. *North American Journal of Fisheries Management* 16:30-40.

Crothers, S. 1988. Individual transferable quotas: The New Zealand experience. *Fisheries* 13(1):10-12.

Crutchfield, J.A. (ed.). 1959. *Biological and Economic Aspects of Fisheries Management.* University of Washington, Seattle.

Crutchfield, J.A., and G. Pontecorvo. 1969. *The Pacific Salmon Fisheries: A Study of Irrational Conservation.* The Johns Hopkins University Press, Baltimore, Maryland.

Crutchfield, J.A., and A. Zellner. 1962. Economic aspects of the Pacific halibut fishery. *Fishery Industrial Research* 1:1-73.

Daníelsson, Á. 1997. Fishery management in Iceland. *Ocean and Coastal Management* 35:121-135.

Davies, N.M. 1992. Fisheries management—A New Zealand perspective. *South Africa Journal of Marine Science* 12:1069-1077.

De Alessi, M. 1998. *Fishing for Solutions.* IEA Studies on the Environment No. 11. The Institute of Economic Affairs, London.

Department of Commerce. 1996. Amendment #9 to the Fishery Management Plan for Atlantic Surfclam and Ocean Quahog Fisheries, April 1996. Mid-Atlantic Fishery Management Council in cooperation with the National Marine Fisheries Service and the New England Fishery Management Council. Silver Spring, Maryland.

Dewees, C.M. 1989. Assessment of the implementation of individual transferable quotas in New Zealand's inshore fishery. *North American Journal of Fisheries Management* 9:131-139.

Dewees, C.M., and E. Ueber. 1990. Effects of different fishery management schemes on bycatch, joint catch, and discards. *Summary of a National Workshop, January 29-31, 1990, San Francisco, Calif.* Report No. T-CSGCP-019. California Sea Grant College Program, La Jolla.

Doeringer, P.B., P.I. Moss, and D.G. Terkla. 1986. *The New England Fishing Economy: Jobs, Income, and Kinship.* University of Massachusetts Press, Amherst.

Durrenberger, P. 1996. *Gulf Coast Soundings: People and Policy in the Mississippi Shrimp Industry.* University Press of Kansas, Lawrence.

Dyer, C.L., and J.R. McGoodwin (eds.). 1994. *Folk Management in the World's Fisheries: Lessons for Modern Fisheries Management.* University Press of Colorado, Niwot, Colorado.

Edney, J.J. 1981. Paradoxes on the commons: Scarcity and the problem of inequality. *Journal of Community Psychology* 9:3-34.

Edwards, S.F. 1994. Ownership of renewable ocean resources. *Marine Resource Economics* 9:253-273.

Ellerman, A.D., R. Schmalensee, P.L. Joskow, J.P. Montero, and E.M. Bailey. 1997. Emissions Trading Under the U.S. Acid Rain Program: Evaluation of Compliance Costs and Allowance Market Performance. MIT Center for Energy and Environmental Policy Research, Cambridge, Massachusetts.

Elliot, G.H. 1973. Problems confronting fishing industries relative to management policies adopted by governments. *Journal of the Fisheries Research Board of Canada* 30(12):2486-2489.

Eythorsson, E. 1996. Coastal communities and ITQ management: The case of Iceland. *Sociologica Ruralis* 36(2):212-223.

Felt, L., B. Neis, and B. McCay. 1997. Co-management. Pp. 185-194 in J.G. Boreman, B.S. Nakashima, H.W. Powles, J.A. Wilson, and R.L. Kendall (eds.). *Northwest Atlantic Groundfish: Perspectives on a Fishery Collapse.* American Fisheries Society, Bethesda, Maryland.

Firth, R. 1959. *Social Change in Tikopia.* Allen & Unwin, London.

Fiske, A.P. 1991. *Structures of Social Life: The Four Elementary Forms of Human Relations.* The Free Press, New York.

Fogarty, M.J., and S.A. Murawski. 1998. Large-scale disturbance and the structure of marine systems: Fishery impacts on Georges Bank. *Ecological Applications* 8(1) Suppl.:S6-S22.

Food and Agriculture Organization (FAO). 1956. R. Turvey and J. Wiseman (eds.), *Proceedings of a Round Table Organized by the International Economic Assoc. Held in Rome, September 1956.* United Nations Food and Agriculture Organization, Rome.

Food and Agriculture Organization (FAO). 1962. *Economic Effects of Fishery Regulation: Report of an FAO Expert Meeting at Ottawa, June 12-17, 1961,* R. Hamlisch (ed.). U.N. Food and Agriculture Organization, Rome.

Fraser, G.A. 1979. Limited entry: Experience of the British Columbia salmon fishery. *Journal of the Fisheries Research Board of Canada* 36:754-763.

Fromm, O., and B. Hansjurgens. 1996. Emission trading in theory and practice: An analysis of RECLAIM in Southern California. *Environment and Planning C—Government and Policy* 14(3):367-384.

Gardner, R., E. Ostrom, and J. Walker. 1990. The nature of common-pool resource problems. *Rationality and Society* 2(3):335-358.

Gatewood, J.B., and B.J. McCay. 1988. Job satisfaction and the culture of fishing: A comparison of six New Jersey fisheries. *MAST/Maritime Anthropological Studies* 1(2):103-128.

Gilroy, H., P.J. Sullivan, S. Lowe, and J.M. Terry. 1996. *Preliminary Assessment of the Halibut and Sablefish IFQ Programs in Terms of Nine Potential Conservation Effects.* International Pacific Halibut Commission, Seattle, Washington, and Alaska Fisheries Science Center, National Marine Fisheries Service, Seattle, Washington.

Ginter, J.J.C. 1995. The Alaska community development quota fisheries management program. *Ocean and Coastal Management* 28(1-3):147-163.

Goodale, H., and M. Raizin. 1992. Clam/Quahog ITQ Administration. Exhibit 5. MacLeod 1992.

Goodenough, W.H. 1951. Pp. 3-5 in *Property, Kin, and Community on Truk.* Yale University Publications in Anthropology, New Haven, Connecticut.

Goodenough, W.H. 1963. *Cooperation in Change: An Anthropological Approach to Community Development.* Russell Sage Foundation, New York.

Gordon, H.S. 1953. An economic approach to the optimum utilization of fishery resources. *Journal of the Fisheries Research Board of Canada* 10:442-457.

Gordon, H.S. 1954. The economic theory of a common property resource: The fishery. *Journal of Political Economy* 62:124-142.

Grafton, R.Q. 1995. Rent capture in a rights-based fishery. *Journal of Environmental Economics and Management* 28(1):48-67.

Graham, M. 1943. *The Fish Gate.* Faber and Faber, London

Granovetter, M. 1985. Economic action and social structure: The problem of embeddedness. *American Journal of Sociology* 91: 481-510.

Greenberg, E.V.C. 1993. Ocean fisheries. In *Sustainable Environmental Law: Integrating Natural Resource and Pollution Abatement Law from Resources to Recovery.* West Environmental Law Institute, St. Paul, Minnesota.

Greenberg, J.A., and M. Herrmann. 1994. Allocative consequences of pot limits in the Bristol Bay red king crab fishery: An economic analysis. *North American Journal of Fisheries Management* 14:307-317.

Grew, P. 1982. A State Perspective on Mining Research Needs of the Forest Service, Conference on Emerging Research Needs and Priorities Concerning Mineral Development on the National Forest System Lands, University of California, Berkeley, April 30, 1982.

Griffith, D. 1997. New immigrants in an old industry: Blue crab processing in Pamlico County, North Carolina. Pp. 153-186 in D. Stull, M. Broadway, and D. Griffith (eds.), *Any Way You Cut It: Meat Processing and Small-Town America.* University Press of Kansas, Lawrence.

Groundfish Development Authority. 1998. *Groundfish Development Authority 1998-1999 Operation Plan.* Department of Fisheries and Oceans, Ottawa, Canada.

Gunderson, D.R. 1984. The great widow rockfish hunt of 1980-82. *North American Journal of Fisheries Management* 4:465-468.

Hahn, R.W., and G.L. Hester. 1989. Where did all the markets go? An analysis of EPA's Emission Trading Program. *Yale Journal of Regulation* 6(1):109-153.

Hanna, S. 1990. The eighteenth century English commons: A model for ocean management. *Ocean and Shoreline Management* 14:155-172.

Hanna, S.S. 1994. Co-management. Pp. 233-242 in K. Gimbel (ed.), *Limited Access Management: A Guidebook for Conservation.* World Wildlife Fund and Center for Marine Conservation, Washington, D.C.

Hanna, S.S. 1995. User participation and fishery management performance within the Pacific Fishery Management Council. *Ocean and Coastal Management* 28(1-3):23-44.

Hanna, S.S. 1997. The new frontier of American fisheries governance. *Ecological Economics* 20:221-233.

Hardin, G. 1968. The tragedy of the commons. *Science* 162:1243-1248

Hausker, K. 1990. Coping with the cap: How auctions can help the allowance market work. *Public Utilities Fortnightly* 125:28-34.

Hausker, K. 1992. The politics and economics of auction design in the market for sulfur dioxide pollution. *Journal of Policy Analysis and Management* 11(4):553-572.

Helgason, A., and G. Pálsson. 1997. Contested commodities: The moral landscape of modernist regimes. *Journal of the Royal Anthropological Institute (incorporating Man)* 3(3):451-471.

Herrmann, M. 1996. Estimating the induced price increase for Canadian Pacific halibut with the introduction of the individual vessel quota program. *Canadian Journal of Agricultural Economics* 44:151-164.

Herrmann, M., J.A. Greenberg, and K.R. Criddle. 1998. Proposed pot limits for the Adak brown king crab fishery: A distinction between open access and common property. *Alaska Fishery Research Bulletin* 5:25-38.

Higgs, R. 1982. Legally induced technical regress in the Washington salmon fishery. *Research in Economic History* 7:55-89.

Hilborn, R., and C.J. Walters. 1992. *Quantitative Fisheries Stock Assessment: Choice, Dynamics, and Uncertainty.* Chapman and Hall, New York.

Hofmann, E.E., and T.M. Powell. 1998. Environmental variability effects on marine fisheries: Four case studies. *Ecological Applications* 8(1) Suppl.:S23-S32.

Homans, F.R. 1993. *Modeling Regulated Open Access Resource Use.* Doctoral Dissertation, Department of Agricultural Economics, University of California, Davis.

Hourston, A. 1980. The decline and recovery of Canada's Pacific herring stocks. *Rapport et Procès-verbaux de Réunions de Conseil International pour l'Exploration de la mer* 177:143-153.

Huppert, D. 1991. Managing the groundfish fisheries of Alaska: History and prospects. *Reviews in Aquatic Sciences* 4(4):339-373.

International Council for the Exploration of the Sea (ICES). 1996. Report by Correspondence of the ICES Study Group on the Management Performance of Individual Transferable Quota (ITQ) Systems. September. ICES C.M. 1996.

International Council for the Exploration of the Sea (ICES). 1997. Report of the ICES Study Group on the Management Performance of Individual Transferable Quota (ITQ) Systems. October. ICES C.M. 1997/H:2.

International Pacific Halibut Commission (IPHC). 1987. *The Pacific Halibut: Biology, Fishery, and Management.* Technical Report 22, International Pacific Halibut Commission, Seattle, Washington.

International Pacific Halibut Commission (IPHC). 1997. *Report of Assessment and Research Activities, 1997.* International Pacific Halibut Commission, Seattle, Washington.

Jarman, C. 1986. The public trust doctrine in the exclusive economic zone. *Oregon Law Review* 65:1-33.

Jarman, C., and J. Archer. 1992. Sovereign rights and responsibilities: Applying public trust principles to the management of EEZ space and resources. *Ocean and Coastal Management* 17(3-4):253-271.

Jentoft, S. 1989. Fisheries co-management: Delegating government responsibility to fishermen's organizations. *Marine Policy* 13(2):137-154.

Jentoft, S., and B.J. McCay. 1995. User participation in fisheries management: Lessons drawn from international experience. *Marine Policy* 19(1):227-246.

Johannes, R.E. 1978. Traditional marine conservation methods in Oceania and their demise. *Annual Review of Ecology and Systematics* 9:349-364.

Johnson, R.N. and G.D. Libecap. 1982. Contracting problems and regulation: The case of the fishery. *American Economic Review* 72:1005-1022.

Johnson, R.W. 1989. Water pollution and the public trust doctrine. *Environmental Law* 19:485-513.

Kalland, A. 1988. *Fishing Villages in Togukawa Japan: The Case of the Fukuoka Domain.* Doctoral dissertation. Department of East Asian Studies, University of Oslo.

Kearney, J.F. 1984. The transformation of the Bay of Fundy herring fisheries 1976-1978: An experiment in fishermen-government co-management. Pp. 185-203 in C. Lamson and A. Hanson, (eds.), *Atlantic Fisheries and Coastal Communities.* Institute of Resource and Environmental Studies, Dalhousie University, Halifax, Nova Scotia.

Kearney, J., A. Bull, M. Recchia, M. Desroches, L. Langille, and G. Cunningham. 1998. Resistance to Privatisation: Community-Based Fisheries Management in an Industrialised Nation. Submitted in Response to a Call for Case Studies for an International Workshop on Community-Based Natural Resource Management, The World Bank, Washington, D.C., May 10-14. Typescript. 5 pp. (Available through John Kearney, Extension Department, St. Francis Xavier University, Antigonish, Nova Scotia, Canada.)

Keen, E.A. 1988. *Ownership and Productivity of Marine Fishery Resources, An Essay on the Resolution of Conflict in the Use of Ocean Pastures.* McDonald and Woodward Publ. Co., Blacksburg, Virginia.

Kinoshita, R.K., A. Grieg, D. Colpo, and J.M. Terry. 1996. *Economic Status of the Groundfish Fisheries off Alaska, 1996.* Resource Ecology and Fisheries Management Division, Alaska Fisheries Science Center, National Marine Fisheries Service, National Oceanic and Atmospheric Administration, Seattle, Washington.

Knapp. G. 1997a. Initial effects of the Alaska halibut and sablefish IFQ program. Some preliminary research results and recommendations. Written testimony provided to the committee.

Knapp, G. 1997b. *Modeling Community Economic Impacts of the Alaska Halibut IFQ Program.* Final Project Report on Saltonstall Kennedy Program award #NA37FD0184. Institute of Social and Economic Research, University of Alaska, Anchorage.

Knapp. G., and D. Hull. 1996. *The First Year of the Alaska IFQ Program: A Survey of Halibut Quota Share Holders.* Alaska Department of Commerce and Economic Development and Alaska Department of Fish and Game, Juneau.

Krueger, A.O. 1974. The political economy of the rent-seeking society. *American Economic Review* 64:291-303.

Larkin, P.A. 1977. An epitaph for the concept of maximum sustained yield. *Transactions of the American Fisheries Society* 106(1):1-11.

Levine, H.B., and M.W. Levine. 1987. *Stewart Island: Anthropological Perspectives on a New Zealand Fishing Community.* Victoria University Occasional Papers in Anthropology, No. 1. Wellington, New Zealand.

Libecap, G. 1989. *Contracting for Property Rights.* Cambridge University Press, New York.

Lieber, M.D. 1994. *More Than a Living: Fishing and the Social Order on a Polynesian Atoll.* Westview Press, Boulder, Colorado.

Lin, B.-H., H.S. Richards, and J.M. Terry. 1988. An analysis of the exvessel demand for Pacific halibut. *Marine Resource Economics* 4:305-314.

Lloyd, W.F. 1968. [1837] Lectures on Population, Value, Poor-Laws, and Rent, Delivered in the University of Oxford During the Years 1832, 1833, 1834, 1835, & 1836. *Reprints of Economic Classics.* Augustus M. Kelley, New York.

Ludwig, D., R. Hilborn, and C. Walters. 1993. Uncertainty, resource exploitation, and conservation: Lessons from history. *Science* 260:17,36.

MacDonald, G. 1992. Subject: Individual Transferable Quota (ITQ) Management System. Exhibit 3. MacLeod 1992.

Macinko, S. 1993. Public or private? United States commercial fisheries management and the public trust doctrine, reciprocal challenges. *Natural Resources Journal* 32:919-955.

MacLeod, E. 1992. Memorandum, Review of the Effectiveness of Our Administrative and Enforcement Obligations Under the Surf Clam/Quahog ITQ Plan. February 25, 1992. Northeast Region, National Marine Fisheries Service, Gloucester, Massachusetts.

Maloney, D.G., and P.H. Pearse. 1979. Quantitative rights as an instrument for regulating commercial fisheries. *Journal of the Fisheries Research Board of Canada* 36:859-866.

Martin, F. 1989. *Common Pool Resources and Collective Action: A Bibliography.* Workshop in Political Theory and Policy Analysis. Indiana University, Bloomington.

Marvin, K.A. 1992. Protecting common property resources through the marketplace: Individual transferable quotas for surf clams and ocean quahogs. *Vermont Law Review* 16:1127-1168.

Matthews, D. 1997. *Beyond IFQ Implementation: A Study of Enforcement Issues in the Alaska Individual Fishing Quota Program.* National Marine Fisheries Service, Silver Spring, Maryland.

Matulich, S.C., R.C. Mittlehammer, and C. Reberte. 1996. Toward a more complete model of individual transferable fishing quotas: Implications of incorporating the processing sector. *Journal of Environmental Economics and Management* 31:112-128.

Matulich, S.C., and M. Sever. 1999. Reconsidering the initial allocation of ITQs: The search for a Pareto safe allocation between fishers and processors. *Land Economics* 75(2).

McCarthy, J.J. 1992. Subject: Surf Clam/Quahog ITQ Review. Exhibit 4. Report from F/EN# (enforcement), Mid-Atlantic Fishery Management Council, Dover, Delaware.

McCay, B.J. 1992. *From the Waterfront: Interviews with New Jersey Fishermen on Marine Safety and Training.* Final Report, Saltonstall/Kennedy Project for Understanding and Teaching Marine Safety. New Jersey Marine Sciences Consortium, Fort Hancock.

McCay, B.J. 1998. *Oyster Wars and the Public Trust.* University of Arizona Press, Tucson.

McCay, B.J. 1999. "That's Not Right": Resistance to Enclosure in a Newfoundland Crab Fishery. Pp. 301-320 in D. Newell and R. Ommer (eds.), *Fishing People, Fishing Places: Issues in Small-Scale Fisheries.* University of Toronto Press, Toronto, Canada.

McCay, B., and J. Acheson (eds.). 1987. *The Question of the Commons: The Culture and Ecology of Communal Resources.* University of Arizona Press, Tucson.

McCay, B.J., R. Apostle, and C.F. Creed. 1998. Individual transferable quotas, co-management, and community: Reflections from Nova Scotia. *Fisheries* 23(4):20-23.

McCay, B.J., R. Apostle, C.F. Creed, A.C. Finlayson, and K. Mikalsen. 1995. Individual transferable quotas (ITQs) in Canadian and U.S. fisheries. *Ocean and Coastal Management* 28 (1-3):85-116.

McCay, B.J., and C.F. Creed. 1987. Crews and labor in the surf clam and ocean quahog fleet of the Mid-Atlantic region. *A Report to the Mid-Atlantic Fisheries Management Council, October 1987.* Mid-Atlantic Fisheries Management Council, Dover, Delaware.

McCay, B.J., and C.F. Creed. 1990. Social structure and debates on fisheries management in the Mid-Atlantic surf clam fishery. *Ocean & Shoreline Management* 13:199-229.

McCay, B.J. and C.F. Creed. 1994. Social impacts of ITQs in the sea clam fishery. *Final Report to the New Jersey Sea Grant College Program.* New Jersey Marine Sciences Consortium, Fort Hancock, New Jersey.

McCay, B.J., J.B. Gatewood, and C.F. Creed. 1989. Labor and the labor process in a limited entry fishery. *Marine Resource Economics* 6:311-330.

McCay, B.J., J.B. Gatewood, and C.F. Creed. 1990. Labor and the labor process in a limited entry fishery. *Marine Resource Economics* 6:311-330.

McCay, B.J., and S. Jentoft. 1998. Market or community failure? Critical perspectives on common property research. *Human Organization* 57(1):21-29.

McEvoy, A.F. 1986. *The Fisherman's Problem: Ecology and Law in the California Fisheries 1850-1980.* Cambridge University Press, New York.

McGoodwin, J.R. 1990. *Crisis in the World's Fisheries.* Stanford University Press, Stanford, California.

McHugh, J.L. 1972. Jeffersonian democracy and the fisheries. Pp. 134-155 in B.J. Rothschild (ed.), *World Fisheries Policy*, University of Washington Press, Seattle.

Menzo, J.E. 1996. Industrial Organizational Impacts of ITQs on the Mid-Atlantic Surf Clam and Ocean Quahog Fishery. Unpublished M.S. Thesis. Agricultural Economics, Rutgers, the State University of New Jersey.

Menzo, J., A. Adelaja, and B. McCay. 1997. Supply Response Behavior Under a Tradable Quota System: The Case of the Mid-Atlantic Surf Clam and Ocean Quahog Fishery. Paper presented at the March 1997 meeting of the Atlantic Economic Society, London.

Mid-Atlantic Fishery Management Council (MAFMC). 1988. Amendment #8 Fishery Management Plan for the Atlantic Surf Clam and Ocean Quahog Fishery. Draft June 30, 1988. Mid-Atlantic Fishery Management Council in cooperation with the National Marine Fisheries Service and the New England Fishery Management Council, Dover, Delaware.

Mid-Atlantic Fishery Management Council (MAFMC). 1990. Amendment #8 Fishery Management Plan for the Atlantic Surf Clam and Ocean Quahog Fishery. June 20, 1990. Mid-Atlantic Fishery Management Council in cooperation with the National Marine Fisheries Service and the New England Fishery Management Council, Dover, Delaware.

Mid-Atlantic Fishery Management Council (MAFMC). 1996. Amendment #9 to the Fishery Management Plan for Atlantic Surf Clam and Ocean Quahog Fisheries, April 1996. Mid-Atlantic Fishery Management Council in cooperation with the National Marine Fisheries Service and the New England Fishery Management Council, Dover, Delaware.

Mid-Atlantic Fishery Management Council (MAFMC). 1997. *Overview of the Surfclam and Ocean Quahog Fisheries and Quota Recommendations for 1998.* Mid-Atlantic Fishery Management Council, Dover, Delaware.

Miller, M.L., and J. van Maanen 1979. "Boats don't fish, people do": Some ethnographic notes on the federal management of fisheries in Gloucester. *Human Organization* 38(4):377-385.

Milliken, W.J. 1994. Individual transferable fishing quotas and antitrust law. *Ocean and Coastal Law Journal* 1:35-58.

Moloney, D.G., and P.H. Pearse. 1979. Quantitative rights as an instrument for regulating commercial fisheries. *Journal of the Fisheries Research Board of Canada* 36:859-866.

MRAG Americas, Inc. 1997. Consolidated Report on the Peer Review of Red Snapper (*Lutjanus campechanus*) Research and Management in the Gulf of Mexico. Prepared for the Office of Science and Technology, National Marine Fisheries Service, Silver Spring, Maryland.

Murawski, S.A., and J.S. Idoine. 1989. Yield sustainability under constant-catch policy and stochastic recruitment. *Transactions of the American Fisheries Society* 118(4):349-367.

Murdock, G.P., C.S. Ford, A.E. Hudson, R. Kennedy, L.W. Simmons, and J.W.M. Whiting. 1945. P. 29 in *Outline of Cultural Materials.* Yale Anthropological Studies II, Yale University, New Haven, Connecticut.

Muse, B., and K Schelle. 1989. *Individual Fishermen's Quotas: A Preliminary Review of Some Recent Programs (CFEC 89-1).* Alaska Commercial Fisheries Entry Commission, Juneau.

Myers, R.A., J. Bridson, and N.J. Barrowman. 1995. Summary of Worldwide Stock and Recruitment Data. *Canadian Technical Report of Fisheries and Aquatic Sciences.* Northwest Atlantic Fisheries Center, St. John's, Newfoundland, Canada.

National Marine Fisheries Service (NMFS). 1996. *Our Living Oceans. Report of the Status of U.S. Living Marine Resources, 1995.* NOAA Technical Memorandum NMFS-F/SPO-19. U.S. Department of Commerce, Washington, D.C.

National Marine Fisheries Service (NMFS). 1997a. Report to the IFQ Review Committee of the National Academy of Sciences, National Research Council, Ocean Studies Board. Restricted Access Management Division, Alaska Region National Marine Fisheries Service, Juneau.

National Marine Fisheries Service (NMFS). 1997b. Central title and lien registry for limited access permits. *Federal Register* 62(44):10249-10252.

National Marine Fisheries Service (NMFS). 1998. *Report to the Fleet.* Restricted Access Management Division, Alaska Region, National Marine Fisheries Service, Juneau.

National Research Council (NRC). 1986. *Proceedings of the Conference on Common Property Resource Management.* National Academy Press, Washington, D.C.

National Research Council (NRC). 1994. *An Assessment of Atlantic Bluefin Tuna.* National Academy Press, Washington, D.C.

National Research Council. 1996. *Upstream: Salmon and Society in the Pacific Northwest.* National Academy Press, Washington, D.C.

National Research Council (NRC). 1997. *Striking A Balance: Improving Stewardship of Marine Areas.* National Academy Press, Washington, D.C.

National Research Council (NRC). 1998a. *Improving Fish Stock Assessments.* National Academy Press, Washington, D.C.

National Research Council (NRC). 1998b. *Review of Northeast Fishery Stock Assessments.* National Academy Press, Washington, D.C.

National Research Council (NRC). 1999a. *The Community Development Quota Program in Alaska and Lessons for the Western Pacific.* National Academy Press, Washington, D.C.

National Research Council (NRC). 1999b. *Sustaining Marine Fisheries.* National Academy Press, Washington, D.C.

Neher, P.A., R. Arnason, and N. Mollett (eds.). 1989. *Rights-Based Fishing.* Kluwer Academic Publishers, Dordrecht, The Netherlands.

Nelson, R.H. 1997. How to reform grazing policy: Creating forage rights on federal rangelands. *Fordham Environmental Law Journal* 8:645-690.

Netboy, A. 1968. *The Atlantic Salmon: A Vanishing Species?* Houghton Mifflin, Boston.

New England Fishery Management Council (NEFMC). 1996. Amendment 7 to the Northeast Multispecies Fishery Management Plan. New England Fishery Management Council, Saugus, Massachusetts.

Newell, D. 1993. *Tangled Webs of History: Indians and the Law in Canada's Pacific Coast Fisheries.* University of Toronto Press, Canada.

New Zealand Fisheries Task Force. 1992. *Sustainable Fisheries. Tiakina nga Taonga a Tangaroa.* Report of the Fisheries Task Force to the Minister of Fisheries on the Review of Fisheries Legislation, Wellington, New Zealand.

Nielsen, L.A. 1976. The evolution of fisheries management philosophy. *Marine Fisheries Review* 38(1):15-23.

Northeast Fishery Science Center (NEFSC). 1995. *Report of the 15th Northeast Regional Stock Assessment Workshop.* NESFC Ref. Doc. 95-0X. Northeast Fishery Science Center, Woods Hole, Massachusetts.

North Pacific Fishery Management Council (NPFMC). 1991a. *Revised Supplement to the Draft Environmental Impact Statement, Regulatory Impact Review, Regulatory Flexibility Analysis for the Groundfish Management Plans for the Gulf of Alaska and the Bering Sea/Aleutian Islands—Longline and Pot Gear Sablefish Management in the Gulf of Alaska and the Bering Sea/Aleutian Islands.* North Pacific Fishery Management Council, Anchorage, Alaska.

North Pacific Fishery Management Council (NPFMC). 1991b. *Environmental Impact Statement, Regulatory Impact Review, Regulatory Flexibility Analysis for the Proposed Individual Fishing Quota Management Alternatives for the Halibut Fisheries in the Gulf of Alaska and Bering Sea/Aleutian Islands.* North Pacific Fishery Management Council, Anchorage, Alaska.

North Pacific Fishery Management Council (NPFMC). 1991c. *North Pacific Fishery Management Council Sablefish and Halibut Fixed Gear Fisheries Individual Fishery Quota System Draft Implementation Plan.* North Pacific Fishery Management Council, Anchorage, Alaska.

North Pacific Fishery Management Council (NPFMC). 1992. IFQs off the port bow. North Pacific Fishery Management Council, Anchorage, Alaska.

North Pacific Fishery Management Council (NPFMC). 1994a. *Faces of the Fisheries.* North Pacific Fishery Management Council, Anchorage, Alaska.

North Pacific Fishery Management Council (NPFMC). 1994b. *Draft Environmental Assessment/ Regulatory Impact Review for License Limitation Alternatives for the Groundfisheries and Crab Fisheries in the Gulf of Alaska and the Bering Sea/Aleutian Islands.* North Pacific Fishery Management Council, Anchorage, Alaska.

North Pacific Fishery Management Council (NPFMC). 1997a. *Stock Assessment and Fishery Evaluation Report for the Groundfish Resources of the Gulf of Alaska.* North Pacific Fishery Management Council, Anchorage, Alaska.

North Pacific Fishery Management Council (NPFMC). 1997b. *Development of the Individual Fishing Quota Program for Sablefish and Halibut Longline Fisheries off Alaska.* North Pacific Fishery Management Council, Anchorage, Alaska.

North Pacific Fishery Management Council (NPFMC). 1998. *Environmental Assessment/Regulatory Impact Review for Plan Amendment 45 for Continuation of the Pollock CDQ Program to the Bering Sea/Aleutian Islands Fishery Management Plan.* North Pacific Fishery Management Council, Anchorage, Alaska.

Nussbaum, B.D. 1992. Phasing down lead in gasoline in the U.S.: Mandates, incentives, trading and banking. Pp. 21-34 in T. Jones and J. Corfee-Morlot (eds.), *Climate Change: Designing a Tradable Permit System.* Organization for Economic Cooperation and Development, Paris.

O'Boyle, R., C. Annand, and L. Brander. 1994. Individual Quotas in the Scotian shelf groundfishery off Nova Scotia, Canada. Pp. 152-168 in K. Gimbel (ed.), *Limiting Access to Marine Fisheries: Keeping the Focus on Conservation.* Center for Marine Conservation and World Wildlife Fund U.S., Washington, D.C.

Organization for Economic Cooperation and Development (OECD). 1997. *Toward Sustainable Fisheries: Economic Aspects of the Management of Living Marine Resources.* Organization for Economic Cooperation and Development, Paris.

Örlygsson, P. 1997. Hver á kvótann? ("Who Owns the Quota?"). Paper presented at a workshop at the Fisheries Research Institute, University of Iceland, November 8.

Ostrom, E. 1990. *Governing the Commons.* Cambridge University Press, Cambridge, U.K.

Ostrom, E., R. Gardner, and J. Walker (eds.). 1994. *Rules, Games, and Common-Pool Resources.* University of Michigan Press, Ann Arbor.

Ostrom, V. and E. Ostrom. 1977. Public goods and public choices. Pp. 7-49 in E.S. Savas (ed.), *Alternatives for Delivering Public Services: Toward Improved Performance.* Westview Press, Boulder, Colorado.

Pacific Fishery Management Council (PFMC). 1993. *Pacific Coast Groundfish Plan.* Pacific Fishery Management Council, Portland, Oregon.

Pacific Fishery Management Council (PFMC). 1997. *Status of the Pacific Coast Groundfish Fishery Through 1997 and Recommended Acceptable Biological Catches For 1998: Stock Assessment and Fishery Evaluation.* Portland, Oregon.

Pacific Fishery Management Council (PFMC). 1998. *Pacific Coast Groundfish Limited Entry Trawl Permit Buyback Business Plan.* Prepared by the Pacific Fishery Management Council's Buyback Committee. Portland, Oregon.

Pálsson, G. 1991. *Coastal Economies, Cultural Accounts: Human Ecology and Icelandic Discourse.* Manchester University Press, New York.

Pálsson, G., and A. Helgason. 1995. Figuring fish and measuring men: The individual transferable quota system in the Icelandic cod fishery. *Ocean and Coastal Management* 28(1-3):117-146.

Pálsson, G., and G. Pétursdóttir (eds.). 1997. *Social Implications of Quota Systems in Fisheries.* Nordic Council of Ministers, Copenhagen.

Parker, G. 1994. *The New Zealand Seafood Industry Economic Review 1993.* New Zealand Fishing Industry Board, Wellington, New Zealand.

Parsons, L.S. 1993. *Management of Marine Fisheries in Canada*. Canadian Bulletin of Fisheries and Aquatic Sciences 225. National Research Council of Canada and Department of Fisheries and Oceans, Ottawa, Canada.

Parsons, T.R. 1996. Taking stock of fisheries management. *Fisheries Oceanography* 5:224-226.

Paulik, G.J., A.S. Hourston, and P.A. Larkin. 1967. Exploitation of multiple stocks by a common fishery. *Journal of the Fisheries Research Board of Canada* 24:2527-2537.

Pautzke, C.G., and C.W. Oliver. 1997. *Development of the Individual Fishing Quota Program for Sablefish and Halibut Longline Fisheries off Alaska*. North Pacific Fishery Management Council, Anchorage, Alaska.

Pearse, P.H. 1991. *Building on Progress—Fisheries Policy Development in New Zealand*. A report prepared for the Minister of Fisheries, MAF Policy, Wellington, New Zealand.

Pinkerton, E. 1987. Intercepting the State: Dramatic processes in the assertion of local co-management rights. Pp. 344-369 in B. McCay and J. Acheson (eds.), *The Question of the Commons*. University of Arizona Press, Tucson.

Pinkerton, E. (ed.). 1989. *Cooperative Management of Local Fisheries: New Directions for Improved Management and Community Development*. University of British Columbia Press, Vancouver, Canada.

Plott, C., and R. Meyer. 1975. The technology of public goods, externalities, and the exclusion principle. Pp. 65-94 in E.S. Mills (ed.), *Economic Analysis of Environmental and Public Problems*. National Bureau of Economic Research, New York.

Rafnsson, V., and H. Gunnarsdóttir. 1992. Fatal accidents among Icelandic seamen: 1966-86. *British Journal of Industrial Medicine* 49:694-699.

Rettig, R.B. 1991. Recent changes in fishery management in developed countries. Paper presented at the IFRS/IIFET Symposium on Fishery Management, August 26-September 3, Tokyo.

Rettig, R.B., and J.C. Ginter. 1978. *Limited Entry as a Fishery Management Tool: Proceedings of a National Conference to Consider Limited Entry as a Tool in Fishery Management*. University of Washington Press, Seattle.

Ricker, W.E. 1958. Maximum sustained yields from fluctuating environments and mixed stocks. *Journal of the Fisheries Research Board of Canada* 15:991-1006.

Rieser, A. 1991. Ecological preservation as a public property right: An emerging doctrine in search of a theory. *Harv. Envtl. L. Rev.* 15:393.

Rieser, A. 1997a. International fisheries law, overfishing, and marine biodiversity. *Georgetown International Environmental Law Review* 9(2):251-279.

Rieser, A. 1997b. Property rights and ecosystem management in U.S. fisheries: Contracting for the commons? *Ecology Law Quarterly* 24(4):813-832.

Roberts, C.M. 1997a. Ecological advice for the global fisheries crisis. *TREE* 12:35-38.

Roberts, C.M. 1997b. Connectivity and management of Caribbean coral reefs. *Science* 278:1454-1457.

Rose, C. 1986. The comedy of the commons: Custom, commerce, and inherently public property. *University of Chicago Law Review* 53(3):711-781.

Rosenberg, A.A., M.J. Fogarty, M.P. Sissenwine, J.R. Beddington, and J.G. Shepherd. 1993. Achieving sustainable use of renewable resources. *Science* 262:828-829.

Ross, R. 1992. Summary: Surf Clam ITQ Implementation-Processor Evaluation. Exhibit 1, MacLeod. 1992.

Royce, W., D. Bevan, J. Crutchfield, G. Paulik, and R. Fletcher. 1963. Salmon Gear Limitation in Northern Washington Waters. University of Washington Publications in Fisheries, New Series 2(1).

Salz, P. 1996. *ITQs in the Netherlands: Twenty Years of Experience*. International Council for Exploration of the Sea, Annual Scientific Meeting, Reykjavík.

Sampson, D.B. 1994. Fishing tactics in a two-species fisheries model: The bioeconomics of bycatch and discarding. *Canadian Journal of Fisheries and Aquatic Sciences* 51:2688-2694.

Sax, J.L. 1970. The public trust doctrine in natural resource law: Effective judicial intervention. *Michigan Law Review* 68:471-475.

Schaefer, M.B. 1957. A study of the dynamics of the fishery for yellowfin tuna in the eastern tropical Pacific Ocean. *Inter-American Tropical Tuna Commission Bulletin* 2:247-268.

Scheiber, H.N., and C. Carr. 1997. The limited entry concept and the pre-history of the ITQ movement in fisheries management. Pp. 235-260 in G. Pálsson and G. Petersdottir (eds.), *Social Implications of Quota Systems in Fisheries*. Nordic Council of Ministers, Copenhagen.

Schlager, E. 1990. *Model Specification and Policy Analysis: The Governance of Coastal Fisheries*. Ph.D. Dissertation, Indiana University, Bloomington.

Schlager, E. 1994. Fishers' institutional responses to common pool resource dilemmas. Pp. 247-265 in E. Ostrom, R. Gardner, and J. Walker (eds.), *Rules, Games, and Common Pool Resources*. University of Michigan Press, Ann Arbor.

Schlager, E., and E. Ostrom. 1992. Property rights regimes and natural resources: A conceptual analysis. *Land Economics* 68(3):249-262.

Scott, A. 1955. The fishery: The objectives of sole ownership. *Journal of Political Economy* 63:116-124.

Scott, A. 1979. Development of the economic theory on fisheries regulation. *Journal of the Fisheries Research Board of Canada* 36:725-741.

Scott, A.D. 1993. *Reminiscences*. Paper presented to Conference on Innovations in Fisheries Management, Norwegian School of Economics and Business Administration, Bergen, May 24-25.

Sen, S., and J.R. Nielsen. 1996. Fisheries co-management: A comparative analysis. *Marine Policy* 20(5):405-418.

Shelley, P., J. Atkinson, E. Dorsey, and P. Brooks. 1994. The New England fisheries crisis: What have we learned? *Tulane Environmental Law Journal* 9(2):211-244.

Sissenwine, M.P., and J.E. Kirkley. 1982. Fishery management techniques: Practical aspects and limitations. *Marine Policy* 6(1):43-57.

Sissenwine, M.P., and P.M. Mace. 1992. ITQs in New Zealand: The era of fixed quota in perpetuity. *Fisheries Bulletin* 90:147-160.

Sissenwine, M.P., and A.A. Rosenberg. 1993. Marine fisheries at a critical juncture. *Fisheries* 18(10):6-14.

Skaptadóttir, U.D. 1996. Housework and wage work: Gender in Icelandic fishing communities. Pp. 87-105 in G. Pálsson and E.P. Durrenberger (eds.), *Images of Contemporary Iceland: Everyday Lives and Global Contexts*. University of Iowa Press, Iowa City.

Smit, W., W.P. Davidse, J. de Jager, C. de Rijter, M.H. Smit, C. Taal, and M.O. van Wijk. 1997. Visserij in Cijfers 1996 (Fisheries in Figures), Periodieke Rapportage 31-96. Netherlands Agricultural Economics Research Institute—Service for Agricultural Research, The Hague.

Smith, A. 1957. Selections from the Wealth of Nations. George J. Stigler (ed.). Harlan Davidson Inc., Arlington Heights, Illinois.

Smith, C.L. 1986. The life cycle of fisheries. *Fisheries* 11(4):20-25.

Solow, A.R. 1994. Detecting change in the composition of a multispecies community. *Biometrics* 50:556-565.

South Atlantic Fishery Management Council (SAFMC). 1991. *Amendment 5 (Wreckfish), Regulatory Impact Review, Initial Regulatory Flexibility Determination and Environmental Assessment for the Fishery Management Plan for the Snapper-Grouper Fishery of the South Atlantic Region*. SAFMC, Charleston, South Carolina.

South Atlantic Fishery Management Council/Gulf Fishery Management Council. 1992. *Regulatory Amendment to the Spiny Lobster Fishery Management Plan for the Gulf of Mexico and South Atlantic*. Gulf of Mexico Fishery Management Council, Tampa, Florida.

Squires, D., H. Campbell, S. Cunningham, C. Dewees, R.Q. Grafton, S.F. Herrick, Jr., J. Kirkley, S. Pascoe, K. Salvanes, B. Shallard, B. Turris, and N. Vestergaard. 1998. Individual transferable quotas in multispecies fisheries. *Marine Policy* 22(2):135-159.

Squires, D., J. Kirkley, and C.A. Tisdell. 1995. Individual transferable quotas as a fisheries management tool. *Reviews in Fisheries Science* 3(2):141-169.

Stephenson R. L., D.E. Lane, D.G. Aldous, and R. Nowak. 1993. Management of the 4WX Atlantic herring (*Clupea harengus*) fishery: An evaluation of recent events. *Canadian Journal of Fisheries and Aquatic Sciences* 50:2742-2757.

Strand, I.E., Jr., J.E. Kirkley, and K.E. McConnell. 1981. Economic analysis and the management of Atlantic surf clams. Pp. 113-288 in L.G. Anderson (ed.), *Economic Analysis for Fisheries Management Plans*. Ann Arbor Science, Ann Arbor, Michigan.

Sullivan, P.J., and S.D. Rebert. 1998. Interpreting Pacific halibut catch statistics in the British Columbia Individual Quota Program. *Canadian Journal of Fisheries and Aquatic Science* 55(1):99-115.

Tang, S.Y. 1992. *Institutions and Collective Action: Self-Governance in Irrigation.* Institute for Contemporary Studies, San Francisco.

Tietenberg, T.H. 1985. *Emissions Trading: An Exercise in Reforming Pollution Policy.* Resources for the Future, Washington, D.C.

Tietenberg, T.H. 1990. Economic instruments for environmental regulation. *Oxford Review of Economic Policy* 6(1):17-33.

Tietenberg, T.H. 1995. Design lessons from existing air pollution control systems: The United States. Pp. 15-32 in S. Hanna and M. Munasinghe (eds.), *Property Rights in a Social and Ecological Context: Case Studies and Design Applications.* The World Bank, Washington, D.C.

Tietenberg, T.H. 1998. Ethical influences on the evolution of the U.S. tradable permit approach to air pollution control. *Ecological Economics* 24(2,3):241-257.

Townsend, R.E. 1995. Fisheries self-governance: Corporate or cooperative structures? *Marine Policy* 19:39-45.

Townsend, R.E. 1997. Corporate management of fisheries. Pp. 195-202 in J. Boreman, B.S. Nakashima, J.A. Wilson, and R.L. Kendall (eds.). *Northwest Atlantic Groundfish: Perspectives on a Fishery Collapse.* American Fisheries Society, Bethesda, Maryland.

Townsend, R.E., and S.G. Pooley. 1995. Distributed governance in fisheries. In S. Hanna and M. Munasinghe (eds.), *Property Rights and the Environment: Social and Ecological Issues.* The World Bank, Washington, D.C.

Trumble, R.J., J.D. Neilson, W.R. Bowering, and D.A. McCaughran. 1993. Atlantic halibut (*Hippoglossus hippoglossus*) and Pacific halibut (*Hippoglossus stenolepis*) and their North American fisheries. *Canadian Bulletin of Fisheries and Aquatic Sciences* 227. National Research Council of Canada, Ottawa.

Turgeon, D.D. 1985. Fishery regulation: Its use under the Magnuson Act and Reaganomics. *Marine Policy* 9:126-133.

Wang, S.D. 1995. The surf clam ITQ management: An evaluation. *Marine Resource Economics* 10:93-98.

Warming, J. 1911. *Grundrente af Fiskegrunde.* Nationaloekonomisk Tidsskrift, Copenhagen.

Weisman, D. 1997. An Economic Analysis of the Mid-Atlantic Surf Clam and Ocean Quahog Fishery Using Logit, Hazard and Survival Rate Functions. Unpublished M.S. thesis, Department of Agricultural Economics and Marketing, Rutgers University, New Brunswick, New Jersey.

Wilen, J.E. 1985. Towards a theory of the regulated fishery. *Marine Resource Economics* 1:369-388.

Wilkinson, C. 1980. The public trust doctrine in public land law. 14 *U.C. Davis Law Review* 269:300-302.

Wilson, J. 1997. The Maine fisheries management initiative. Pp. 335-353 in G. Pálsson and G. Pétursdottir (eds.), *Social Implications of Quota Systems in Fisheries*. The Nordic Council of Ministers, Copenhagen.

Wilson, J.A., J.M. Acheson, M. Metcalfe, and P. Kleban. 1994. Chaos, complexity and community management of fisheries. *Marine Policy* 18(4):291-305.

Young, H.P. 1994. *Equity: In Theory and Practice*. Princeton University Press, Princeton, University. Pp. 164-167.

Young, M.D. 1995. The design of fishing-right systems: The New South Wales experience. *Ocean and Coastal Management* 28:54-61.

Young, M.D., and B.J. McCay. 1995. Building equity, stewardship, and resilience into market-based property-rights systems. Pp. 87-102 in S. Hanna and M. Munasinghe (eds.), *Property Rights and the Environment: Social and Ecological Issues*. The Beijer International Institute of Ecological Economics and The World Bank, Washington, D.C.

APPENDIXES

APPENDIX

A

Relevant Sections of the Magnuson-Stevens Fishery Conservation and Management Act

SEC. 108(f)
INDIVIDUAL FISHING QUOTA REPORT[1]

(1) Not later than October 1, 1998, the National Academy of Sciences, in consultation with the Secretary of Commerce and the Regional Fishery Management Councils, shall submit to the Congress a comprehensive final report on individual fishing quotas, which shall include recommendations to implement a national policy with respect to individual fishing quotas. The report shall address all aspects of such quotas, including an analysis of—

(A) the effects of limiting or prohibiting the transferability of such quotas;

(B) mechanisms to prevent foreign control of the harvest of United States fisheries under individual fishing quota programs, including mechanisms to prohibit persons who are not eligible to be deemed a citizen of the United States for the purpose of operating a vessel in the coastwise trade under section 2(a) and section 2(c) of the Shipping Act, 1916 (46 U.S.C. 802 (a) and (c)) from holding individual fishing quotas;

(C) the impact of limiting the duration of individual fishing quota programs;

(D) the impact of authorizing Federal permits to process a quantity of fish that correspond to individual fishing quotas, and of the value created for recipients of any such permits, including a comparison of such value to the value of the corresponding individual fishing quotas;

(E) mechanisms to provide for diversity and to minimize adverse social and

---

[1] Legislative mandate from the Sustainable Fisheries Act of 1996.

economic impacts on fishing communities, other fisheries affected by the displacement of vessels, and any impacts associated with the shifting of capital value from fishing vessels to individual fishing quotas, as well as the use of capital construction funds to purchase individual fishing quotas;

(F) mechanisms to provide for effective monitoring and enforcement, including the inspection of fish harvested and incentives to reduce bycatch, and in particular economic discards;

(G) threshold criteria for determining whether a fishery may be considered for individual fishing quota management, including criteria related to the geographical range, population dynamics and condition of a fish stock, the socioeconomic characteristics of a fishery (including participants' involvement in multiple fisheries in the region), and participation by commercial, charter, and recreational fishing sectors in the fishery;

(H) mechanisms to ensure that vessel owners, vessel masters, crew members, and United States fish processors are treated fairly and equitably in initial allocations, to require persons holding individual fishing quotas to be on board the vessel using such quotas, and to facilitate new entry under individual fishing quota programs;

(I) potential social and economic costs and benefits to the nation, individual fishing quota recipients, and any recipients of Federal permits described in subparagraph (D) under individual fishing quota programs, including from capital gains revenue, the allocation of such quotas or permits through Federal auctions, annual fees and transfer fees at various levels, or other measures;

(J) the value created for recipients of individual fishing quotas, including a comparison of such value to the value of the fish harvested under such quotas and to the value of permits created by other types of limited access systems, and the effects of creating such value on fishery management and conservation; and

(K) such other matters as the National Academy of Sciences deems appropriate.

(2) The report shall include a detailed analysis of individual fishing quota programs already implemented in the United States, including the impacts: of any limits on transferability, on past and present participants, on fishing communities, on the rate and total amount of bycatch (including economic and regulatory discards) in the fishery, on the safety of life and vessels in the fishery, on any excess harvesting or processing capacity in the fishery, on any gear conflicts in the fishery, on product quality from the fishery, on the effectiveness of enforcement in the fishery, on the size and composition of fishing vessel fleets, on the economic value created by individual fishing quotas for initial recipients and non-recipients, on conservation of the fishery resource, on fishermen who rely on participation in several fisheries, on the success in meeting any fishery management plan goals, and the fairness and effectiveness of the methods used for allocating quotas and control-

ling transferability. The report shall also include any information about individual fishing quota programs in other countries that may be useful.

(3) The report shall identify and analyze alternative conservation and management measures, including other limited access systems such as individual transferable effort systems, that could accomplish the same objectives as individual fishing quota programs, as well as characteristics that are unique to individual fishing quota programs.

(4) The Secretary of Commerce shall, in consultation with the National Academy of Sciences, the Councils, the fishing industry, affected States, conservation organizations and other interested persons, establish two individual fishing quota review groups to assist in the preparation of the report, which shall represent:
(A) Alaska, Hawaii, and the other Pacific coastal States; and
(B) Atlantic coastal States and the Gulf of Mexico coastal States.
The Secretary shall, to the extent practicable, achieve a balanced representation of viewpoints among the individuals on each review group. The review groups shall be deemed to be advisory panels under section 302(g) of the Magnuson Fishery Conservation and Management Act, as amended by this Act.

(5) The Secretary of Commerce, in consultation with the National Academy of Sciences and the Councils, shall conduct public hearings in each Council region to obtain comments on individual fishing quotas for use by the National Academy of Sciences in preparing the report required by this subsection. The National Academy of Sciences shall submit a draft report to the Secretary of Commerce by January 1, 1998. The Secretary of Commerce shall publish in the Federal Register a notice and opportunity for public comment on the draft of the report, or any revision thereof. A detailed summary of comments received and views presented at the hearings, including any dissenting views, shall be included by the National Academy of Sciences in the final report.

OTHER SECTIONS OF THE MAGNUSON-STEVENS ACT RELEVANT TO IFQS

SEC. 3 (16 U.S.C. 1802)

DEFINITIONS.—

(21) The term "individual fishing quota" means a Federal permit under a limited access system to harvest a quantity of fish, expressed by a unit or units representing a percentage of the total allowable catch of a fishery that may be

received or held for exclusive use by a person. Such term does not include community development quotas as described in section 305(i).

SEC. 108(g) (Magnuson-Stevens Act [uncodified])

NORTH PACIFIC LOAN PROGRAM.—

(1) By not later than October 1, 1997 the North Pacific Fishery Management Council shall recommend to the Secretary of Commerce a program which uses the full amount of fees authorized to be used under section 303(d)(4) of the Magnuson Fishery Conservation and Management Act, as amended by this Act, in the halibut and sablefish fisheries off Alaska to guarantee obligations in accordance with such section.

(2)(A) For the purposes of this subsection, the phrase 'fishermen who fish from small vessels' in section 303(d)(4)(A)(i) of such Act shall mean fishermen wishing to purchase individual fishing quotas for use from Category B, Category C, or Category D vessels, as defined in part 676.20(c) of title 50, Code of Federal Regulations (as revised as of October 1, 1995), whose aggregate ownership of individual fishing quotas will not exceed the equivalent of a total of 50,000 pounds of halibut and sablefish harvested in the fishing year in which a guarantee application is made if the guarantee is approved, who will participate aboard the fishing vessel in the harvest of fish caught under such quotas, who have at least 150 days of experience working as part of the harvesting crew in any United States commercial fishery, and who do not own in whole or in part any Category A or Category B vessel, as defined in such part and title of the Code of Federal Regulations.

(B) For the purposes of this subsection, the phrase "entry level fishermen" in section 303(d)(4)(A)(ii) of such Act shall mean fishermen who do not own any individual fishing quotas, who wish to obtain the equivalent of not more than a total of 8,000 pounds of halibut and sablefish harvested in the fishing year in which a guarantee application is made, and who will participate aboard the fishing vessel in the harvest of fish caught under such quotas.

SEC. 108(i) (Magnuson-Stevens Act [uncodified])

EXISTING QUOTA PLANS.—Nothing in this Act (Magnuson-Stevens Act) or the amendments made by this Act shall be construed to require a reallocation of individual fishing quotas under any individual fishing quota program approved by the Secretary before January 4, 1995.

SEC. 303 (16 U.S.C. 1853)

(d) INDIVIDUAL FISHING QUOTAS.—

(1) (A) A Council may not submit and the Secretary may not approve or implement before October 1, 2000, any fishery management plan, plan amendment, or regulation under this Act which creates a new individual fishing quota program.

(B) Any fishery management plan, plan amendment, or regulation approved by the Secretary on or after January 4, 1995, which creates any new individual fishing quota program shall be repealed and immediately returned by the Secretary to the appropriate Council and shall not be resubmitted, reapproved, or implemented during the moratorium set forth in subparagraph (A).

(2) (A) No provision of law shall be construed to limit the authority of a Council to submit and the Secretary to approve the termination or limitation, without compensation to holders of any limited access system permits, of a fishery management plan, plan amendment, or regulation that provides for a limited access system, including an individual fishing quota program.

(B) This subsection shall not be construed to prohibit a Council from submitting, or the Secretary from approving and implementing, amendments to the North Pacific halibut and sablefish, South Atlantic wreckfish, or Mid-Atlantic surf clam and ocean (including mahogany) quahog individual fishing quota programs.

(3) An individual fishing quota or other limited access system authorization—

(A) shall be considered a permit for the purposes of sections 307, 308, and 309;

(B) may be revoked or limited at any time in accordance with this Act;

(C) shall not confer any right of compensation to the holder of such individual fishing quota or other such limited access system authorization if it is revoked or limited; and

(D) shall not create, or be construed to create, any right, title, or interest in or to any fish before the fish is harvested.

(4) (A) A Council may submit, and the Secretary may approve and implement, a program which reserves up to 25 percent of any fees collected from a fishery under section 304(d)(2) to be used, pursuant to section 1104A(a)(7) of the Merchant Marine Act, 1936 (46 U.S.C. App. 1274(a)(7)), to issue obligations that aid in financing the—

(i) purchase of individual fishing quotas in that fishery by fishermen who fish from small vessels; and

(ii) first-time purchase of individual fishing quotas in that fishery by entry level fishermen.

(B) A Council making a submission under subparagraph (A) shall recommend criteria, consistent with the provisions of this Act, that a fisherman must meet to qualify for guarantees under clauses (i) and (ii) of subparagraph (A) and the portion of funds to be allocated for guarantees under each clause.

(5) In submitting and approving any new individual fishing quota program on or after October 1, 2000, the Councils and the Secretary shall consider the report of the National Academy of Sciences required under section 108(f) of the Sustainable Fisheries Act, and any recommendations contained in such report, and shall ensure that any such program—

(A) establishes procedures and requirements for the review and revision of the terms of any such program (including any revisions that may be necessary once a national policy with respect to individual fishing quota programs is implemented), and, if appropriate, for the renewal, reallocation, or reissuance of individual fishing quotas;

(B) provides for the effective enforcement and management of any such program, including adequate observer coverage, and for fees under section 304(d)(2) to recover actual costs directly related to such enforcement and management; and

(C) provides for a fair and equitable initial allocation of individual fishing quotas, prevents any person from acquiring an excessive share of the individual fishing quotas issued, and considers the allocation of a portion of the annual harvest in the fishery for entry-level fishermen, small vessel owners, and crew members who do not hold or qualify for individual fishing quotas.

SEC. 304 (16 U.S.C. 1854)

(c) PREPARATION AND REVIEW OF SECRETARIAL PLANS.—

(1) The Secretary may prepare a fishery management plan, with respect to any fishery, or any amendment to any such plan, in accordance with the national standards, the other provisions of this Act, and any other applicable law, if—

(A) the appropriate Council fails to develop and submit to the Secretary, after a reasonable period of time, a fishery management plan for such fishery, or any necessary amendment to such a plan, if such fishery requires conservation and management;

(B) the Secretary disapproves or partially disapproves any such plan or amendment, or disapproves a revised plan or amendment, and the Council involved fails to submit a revised or further revised plan or amendment; or

(C) the Secretary is given authority to prepare such plan or amendment under this section. In preparing any such plan or amendment, the Secretary shall consult with the Secretary of State with respect to foreign fishing and with the Secretary of the department in which the Coast Guard is operating with respect to enforcement at sea. The Secretary shall also prepare such proposed regulations as he deems necessary or appropriate to carry out each plan or amendment prepared by him under this paragraph.

(2) In preparing any plan or amendment under this subsection, the Secretary shall—

(A) conduct public hearings, at appropriate times and locations in the geographical areas concerned, so as to allow interested persons an opportunity to be heard in the preparation and amendment of the plan and any regulations implementing the plan; and

(B) consult with the Secretary of State with respect to foreign fishing and with the Secretary of the department in which the Coast Guard is operating with respect to enforcement at sea.

(3) Notwithstanding paragraph (1) for a fishery under the authority of a Council, the Secretary may not include in any fishery management plan, or any amendment to any such plan, prepared by him, a provision establishing a limited access system, including any individual fishing quota program unless such system is first approved by a majority of the voting members, present and voting, of each appropriate Council.

(d) ESTABLISHMENT OF FEES.—

(1) The Secretary shall by regulation establish the level of any fees which are authorized to be charged pursuant to section 303(b)(1). The Secretary may enter into a cooperative agreement with the States concerned under which the States administer the permit system and the agreement may provide that all or part of the fees collected under the system shall accrue to the States. The level of fees charged under this subsection shall not exceed the administrative costs incurred in issuing the permits.

(2)(A) Notwithstanding paragraph (1), the Secretary is authorized and shall collect a fee to recover the actual costs directly related to the management and enforcement of any—

(i) individual fishing quota program; and

(ii) community development quota program that allocates a percentage of the total allowable catch of a fishery to such program.

(B) Such fee shall not exceed 3 percent of the exvessel value of fish harvested under any such program, and shall be collected at either the time of the landing, filing of a landing report, or sale of such fish during a fishing season or in the last quarter of the calendar year in which the fish is harvested.

(C) (i) Fees collected under this paragraph shall be in addition to any other fees charged under this Act and shall be deposited in the Limited Access System Administration Fund established under section 305(h)(5)(B), except that the portion of any such fees reserved under section 303(d)(4)(A) shall be deposited in the Treasury and available, subject to annual appropriations, to cover the costs of new direct loan obligations and new loan guarantee commitments as required by section 504(b)(1) of the Federal Credit Reform Act (2 U.S.C. 661c(b)(1)).

(ii) Upon application by a State, the Secretary shall transfer to such State up to 33 percent of any fee collected pursuant to subparagraph (A) under a community development quota program and deposited in the Limited Access System Administration Fund in order to reimburse such State for actual costs directly incurred in the management and enforcement of such program.

SEC. 305 (16 U.S.C. 1855)

(h) CENTRAL REGISTRY SYSTEM FOR LIMITED ACCESS SYSTEM PERMITS.—

(1) Within 6 months after the date of enactment of the Sustainable Fisheries Act, the Secretary shall establish an exclusive central registry system (which may be administered on a regional basis) for limited access system permits established under section 303(b)(6) or other Federal law, including individual fishing quotas, which shall provide for the registration of title to, and interests in, such permits, as well as for procedures for changes in the registration of title to such permits upon the occurrence of involuntary transfers, judicial or nonjudicial foreclosure of interests, enforcement of judgments thereon, and related matters deemed appropriate by the Secretary. Such registry system shall—

(A) provide a mechanism for filing notice of a nonjudicial foreclosure or enforcement of a judgment by which the holder of a senior security interest acquires or conveys ownership of a permit, and in the event of a nonjudicial foreclosure, by which the interests of the holders of junior security interests are released when the permit is transferred;

(B) provide for public access to the information filed under such system, notwithstanding section 402(b); and

(C) provide such notice and other requirements of applicable law that the Secretary deems necessary for an effective registry system.

(2) The Secretary shall promulgate such regulations as may be necessary to carry out this subsection, after consulting with the Councils and providing an opportunity for public comment. The Secretary is authorized to contract with non-Federal entities to administer the central registry system.

(3) To be effective and perfected against any person except the transferor, its heirs and devisees, and persons having actual notice thereof, all security inter-

ests, and all sales and other transfers of permits described in paragraph (1), shall be registered in compliance with the regulations promulgated under paragraph (2). Such registration shall constitute the exclusive means of perfection of title to, and security interests in, such permits, except for Federal tax liens thereon, which shall be perfected exclusively in accordance with the Internal Revenue Code of 1986 (26 U.S.C. 1 et seq.). The Secretary shall notify both the buyer and seller of a permit if a lien has been filed by the Secretary of the Treasury against the permit before collecting any transfer fee under paragraph (5) of this subsection.

(4) The priority of security interests shall be determined in order of filing, the first filed having the highest priority. A validly-filed security interest shall remain valid and perfected notwithstanding a change in residence or place of business of the owner of record. For the purposes of this subsection, "security interest" shall include security interests, assignments, liens and other encumbrances of whatever kind.

(5) (A) Notwithstanding section 304(d)(1), the Secretary shall collect a reasonable fee of not more than one-half of one percent of the value of a limited access system permit upon registration of the title to such permit with the central registry system and upon the transfer of such registered title. Any such fee collected shall be deposited in the Limited Access System Administration Fund established under subparagraph (B).

(B) There is established in the Treasury a Limited Access System Administration Fund. The Fund shall be available, without appropriation or fiscal year limitation, only to the Secretary for the purposes of—

(i) administering the central registry system; and

(ii) administering and implementing this Act in the fishery in which the fees were collected. Sums in the Fund that are not currently needed for these purposes shall be kept on deposit or invested in obligations of, or guaranteed by, the United States.

SEC. 402. (16 U.S.C. 1881a)

(b) CONFIDENTIALITY OF INFORMATION.—

(1) Any information submitted to the Secretary by any person in compliance with any requirement under this Act shall be confidential and shall not be disclosed, except—

(A) to Federal employees and Council employees who are responsible for fishery management plan development and monitoring;

(B) to State or Marine Fisheries Commission employees pursuant to an agreement with the Secretary that prevents public disclosure of the identity or business of any person;

(C) when required by court order;

(D) when such information is used to verify catch under an individual fishing quota program;

(E) that observer information collected in fisheries under the authority of the North Pacific Council may be released to the public as specified in a fishery management plan or regulation for weekly summary bycatch information identified by vessel, and for haul-specific bycatch information without vessel identification; or

(F) when the Secretary has obtained written authorization from the person submitting such information to release such information to persons for reasons not otherwise provided for in this subsection, and such release does not violate other requirements of this Act.

SEC. 407. (16 U.S.C. 1883)

GULF OF MEXICO RED SNAPPER RESEARCH

(b) PROHIBITION.— In addition to the restrictions under section 303(d)(1)(A), the Gulf Council may not, prior to October 1, 2000, undertake or continue the preparation of any fishery management plan, plan amendment or regulation under this Act for the Gulf of Mexico commercial red snapper fishery that creates an individual fishing quota program or that authorizes the consolidation of licenses, permits, or endorsements that result in different trip limits for vessels in the same class.

(c) REFERENDUM.—

(1) On or after October 1, 2000, the Gulf Council may prepare and submit a fishery management plan, plan amendment, or regulation for the Gulf of Mexico commercial red snapper fishery that creates an individual fishing quota program or that authorizes the consolidation of licenses, permits, or endorsements that result in different trip limits for vessels in the same class, only if the preparation of such plan, amendment, or regulation is approved in a referendum conducted under paragraph (2) and only if the submission to the Secretary of such plan, amendment, or regulation is approved in a subsequent referendum conducted under paragraph (2).

(2) The Secretary, at the request of the Gulf Council, shall conduct referendums under this subsection. Only a person who held an annual vessel permit with a red snapper endorsement for such permit on September 1, 1996 (or any person to whom such permit with such endorsement was transferred after such date) and vessel captains who harvested red snapper in a commercial fishery using such endorsement in each red snapper fishing season occurring between

January 1, 1993, and such date may vote in a referendum under this subsection. The referendum shall be decided by a majority of the votes cast. The Secretary shall develop a formula to weigh votes based on the proportional harvest under each such permit and endorsement and by each such captain in the fishery between January 1, 1993, and September 1, 1996. Prior to each referendum, the Secretary, in consultation with the Council, shall—

(A) identify and notify all such persons holding permits with red snapper endorsements and all such vessel captains; and

(B) make available to all such persons and vessel captains information about the schedule, procedures, and eligibility requirements for the referendum and the proposed individual fishing quota program.

Appendix

B

Committee Biographies

Jan S. Stevens chaired the Committee to Review Individual Fishing Quotas. He earned an LL.B. from the University of California, Berkeley, and recently retired as an assistant attorney general for the State of California. Mr. Stevens managed the Lands Law Section of the attorney general's office, which advises the California Coastal Commission; the State Lands Commission; and the Lake Tahoe, Coastal, Santa Monica Mountains, and Coachella Valley Conservancies. He has taught environmental law at the University of California and published articles in the areas of lands, natural resources, and the public trust doctrine.

John H. Annala earned a Ph.D. in marine biology from the University of New Hampshire in 1974. Dr. Annala currently serves as the manager of science policy for the Ministry of Fisheries in New Zealand. His research interests include rock lobsters, inshore and deepwater finfish, stock assessment, fisheries management, and management of research.

James H. Cowan, Jr., earned a Ph.D. in marine sciences and experimental statistics from Louisiana State University in 1985. Dr. Cowan currently serves as an associate professor for the University of South Alabama's Department of Marine Sciences. His research interests include fisheries ecology, biological and fisheries oceanography, predation, and feeding ecology and recruitment variability of early life stages of fishes.

Keith R. Criddle earned a Ph.D. in agricultural economics from the University of California, Davis, in 1989. Dr. Criddle currently serves as the Economics

Department head at Utah State University. His areas of research have included the economic impacts of potential policy changes affecting the total allowable catch for walleye pollock and predicting the consequences of alternative harvest regulations in a sequential fishery.

Ward H. Goodenough earned a Ph.D. in anthropology from Yale University in 1949. Dr. Goodenough is presently a professor emeritus at the University of Pennsylvania. His research interests include cultures and languages of the Pacific, social organization and land tenure, religion, ethnographic methods, formal analysis of ethnographic data, and culture theory.

Susan S. Hanna earned a Ph.D. in agricultural and resource economics at Oregon State University in 1981. Dr. Hanna currently serves as a professor in the Department of Agricultural and Resource Economics at Oregon State University and is a former member of the Ocean Studies Board. Her research interests include marine economics, resource allocation and property rights, fisheries management, institutional economics, resource use under uncertainty, and economic history of natural resources.

Rögnvaldur Hannesson earned a Ph.D. in economics from the University of Lund, Sweden, in 1974. Dr. Hannesson has served as a professor of fisheries economics at the Norwegian School of Economics and Business Administration since 1983. His research interests include fisheries management, the economics of fish resources, and extended fishing limits.

Bonnie J. McCay earned a Ph.D. in anthropology from Columbia University in 1976. Dr. McCay is a professor in the Department of Human Ecology at Cook College of Rutgers University. Her research interests include common property issues, participatory democracy in fisheries management, and the sustainability of resource-dependent coastal communities.

Michael K. Orbach earned a Ph.D. in cultural anthropology from the University of California, San Diego, in 1975. Dr. Orbach is presently a professor in marine affairs and policy and director of the Duke University Marine Laboratory. His research interests include fisheries management, modernization and marine fisheries policy, and environmental planning.

Gísli Pálsson earned a Ph.D. in social anthropology from the University of Manchester in 1982. Dr. Pálsson currently serves as the director of the Institute of Anthropology and is also a professor in the Department of Anthropology for the Faculty of Social Science at the University of Iceland. His current research is focused on evaluating the social implications and development of the quota system in the Icelandic cod fishery and comparing the ecological knowledge of fishermen and professional marine biologists.

Alison Rieser earned a J.D. from the George Washington University in 1976 and an LL.M. from Yale Law School in 1990. Since 1993, Professor Rieser has served as a professor of law and director of the Marine Law Institute for the University of Maine School of Law. Her research interests include natural resources law, fisheries law, protection of marine biodiversity, and law of the sea.

David B. Sampson earned a Ph.D. in environmental technology from the Imperial College of Science and Technology at the University of London in 1989. Dr. Sampson currently serves as an associate professor of fisheries with the Coastal Oregon Marine Experiment Station and the Department of Fisheries and Wildlife at Oregon State University. His research focuses on the dynamics of fishery systems, the response of fishermen to changing conditions within the fisheries, and fish stock assessment.

Edella C. Schlager earned a Ph.D. in political science from Indiana University in 1990. Dr. Schlager is currently an associate professor for the School of Public Administration and Policy at the University of Arizona. She studies the emergence and evolution of institutional arrangements devised by communities to govern natural resources on which they are economically dependent. Her research focuses on coastal fisheries and water.

Richard E. Stroble earned a B.A. in finance from the University of Washington in 1970. Mr. Stroble is currently the chief executive officer of Merrill and Ring Inc., a family-owned corporation that has held forest lands in Washington State and British Columbia for more than 100 years. The company is active in professional forestry issues and public policy, but has no ties to fisheries.

Thomas H. Tietenberg earned his Ph.D. in economics from the University of Wisconsin in 1971. A former president of the Association of Environmental and Resource Economists, Dr. Tietenberg currently holds the Mitchell Family Chair in Economics at Colby College. His research focuses on economics and environmental policy, economics of global warming, and pollution emissions trading.

Appendix
C

Relevant Section From the Shipping Act of 1916[1]

Sec. 802. Corporation, partnership, or association as citizen

(a) Ownership of controlling interest

Within the meaning of this chapter no corporation, partnership, or association shall be deemed a citizen of the United States unless the controlling interest therein is owned by citizens of the United States, and, in the case of a corporation, unless its president or other chief executive officer and the chairman of its board of directors are citizens of the United States and unless no more of its directors than a minority of the number necessary to constitute a quorum are noncitizens and the corporation itself is organized under the laws of the United States or of a state, Territory, District, or possession thereof, but in the case of a corporation, association, or partnership operating any vessel in the coastwise trade the amount of interest required to be owned by citizens of the United States shall be 75 per centum.

(c) Determination of seventy-five per centum of interest

Seventy-five per centum of the interest in a corporation shall not be deemed to be owned by citizens of the United States (a) if the title to 75 per centum of its stock is not vested in such citizens free from any trust or fiduciary obligation in favor of any person not a citizen of the United States; or (b) if 75 per centum of the voting power in such corporation is not vested in citizens of the United States; or (c) if, through any contract or understanding, it is so arranged that more than 25 per centum of the voting power in such corporation may be exercised, directly or indirectly, in behalf of any person who is not a citizen of the United States; or (d) if by any other means whatsoever control of any interest in the corporation in excess of 25 per centum is conferred upon or permitted to be exercised by any person who is not a citizen of the United States.

---

[1] 46 U.S.C. Appendix—Shipping Sec. 802.

Appendix

D

National Standards in the Magnuson-Stevens Fishery Conservation and Management Act[1]

(a) IN GENERAL—Any fishery management plan prepared, and any regulation promulgated to implement any such plan, pursuant to this title shall be consistent with the following national standards for fishery conservation and management:

> (1) Conservation and management measures shall prevent overfishing while achieving, on a continuing basis, the optimum yield from each fishery for the United States fishing industry.

> (2) Conservation and management measures shall be based on the best scientific information available.

> (3) To the extent practicable, an individual stock of fish shall be managed as a unit throughout its range, and interrelated stocks of fish shall be managed as a unit or in close coordination.

> (4) Conservation and management measures shall not discriminate between residents of different States. If it becomes necessary to allocate or assign fishing privileges among various United States fishermen, such allocation shall be (A) fair and equitable to all such fishermen; (B) reasonably calculated to promote conservation; and (C) carried out in such

---

[1] 16 U.S.C. 1851, Sec. 201.

manner that no particular individual, corporation, or other entity acquires an excessive share of such privileges.

(5) Conservation and management measures shall, where practicable, consider efficiency in the utilization of fishery resources; except that no such measure shall have economic allocation as its sole purpose.

(6) Conservation and management measures shall take into account and allow for variations among, and contingencies in, fisheries, fishery resources, and catches.

(7) Conservation and management measures shall, where practicable, minimize cost and avoid unnecessary duplication.

(8) Conservation and management measures shall, consistent with the conservation requirements of this Act (including the prevention of overfishing and rebuilding of overfished stocks), take into account the importance of fishery resources to fishing communities in order to (A) provide for the sustained participation of such communities, and (B) to the extent practicable, minimize adverse economic impacts in such communities. [Added in 1996]

(9) Conservation and management measures shall, to the extent practicable, (A) minimize bycatch and (B) to the extent bycatch cannot be avoided, minimize the mortality of such bycatch. [Added in 1996]

(10) Conservation and management measures shall, to the extent practicable, promote the safety of human life at sea. [Added in 1996]

APPENDIX
E

History of Changes to the Magnuson-Stevens Fishery Conservation and Management Act[1]

Public Law No.	Name and Major Provisions
95-6 (1977)	"Fishery Conservation Zone Transition Act." Technical amendments.
95-354 (1978)	"Processor preference" amendment. Foreign vessels could receive U.S.-harvested fish only if surplus to needs of U.S. processors.
96-61 (1979)	"Packwood-Magnuson Amendment." Provided decreased allocations for nations diminishing the effectiveness of the International Whaling Convention.
96-118 (1979)	Technical amendments.
96-470 (1980)	Technical amendments.
96-561 (1980)	"American Fisheries Promotion Act." Changed name of statute to "Magnuson Fishery Conservation and Management Act."

---

[1] Source: NOAA's Office of General Counsel for Fisheries, 1998.

Added provision for 100% observer coverage in foreign fisheries.

Significantly raised foreign fishing fees.

Established pro-development criteria for use in allocating surplus fishery resources ("fish and chips").

97-191 (1982)   Allowed foreign processing in internal waters if approved by State Governor.

97-453 (1983)   Added a provision on foreign recreational fishing.

Established a supplemental observer program.

Changed composition of Western Pacific Council.

Modified process for Council appointments.

Removed Councils from Federal Advisory Committee Act requirements.

Required submission of proposed regulations along with a fishery management plan (FMP) or amendment.

Added data collection program if requested by Council.

Expedited Secretarial review by placing a deadline on the decision to approve or disapprove an FMP amendment.

Allowed States to share permit fees if they administer the permitting system.

Required Secretary to promulgate regulations implementing an FMP or amendment within 10 days of receipt.

Changed circumstances under which the Secretary may issue emergency regulations; lengthened effective time periods; required Secretary to issue emergency regulations if unanimous vote of Council.

Directed the Secretary to comply with the Paperwork Reduction Act, the Regulatory Flexibility Act, and E.O. 12291 within the time limits set for FMP implementation.

Extended State jurisdiction to certain "pockets of water" within the exclusive economic zone (EEZ).

Gave Secretary subpoena power in civil penalty hearings.

Removed imprisonment as a sanction for foreign fishing violations.

Gave general arrest authority to enforcement officers.

98-623 (1984)   Added more development criteria for foreign allocation decisions.

Specified that national standard guidelines are advisory.

Extended State jurisdiction to certain "pockets of water" within the EEZ.

99-386 (1986)   Technical amendments.

99-659 (1986)   Asserted sovereign rights over fish in the newly acclaimed
                        exclusive economic zone.
                      Required foreign fishing vessels to meet health and safety
                        standards.
                      Clarified the annual nature of a foreign fishing permit and
                        specified procedures for temporary suspension of permit.
                      Added a surcharge to foreign fishing fees for nations har-
                        vesting unacceptable levels of U.S.-origin anadromous
                        species.
                      Directed Secretary to make appointments so fishery partici-
                        pants are fairly represented.
                      Required Federal agencies to respond to Council comments
                        on activities affecting fishery habitat; made habitat infor-
                        mation a mandatory part of FMPs.
                      Established financial disclosure requirement for Council
                        members and executive directors.
                      Directed Councils to consider temporary adjustments regard-
                        ing access to fisheries for those affected by weather.
                      Revised procedure for preliminary review of FMPs and
                        amendments and shortened schedule for publication of
                        proposed rules.
                      Modified fishery research provision to ensure more partici-
                        pation by Councils.
                      Made technical amendments to civil penalty and forfeiture
                        provisions.
                      Prohibited, with criminal sanctions, submission of false in-
                        formation concerning processing capacity or other mat-
                        ters.
                      Allowed sums from penalties and forfeitures to be used to
                        pay storage costs and rewards.

100-66 (1988)   Technical amendment.

100-239 (1988)   Redefined "vessel of the United States."

100-629 (1988)   Required transponders on foreign fishing vessels.
                      Prohibited foreign fishing vessels not authorized to "engage
                        in fishing" from operating in the EEZ without their gear
                        stowed below deck or otherwise made unusable for
                        fishing.

101-627 (1990) "Fishery Conservation Amendments of 1990."

Provided management authority over tuna as of January 1, 1992.

Restored original provision for setting total allowable level of foreign fishing.

Limited Council members to no more than three consecutive terms and decreased their compensation from GS-18 to GS-16.

Allowed Councils to meet outside their areas of authority but within their constituent states; required Regional Director to submit written minority statement.

Required each Council to establish a fishing industry advisory committee and to comment on federal and state activities that are likely to affect habitat of anadromous fish.

Added to required contents of FMPs, including a "fishery impact statement," and added to discretionary contents of FMPs permits and fees for vessel operators and fish processors.

Clarified that state employees may receive confidential data pursuant to an agreement.

Directed the Secretary to promulgate regulations restricting use of information collected by voluntary observers.

Required publication of a strategic plan for fisheries research every three years.

Gave the Secretary management authority over highly migratory species (tuna and other pelagics) in the Atlantic, Gulf of Mexico, and Caribbean.

Directed the Secretary to conduct research on incidental harvest in shrimp trawl fishery and restricted until January 1, 1994, measures to reduce mortality of nontarget resources in that fishery.

Facilitated judicial review of "framework actions" and expedited all review proceedings.

Added to prohibited acts the theft of fish and fishing gear, forcible assault on an observer, large-scale driftnet fishing under U.S. jurisdiction, roe stripping, and violation of an international fishery agreement or regulations implementing it.

Increased maximum civil penalty from $25,000.00 to $100,000.00; specified the Secretary may suspend or revoke domestic permits and made other changes to permit provisions.

Added assault on an observer as a criminal offense; increased maximum fines from $50,000.00 to $100,000.00 and (for aggravated offenses) from $100,000.00 to $200,000.00.

Established new uses for the fund from penalties and forfeitures.

Provided for a North Pacific Fisheries Research Plan, financing observers through fees paid by all fishery participants.

102-251 (1992)   Implemented U.S.-U.S.S.R. maritime boundary agreement by adding three "special areas" to management authority. The amendments will not take effect until the agreement enters into force for the United States.

Authorized State Department to negotiate a three-year fishery agreement with Russia for access to "special areas."

102-567 (1992)   Restored authority to conduct fisheries research, deleted by P.L. 101-627.

Provided for agreements between the Secretary and New England States for enforcement of the Multispecies FMP and for Coast Guard-industry working group; ear-marked penalties to enforce that FMP.

Established Northwest Atlantic Ocean Fisheries Reinvestment Program.

102-582 (1992)   Lifted three-term limit on Council appointments for 1993 cycle.

Increased limit on fees charged under North Pacific Fisheries Research Plan from 1% to 2%.

APPENDIX
F

Acronyms and Glossary

ACRONYMS

ABC	allowable biological catch
ACCSP	Atlantic Coastal Cooperative Statistics Program
ACE	annual catch entitlements (New Zealand)
ADF&G	Alaska Department of Fish and Game
BSAI	Bering Sea-Aleutian Islands
CDQ	community development quota
CFEC	Commercial Fisheries Entry Commission (Alaska)
CFMC	Caribbean Fishery Management Council
CFQ	community fishing quota
CFR	Code of Federal Regulations
CPUE	catch per unit effort
DFO	Department of Fisheries and Oceans (Canada)
DMP	dockside monitoring program
EEZ	exclusive economic zone
ENSO	El Niño-Southern Oscillation
F	fishing mortality
FAO	Food and Agriculture Organization, United Nations

FCMA	Fishery Conservation and Management Act
FLPMA	Federal Land Policy Management Act
FMFC	Florida Marine Fisheries Commission
FMP	fishery management plan
GDA	Groundfish Development Authority (Canada)
GDP	gross domestic product
GFMC	Gulf Fishery Management Council
GOA	Gulf of Alaska
GRT	gross register ton
IBQ	individual bycatch quota
ICES	International Council for the Exploration of the Sea
ICNAF	International Commission for Northwest Atlantic Fisheries
IFQ	individual fishing quota
IPHC	International Pacific Halibut Commission
ITQ	individual transferable quota
IVBQ	individual vessel bycatch quota
IVQ	individual vessel quota
LATE	Local Authority Trading Enterprise (New Zealand)
LPUE	landings per unit effort
MAFMC	Mid-Atlantic Fishery Management Council
MEY	maximum economic yield
MRFSS	Marine Recreational Fisheries Statistics Survey
MSFCMA	Magnuson-Stevens Fishery Conservation and Management Act
MSY	maximum sustainable yield
NAFO	Northwest Atlantic Fisheries Organization
NAS	National Academy of Sciences
NEFMC	New England Fishery Management Council
NEFSC	Northeast Fishery Science Center (NMFS)
NMFS	National Marine Fisheries Service
NOAA	National Oceanic and Atmospheric Administration
NPFMC	North Pacific Fishery Management Council
NRC	National Research Council
OAA	Office of Administrative Appeals (NMFS)
OAE	open access equilibrium
OECD	Organization for Economic Cooperation and Development
OFL	overfishing limit
OSB	Ocean Studies Board

OY   optimum yield

PFMC   Pacific Fishery Management Council
PSP   paralytic shellfish poisoning

QMA   quota management area (New Zealand)
QMS   quota management system (New Zealand)
QS   quota share

R recruitment
RAM   Restricted Access Management Division

SAFMC   South Atlantic Fishery Management Council
SAR   search and rescue
SCOQ   surf clam/ocean quahog
SFA   Sustainable Fisheries Act of 1996
SSC   scientific and statistical committee (of a regional fishery
                        management council)
SURFs   stock-use rights in fisheries

TAC   total allowable catch
TURFs   territorial use rights in fisheries

USCG   U.S. Coast Guard

WPFMC   Western Pacific Fishery Management Council

GLOSSARY

allowable biological catch (ABC): Maximum amount of fish stock that could be
    harvested without adversely affecting recruitment or other biological com-
    ponents of the stock. The ABC level is typically higher than the total allow-
    able catch, leaving a buffer between the two.

Australian "drop-through" system: Approach developed in the New South Wales
    fishery that establishes a cascade of fixed-term entitlements for quota share-
    holders to allow the introduction of new management measures. Under this
    scheme, initial entitlements of quota share are defined for a finite period, but
    one long enough to encourage investments. Periodically, a comprehensive
    review is undertaken to develop a new set of entitlements. These entitle-
    ments would confer a similar, but not necessarily identical, set of rights and

obligations on the quota holders. This process would continue until such time as it appeared that no more modifications were necessary.

biomass: Amount or mass of some organism, such as fish.

blocked quota: Quota shares in the Alaskan halibut and sablefish IFQ program that are not allowed to be subdivided when transferred. There are limits on the size of the blocked quota and on the number of blocks that an individual may own in a given area. This is intended to ensure the availability of small units of quota for purchase by new entrants.

bycatch: Fish other than the primary target species that are caught incidental to the harvest of the primary target species. Bycatch may be retained or discarded. Discards may occur for regulatory or economic reasons.

capital stuffing: Investing in gear, technology, engines, processing lines, and other capital components of a fishing operation in order to maximize the ability of a vessel or processing facility to harvest or process fish. These investments are made so that the vessel or processing facility can harvest and process fish as rapidly as possible under a derby fishery or in a race for fish.

Caribbean Fishery Management Council (CFMC): One of eight regional councils mandated in the MSFCMA to develop management plans for fisheries in federal waters. The CFMC develops fishery management plans for fisheries off the coast of the U.S. Virgin Islands and the Commonwealth of Puerto Rico.

catcher vessel: Vessel that harvests fish but does not have onboard processing capacity.

catcher-processor vessel: Vessel that can both catch fish and process the catch on-board. Also referred to as factory trawlers or freezer-longliners.

catch per unit effort (CPUE): Weight of fish harvested for each unit of effort expended by vessels in the fishery. CPUE can be expressed as weight of fish captured per fishing trip, per hour spent at sea, or through other standardized measures.

charterboat: Boat designed for carrying for hire a group of passengers who are engaged in recreational fishing.

coefficient of variation: Standard deviation divided by the mean, showing standard deviation as a percentage of the mean. A high coefficient of variation is indicative of wide variation in the data being analyzed.

cohort: Fish born in a given year. (See year class.)

Commercial Fisheries Entry Commission (CFEC): Agency responsible for tracking and approving the transfer of permits in Alaska's limited entry fisheries. Although primarily responsible for Alaska's salmon and herring limited entry programs, CFEC has participated in evaluating the effects of the Alaskan halibut and sablefish IFQ programs.

common-pool resources: Resources such as groundwater, open-access fisheries, or public grazing lands that are held for public use. Common-pool resources have features that make it difficult to exclude others from their use, and one person's use can affect what is available to another person.

common property: Form of resource ownership with a set of well-defined users capable of excluding other potential users and having well-understood rules regarding their rights and obligations with respect to other users and the resource.

community development quota: Program in Western Alaska under which a percentage of the TAC of Bering Sea commercial fisheries is allocated to specific communities. Communities eligible for this program must be located within 50 miles of the Bering Sea coast, or on an island within the Bering Sea; meet criteria established by the State of Alaska; be a village certified by the Secretary of the Interior pursuant to the Alaska Native Claims Settlement Act; and consist of residents who conduct more than half of their current commercial or subsistence fishing in the Bering Sea or waters surrounding the Aleutian Islands. These communities cannot have previously developed harvesting or processing capable of substantial participation in the Bering Sea fisheries in order to qualify for the program. Currently, 7.5% of the total allowable catch in the pollock, halibut and sablefish, crab, and groundfish fisheries is allocated to the CDQ program (see Box 4.3).

control date: Date established for defining the pool of potential participants in a given management program. Control dates can establish a range of years during which a potential participant must have been active in a fishery in order to qualify for quota share.

data fouling: Process whereby improper data reporting and collection procedures from a fishery can result in unrepresentative samples of what is actually being harvested in the fishery (e.g., misreporting of highgrading or bycatch rates). Based on these samples, incorrect inferences may be drawn about the true biological, economic, or social components of the fishery.

Department of Fisheries and Oceans, Canada (DFO): Federal agency in Canada responsible for management of fisheries in Canadian federal waters.

derby: Fishery in which the TAC is fixed and participants in the fishery do not have individual quotas. The fishery is closed once the TAC is reached, and participants attempt to maximize their harvests as quickly as possible. Derby fisheries can result in capital stuffing and a race for fish.

discard: Fish that are not retained for market.

economic overfishing: Condition in which a reduction in fishing effort results in an improvement in net revenue from the harvest.

economic rent: Difference between total revenue and all necessary costs of production, including a normal return on invested capital. This difference will prevail in a successfully managed fishery because fish stocks cannot be replicated on any scale desired, in contrast to automobile factories and other manufacturing industries. The rent in the fishery reflects the scarcity value of the fish stocks.

El Niño-Southern Oscillation (ENSO): Coupled oceanographic-atmospheric phenomenon resulting in a shift of sea surface temperatures beginning in the tropical Pacific. ENSO has widespread effects on oceanographic and atmospheric phenomena throughout the entire Pacific region and affects Pacific fisheries.

exclusive economic zone (EEZ): Zone extending from the shoreline out to 200 nautical miles in which the country owning the shoreline has the exclusive right to conduct certain activities such as fishing. In the United States, the EEZ is split into state waters (typically from the shoreline out to 3 nautical miles) and federal waters (typically from 3 to 200 nautical miles).

exploitation rate: Amount of fish harvested from a stock relative to the size of the stock, often expressed as a percentage.

externalities: Occur when the costs or benefits of a resource user's actions are not borne fully by the individual user; other resource users share the costs or benefits. Because of the common-pool nature of fisheries (see Box 2.1), fishermen impose externalities on one another. Such externalities occur through highgrading, as well as when fishermen are racing to use up their time allocated to fish.

factory trawler: (See catcher-processor.)

finfish: Vertebrate and cartilaginous fishery species, not including crustaceans, cephalopods, or other mollusks.

fishery management plan (FMP): Management plan for fisheries operating in the federal EEZ produced by regional fishery management councils and submitted to the Secretary of Commerce for approval. These plans must meet certain mandatory requirements in the MSFCMA before they can be approved or implemented.

fishing effort: In casual usage, this term refers to the amount of fishing. Depending on the context, fishing effort may refer to the number of fishing vessels, the amount of fishing gear (nets, traps, hooks), or the total amount of time that vessels and gear are actively engaged in fishing. Fishery economists often use the term to describe the quantity of productive inputs (e.g., labor, capital, fuel, ice) that are applied in fishing activities. Fishery scientists sometime distinguish between nominal fishing effort, which is the aggregate amount of time spent fishing, and standardized fishing effort, which is the amount of time spent fishing after adjustments are made for differences in fishing power among vessels and gear types.

fishing mortality (F): Deaths of fish that result from the fishing process (because fish are caught and retained, because they are discarded and subsequently die, or because they are caught in the gear and escape but subsequently die). Deaths that are not attributable to fishing activities are described as natural mortality. Fishery scientists often measure fishing mortality as an instantaneous rate, which is related mathematically to the exploitation rate and to standardized fishing effort.

fishing power: Measure of the relative ability of a fishing vessel (and its gear and crew) to catch fish, in reference to some standard vessel, given that both vessels are fishing under identical conditions (e.g., simultaneously on the same fishing grounds).

Food and Agriculture Organization (FAO): United Nations organization founded in 1945 with a mandate to raise levels of nutrition and standards of living, to improve agricultural productivity, and to better the condition of rural populations. FAO is active in land and water development, plant and animal production, forestry, fisheries, economic and social policy, investment, nutrition, food standards, and commodities and trade.

gear restrictions: Limits placed on the type, amount, number, or techniques allowed for a given type of fishing gear.

ghost fishing: Incidental capture of fish caused by gear that is lost or abandoned at sea.

groundfish: Collective term loosely applied to most commercially harvested marine fish other than salmonids, scombrids, and clupeids. Although many groundfish species are demersal (e.g., yellowtail flounder, yellowfin sole), other species are semidemersal or pelagic (e.g., pollock, cod, haddock, Atka mackerel).

growth overfishing: Condition in which the total weight of the harvest from a fishery is improved when fishing effort is reduced, and this improvement in harvest is due to an increase in the average weight of harvested fish.

gross register ton (GRT): A unit of the internal volume of a ship, equal to 100 cubic feet. Gross registered tonnage is the total volume or capacity of a vessel.

Gulf Fishery Management Council (GFMC): One of eight regional councils mandated in the MSFCMA to develop management plans for fisheries in federal waters. The GFMC develops fishery management plans for fisheries off the coast of the states of Texas, Louisiana, Mississippi, Alabama, and the west coast of Florida.

Gulf of Alaska (GOA): Region of the EEZ off the shore of Alaska extending from the southeastern edge of Alaska to the eastern side of the Aleutian Island chain.

highgrading: Form of selective sorting of fish in which higher value, more marketable fish are retained and fish that could be legally retained, but are less marketable, are discarded.

individual fishing quota (IFQ): Fishery management tool used in the Alaska halibut and sablefish, wreckfish, and SCOQ fisheries in the United States, and other fisheries throughout the world, that allocates a certain portion of the TAC to individual vessels, fishermen, or other eligible recipients based on initial qualifying criteria.

individual transferable quota (ITQ): Individual fishing quota that is transferable.

input controls: Fishery management measures that seek to limit the amount or effectiveness of effort in a fishery. These include limited licenses that restrict the number of fishermen, gear restrictions that limit the type or amount of gear that may be used, and effort quotas that restrict the amount of effort or time that is allowed in fishing activities.

International Commission for Northwest Atlantic Fisheries (ICNAF): Fishery management organization founded by the United States and Canada in 1949 for joint scientific and management measures affecting certain groundfish stocks. ICNAF later evolved into NAFO.

International Council for the Exploration of the Sea (ICES): International body established in 1902, ICES is a scientific forum for the exchange of information and ideas on the sea and its living resources and for the promotion and coordination of marine research by scientists in its member countries. Membership has increased from the original 7 countries in 1902 to the present 19 countries.

International Pacific Halibut Commission (IPHC): International management and advisory body established in 1923 to oversee the management of halibut in the North Pacific region. Member states of the commission include the United States and Canada. The IPHC is responsible for conducting stock assessments and providing recommendations on the appropriate level of harvest and other regulations to managers in Canada and the United States.

landings per unit effort (LPUE): Means of quantifying the CPUE. LPUE is the amount, or biomass, of fish landed per given unit of measure, typically measured on a per-trip or per-day basis.

longline: Fishing method using a horizontal mainline to which weights and baited hooks are attached at regular intervals. The horizontal mainline is connected to the surface by floats. The mainline can extend from several hundred yards to several miles and may contain several hundred to several thousand baited hooks.

longliner: Vessel specifically designed to catch fish using the longline fishing method.

Magnuson-Stevens Fishery Conservation and Management Act (MSFCMA): Federal legislation responsible for establishing the fishery management councils and the mandatory and discretionary guidelines for federal fishery management plans. This legislation was originally enacted in 1976 as the Fishery Management and Conservation Act; its name was changed to the Magnuson Fishery Conservation and Management Act in 1980, and in 1996 it was renamed the Magnuson-Stevens Fishery Conservation and Management Act.

Marine Recreational Fisheries Statistics Survey (MRFSS): Primary source of marine recreational data. MRFSS is operated by NMFS with the cooperation of coastal states. MRFSS is a design-based survey that produces estimates of total effort and catch in directed recreational fisheries.

maximum sustainable yield (MSY): Largest average catch that can be harvested on a sustainable basis from a stock under existing environmental conditions. MSY is a deterministic single-species construct that may have difficulty reflecting the stochastic nature of stock dynamics.

metric ton (mt): 2,000 kilograms (equivalent to 2,206 pounds).

Mid-Atlantic Fishery Management Council (MAFMC): One of eight regional councils mandated in the MSFCMA to develop management plans for fisheries in federal waters. The MAFMC develops fishery management plans for fisheries off the coasts of New York, New Jersey, Delaware, Maryland, Virginia, and North Carolina. Pennsylvania is also represented on the council.

multispecies fishery: Fishery in which more than one species is caught at the same time. Because of the imperfect selectivity of most fishing gear, most fisheries are "multispecies." Term is often used to refer to fisheries where more than one species is intentionally sought and retained.

National Academy of Sciences (NAS): Private nonprofit, self-perpetuating society of scientists. The NAS was granted a charter by Congress in 1863 that requires it to advise the federal government on scientific and technical matters.

National Marine Fisheries Service (NMFS): Federal agency within NOAA responsible for overseeing fisheries science and regulation.

National Oceanic and Atmospheric Administration (NOAA): Agency within the Department of Commerce responsible for ocean and coastal management.

National Research Council (NRC): Operating arm of the National Academy of Sciences.

New England Fishery Management Council (NEFMC): One of eight regional councils mandated in the MSFCMA to develop management plans for fisheries in federal waters. The NEFMC develops fishery management plans for fisheries off the coasts of Maine, New Hampshire, Massachusetts, Connecticut, and Rhode Island.

North Pacific Fishery Management Council (NPFMC): One of eight regional councils mandated in the MSFCMA to develop management plans for fisheries in federal waters. The NPFMC develops fishery management plans for fisheries off the coast of Alaska. It is comprised of voting members from Alaska, Washington, and Oregon.

open access: Condition in which access to a fishery is not restricted (i.e., no license limitation, quotas, or other measures that would limit the amount of fish that an individual fisherman can harvest).

optimum yield (OY): Term defined in the MSFCMA as the amount of fish providing the greatest overall benefit to the nation based on the MSY from the fishery as reduced by any relevant economic, social, or ecological factors.

Organization for Economic Cooperation and Development (OECD): International organization formed in 1961 of member nations in North America, Europe, Asia, and Australia to sustain economic growth and improve international trade.

output controls: Fishery management measures designed to limit the amount of catch or harvest in a fishery. These measures include catch quotas such as the TAC, IFQs, or IVQs.

overfishing: Harvesting at a rate greater than necessary to meet economic or biological goals for fishery. Overfishing is defined in the Magnuson-Stevens Act.

overfishing limit (OFL): Point at which fishing seriously compromises a fishery's continued, sustained productivity. Overfishing limits may be set based on standardized biological criteria established for a particular fishery. Overfishing limits may also incorporate economic and social considerations relevant to a particular fishery.

Pacific Fishery Management Council (PFMC): One of eight regional councils mandated in the MSFCMA to develop management plans for fisheries in federal waters. The PFMC develops fishery management plans for fisheries off the coasts of Washington, Oregon, and California. Idaho is also represented on the council.

paralytic shellfish poisoning (PSP): Condition in humans caused by the ingestion of bivalve mollusks that have accumulated dangerous levels of neurotoxins from phytoplankton.

pelagic: Referring to the open ocean.

poaching: Catching fish for which no quota is held. Illegally harvesting fish.

purse seine fishing: Fishing for certain species (e.g., tuna, herring, salmon) in which the school of fish is encircled with a large vertical net. The fish are trapped by "pursing" (closing) the bottom of the net by pulling it up from the center.

quota: Percentage or amount of fish that can be harvested.

quota busting: Harvesting fish in excess of the amount allowable for an individual's quota share.

quota management area (QMA): Geographic area used in the management of New Zealand fisheries. There are 10 QMAs.

quota management system (QMS): Overall management system used in the New Zealand fisheries managed by IFQs.

quota share (QS): Amount of quota, translated into pounds or number of fish, that a particular individual or corporation is allowed to harvest or process.

recruitment (R): Number, or percentage, of fish that survive from birth to a specific age or size. The specific size or age at which recruitment is measured may correspond to when the fish first become vulnerable to capture in a fishery or when the number of fish in a cohort can be estimated reliably by stock assessment techniques.

race for fish: Situation that can result in a fishery having a TAC without any limitation on fishing by the individual fisherman. This situation provides incentives for all participants in the fishery to harvest the TAC as quickly as possible before the fishery is closed. It typically leads to excessive fleet capacity and fishing effort (capital stuffing) and increasingly shorter fishing seasons.

recreational fishing: Fishing whose primary intent is for sport and pleasure, not for the sale, barter, or trade of fish.

recruitment overfishing: This condition results from fishing at a high enough level to reduce the biomass of reproductively mature fish (spawning biomass) to a level at which future recruitment is reduced. Recruitment overfishing is characterized by a decreasing proportion of older fish in the fishery and consistently low average recruitment over time.

regional fishery management council: Eight regional fishery management councils are mandated in the MSFCMA to be responsible for developing fishery management plans for fisheries in federal waters. Councils are composed of voting members from NMFS, state fishery managers, and individuals selected by governors of the coastal states. Nonvoting members include individuals from the U.S. Coast Guard, the U.S. Fish and Wildlife Service, and other federal officials. Regional councils exist for the Caribbean, Gulf of Mexico, Mid-Atlantic, New England, North Pacific, Pacific, South Atlantic, and Western Pacific regions.

riparian: Living on or near the bank of a river or lake.

scientific and statistical committee (SSC): Fishery management advisory body composed of federal, state, and academic scientists that provides scientific advice to a fishery management council.

South Atlantic Fishery Management Council (SAFMC): One of eight regional councils mandated in the MSFCMA to develop management plans for fisheries in federal waters. The SAFMC develops fishery management plans for fisheries off the coasts of North Carolina, South Carolina, Georgia, and the east coast of Florida.

Superexclusive area registration: A management tool used by the Alaska Department of Fish and Game for some (small) Bering Sea crab fisheries. Participation is open to any vessel, provided that the vessel agrees not to participate in any other crab fishery. Because this tool is applied to crab fisheries with low guideline harvest levels and because participation in the fishery precludes participation in any other Bering Sea crab fisheries, few vessels will choose to participate. Most of the vessels that do choose to participate will be small and local to the fishery.

surf clam/ocean quahog (SCOQ): Surf clam and ocean quahog IFQ fishery managed by the MAFMC.

surimi: Protein paste derived from processing raw fish, primarily pollock and whiting. Surimi can be combined with flavoring agents and other substances and extruded to create marketable foodstuffs (e.g., imitation crab meat).

total allowable catch (TAC): Total catch permitted to be caught from a stock in a given period, typically a year. In the United States, this limit is determined by regional fishery management councils in consultation with NMFS and scientific and statistical committees where they are used.

transshipment: Transfer of product from one ship to another at sea for its further transport.

trawling: Fishing technique in which a net is dragged behind the vessel and retrieved when full of fish. This technique is used extensively in the harvest of pollock, cod, and other species in North Pacific fisheries. It includes bottom- and midwater fishing activities.

trolling: Fishing technique in which a lure is attached to a line dragged through the water. This technique is used in fishing for tuna and other pelagic species.

two-fee system: System for recovering administrative costs and collecting additional fees using two different methods of fee collection. Monitoring, enforcement, and administrative costs would be covered by a per-unit fee levied on quota share determined by the magnitude of the costs it is designed to cover. Fees above administrative costs could be funded by a per-unit fee on actual catch. This fee would be relatively stable, although it could be indexed to some measure of inflation to ensure that its real value did not decline over time.

two-pie system: Form of quota allocation in which both harvesters and processors are allocated shares of quota. The harvester and processor allocations would be transferable within but not between each category.

unblocked quota: Quota shares in the Alaskan halibut and sablefish IFQ programs that are allowed to be subdivided when transferred. There are limits on the total number of unblocked quota shares that an individual may own.

underexploited: Fish species that are not exploited to the optimum yield or maximum sustainable yield.

Western Pacific Fishery Management Council (WPFMC): One of eight regional councils mandated in the MSFCMA to develop management plans for fisheries in federal waters. The WPFMC develops fishery management plans for fisheries off the coasts of Hawaii, Guam, the Commonwealth of the Northern Mariana Islands, American Samoa, and uninhabited U.S. territories in the Western Pacific.

year class: Fish of a given species spawned or hatched in a given year; a three-year-old fish caught in 1998 would be a member of the 1995 year class (See cohort).

zero-revenue auction: Form of auction used in the Acid Rain Program to control sulfur emissions. Under this system, the government takes back some proportion of the allocation each year for sale in an auction. Quota holders are allowed to buy back the quota they put up to bid, but they will succeed only if they are the highest bidder. Quota shares are auctioned to the highest bidders, and the revenue is returned to the holders of the auctioned quota shares. In principle, the auction could involve either quota shares or annual quota in fisheries. Significantly, all components of auction transactions (e.g., price, identification of buyers, quantities transacted) are public information.

APPENDIX

G

Individual Fishing Quota Case Studies

This appendix presents data and information on the currently implemented federal individual fishing quota (IFQ) and transferable trap certificate programs in the United States and on selected systems from other nations. Its intent is not to evaluate these systems with respect to their desirability or lack thereof but to present empirical data on their genesis, characteristics, and effects. Although the committee gathered information on several other systems that are in various stages of development, the focus here is on those systems that actually have been implemented and from which some documented results are available.

The appendix is organized in two sections: (1) the U.S. federal experience and (2) selected foreign experiences. The common characteristics of these fisheries and their IFQ systems and the lessons learned from their experiences can be found in Chapter 3. The U.S. federal experiences that form the core of the analysis are summarized according to eight topics that are described for each case:

1. Prior regulatory conditions in the fishery;
2. Prior biological and ecological conditions in the fishery;
3. Prior economic and social conditions in the fishery;
4. Problems and issues that led to the consideration of an IFQ program;
5. Objectives of the IFQ program;
6. IFQ program development process and the transition to IFQs;
7. The IFQ program; and
8. Outcomes of the IFQ program.

The foreign experiences follow the same general format, although any comparison among the cases must be made carefully because of the different policy and management frameworks and political, social, and economic conditions under which these systems were developed.

SUMMARY OF U.S. EXPERIENCE

Fishery managers in the United States have gained substantial experience with individual fishing quotas and related systems in the past eight years. In this section, three of the four existing IFQ programs (surf clam/ocean quahog, halibut, and sablefish), plus the spiny lobster transferable trap certificate program, are discussed in the order of their implementation. The wreckfish program is summarized in Chapter 3.

Each section describes the conditions that existed in the fishery prior to IFQs, including the factors that most directly led to IFQs (if implemented), and characteristics and outcomes of the program.

Surf Clam and Ocean Quahog (SCOQ) ITQ Case Study

Surf clams (SC: *Spisula solidissima*) and ocean quahogs (OQ: *Arctica islandica*) are bivalve mollusks that occur along the U.S. East Coast, primarily from Maine to Virginia. Commercial concentrations of surf clams are found primarily off the Mid-Atlantic coast. In this region, they are found from the beach zone to a depth of about 60 m. Ocean quahogs have a similar distribution, overlapping considerably, but they are also found in deeper waters, from 8 to 256 m.

These two closely related fisheries are largely (but not entirely) conducted by the same vessels, in the range of 40-110 gross register tons (GRT), which employ hydraulic clam dredges. Most of the catch is shucked and processed into a variety of clam products (minced clams, clam strips, juice, sauce, chowder). Apart from a small bait fishery, the recreational fishery is insignificant. Surf clam fishing began in the 1940s; ocean quahog fishing began in the 1970s. In addition, a small fishery for ocean quahogs found in shallow waters in the Gulf of Maine began in the 1980s; its market is for fresh in-shell product.

The SCOQ fishery was the first to be managed under the Magnuson-Stevens Act in 1977; the first limited access fishery in the exclusive economic zone (EEZ), through the moratorium created in 1977; and the first IFQ fishery in the EEZ, in 1990. Like the New Zealand IFQ program, but even more so, it is designed according to the prescriptions of free-market liberalism: there are few constraints on ownership eligibility, transfer, and other features, as described below.

Several features of the SCOQ fishery make it a relatively simple case for IFQ management. There is little competition for its product, although this is changing with the advent of clams from Iceland and elsewhere. The geographic range is relatively small; the number of vessels has never exceeded 140, and is now less than 50, and the number of vessel owners is much smaller; because of the tight linkages between harvesting and processing, the number of landing sites is also small. Moreover, the fishery is highly specialized: the vessels are not easily used for other purposes when outfitted with hydraulic clam dredges; to date bycatches have not been discussed as a problem. Finally, the commodity orientation of IFQs is appropriate because there have been no recreational or environmentalist claims for other values.

Prior Regulatory Conditions in the Fishery

In 1990, prior to IFQs, different regulations were applied to surf clams and ocean quahogs, to restrict the harvest of surf clams and encourage development of the ocean quahog fishery.

1. Quota setting and catch limits—Quota setting for both species became an annual process of the Mid-Atlantic Fishery Management Council (MAFMC), within a framework plan that establishes the optimum yield (OY) within a range of bushels.

With the beginning of EEZ surf clam management in 1977, a total allowable catch (TAC) was estimated for the Mid-Atlantic surf clam fishery and divided into quarterly quotas. Fishing time limits per fishing vessel were set to help spread catch over time, so as to stabilize product input to processors. Conservative TACs were set. The policy was to set the TAC to allow a 10-year supply horizon, or at least 10 years of harvest on the present standing stock. There was a separate TAC for the smaller fishery in the New England region. At the same time, the state of New Jersey also began to regulate surf clam harvests within 3 nautical miles of its shores.

A TAC for ocean quahogs was also set in 1977 but there were no time restrictions. The TAC was set high to spur development of this fishery and take pressure off the surf clam stock. The TAC was never met. Concern about the longevity and lack of recruitment of ocean quahogs, however, led to the adoption of a 30-year supply horizon.

2. Reporting requirements—In the Mid-Atlantic and New England regions, all SCOQ-harvesting vessels were required to report their catches in detailed logs. Processors also had to report how much product they accepted and from whom. This created a record of individual vessel performance.

3. Access restrictions—Access in the ocean quahog and New England surf clam fisheries was essentially unrestricted. The ability to restrict entry was allowed in the SCOQ fishery management plan (FMP) but was among the many provisions directed toward ocean quahogs that were never put in place (Brandt, 1994-1995). A permit and logbook reporting were all that was required. In the

State of Maine a small-scale ocean quahog ("mahogany clam") fishery developed in the late 1980s. It was open access, although regulated by the state because of concerns about the toxin that causes paralytic shellfish poisoning (PSP). This state fishery overlapped with the federal fishery in the EEZ but was not considered part of the larger management regime until the advent of IFQs.

The Mid-Atlantic surf clam fishery was the first EEZ fishery in the nation to be managed with limited access. The commercial fishery for surf clams began after World War II. It was an open-access, boom-and-bust fishery until 1977. The State-Federal Surf Clam Project depended on states to enact regulations, but most of the fishery took place beyond 3 nautical miles from shore.

In 1978, a vessel moratorium was established, grandfathering all vessels in or being built for the surf clam fishery in 1977 and/or fishing in 1978 (184 vessels were included at first, but some were dropped because of inactivity, leaving 142 (MAFMC, 1990). Thereafter, access was contingent on owning one of the original boats or its replacement. There were no restrictions on sale or purchase of these vessels, and capitalized values of moratorium permits were very high (estimated at $50,000-$150,000) (MAFMC, 1990). The moratorium lasted until 1990.

Prior Biological and Ecological Conditions in the Fishery

The population ecology of surf clams and ocean quahogs is distinctive, leading to "mining" rather than "sustainable resource" management strategies (cf. Murawski and Idoine, 1989). Their biomass is dominated by a few large year classes. Year-to-year recruitment variability is very high. They have erratic sets and few year classes that make it to "recruitment" size. For surf clams, recruitment to harvestable size is achieved in 6 to 7 years; ocean quahog recruitment is more difficult to determine, the majority of individuals found in the Mid-Atlantic region being very old, far beyond 20 years. Adult clams grow very slowly and may live a long time, particularly ocean quahogs, one of which is believed to have lived for 225 years (Brownlow and Ropes, 1985). Accordingly, the major management decision has been how long the present standing stock should last.

The major goal of the surf clam FMP was to restore depleted populations. Surf clams were subject to heavy fishing pressure from the late 1960s to the mid-1970s; localized stocks were depleted and the fishing fleet moved to new grounds. In 1976, a period of low dissolved oxygen in waters near the seafloor off the coast of New Jersey killed a large portion of the surf clam stock. This event prompted action, first from New Jersey and subsequently from the new MAFMC to try to prevent an unregulated industry from reducing the remaining clams to economic extinction.

Ocean quahogs are found over a much broader range of the North Atlantic region and in deeper waters than surf clams. Their life-cycle characteristics are similar to those of surf clams (erratic sets, few successful year classes), but recruitment, growth, and maturity take longer. As one scientist said, "Ocean quahogs are like a 'living rock'" (Jeff Weinberg, meeting of the Surf Clam and Science and Statistical Committees, September 4, 1996).

There was a high level of scientific uncertainty about the population dynamics of both species. However, harvesters knew where to find these clams because they are sedentary creatures. With hydraulic dredging gear, they are easy to harvest.

For both clam species, there is no discernible relationship between the size of the spawning stock and the number of clams recruited, and harvesters rely substantially on a few large year classes to buffer interannual variability, leading to analogies to mining when talking about management strategies. This is why the TAC was set conservatively, using the figure of 10 years' sustainability or "supply years" for surf clams and 30 years for ocean quahogs in setting annual TACs. An important effect of the reliance on occasionally large year classes during the moratorium period (1978-1989) was the creation of excess harvesting capacity. As the 1976 year class in the New Jersey area and the 1977 year class in the Delmarva area grew large enough to harvest, this created a bonanza that was easy to harvest but, within the context of a TAC, fixed for the long term. The result was further restrictions on fishing time, so that by 1987, surf clam boats were allowed to fish for only eight hours every month (see Marvin, 1992), even though the annual quota had increased greatly.

Prior Economic and Social Conditions in the Fishery

From 1977 to 1989, the moratorium on new entries created a situation in which the harvesting sector of the industry retained virtually the same number of vessels (about 144 vessels with surf clam permits in the Mid-Atlantic region), although the participation of these vessels varied from year to year and with the prices of clams.

Although the number of vessels in the surf clam fishery remained virtually unchanged during the moratorium period, liberal interpretations of the replacement policy on the part of the Northeast Region of the National Marine Fisheries Service (NMFS) allowed changes in total fleet capacity. The number of small vessels (class 1) decreased from 14 to 8 between 1980 and 1987, while the number of large vessels (class 3) increased from 59 to 75 in that period. The number of class 2 vessels decreased from 54 to 50 (MAFMC, 1990). This increase in capacity contributed to rising catch per unit effort (CPUE), as did the growth of the 1977 and 1978 year classes of surf clams and industry changes in harvesting gear. The MAFMC staff computed estimates of revenues versus costs for different classes of the fleet, and estimated that a loss of more than $3 million must have occurred during the moratorium, given the costs of catching clams (MAFMC, 1990).

Detailed data on the processing sector are not available. Clams are processed for canned chowder, canned whole and minced clams, and breaded strips.

Ocean quahogs are partly substitutable for surf clams, but the latter are definitely preferred for technological and quality reasons (ocean quahogs are tougher and high in iodine). As early as 1980, concentrated market power was evident in the processing sector (Strand et al., 1981); this remained true throughout the decade. During this period the industry structure that had existed before the moratorium remained: a few large, vertically integrated firms dominated the industry in their dealings with numerous smaller processors and "independent" vessel owners (including a few who amassed large fleets during the moratorium).

The year in which much of the politicking about ITQs occurred (1987) was also the year of the lowest recorded average prices for surf clams and ocean quahogs (see MAFMC, 1990).

Crew employment declined during the moratorium period, as vessel owners adapted to time restrictions by using the same crew members on more than one vessel (McCay and Creed, 1987, 1990; McCay et al., 1989).

Many of the clam vessels were unionized prior to 1979; after that time, when one of the processing firms was relocated and its boats were sold, mostly to their captains, unionization ended, and no associations arose to represent the interests of captains and crew in the fishery management process. However, vessel owners and processors were very active in this process, and several organizations appeared from time to time to help galvanize industry efforts to cooperate with the MAFMC in managing this fishery. There was a strong spirit of "co-management" from the outset (Turgeon, 1985). A job satisfaction study done in New Jersey (Gatewood and McCay, 1988) showed that in comparison with other types of commercial fishermen, crew members who worked on clam vessels received higher incomes and were less likely to see fishing as a challenge and adventure; there was a somewhat lower degree of commitment to and dependence on clam fishing than other types of fishing (i.e., dragging or longlining). This did not hold true for captains on clam vessels, most of whom had little experience in other occupations.

Fishing ports and processor locations for clams are spread throughout the Mid-Atlantic region and into New England. Most of the processors are found at seaport communities, but a few large ones have facilities inland as well, where fruits and vegetables are processed. The labor force in clam processing tends to be much the same as in poultry, and fruit and vegetable processing; it is dominated by ethnic and racial minorities, and in places dependent on immigrants, in some cases bused from the inner cities. No research has been done on the relationships between changes in the clam fisheries and the fortunes of either the processing firms or their employees (but see Griffith, 1997), much less on how such changes affect the communities in which the firms are located or the employees live.

Similarly, no research has focused on the community aspects of the harvesting sector of the clam fisheries. The fishing fleets move around quite a bit over time, following clams or clam buyers; hence many crew members are long-distance commuters (e.g., between New Bedford, Massachusetts, and Cape May, New Jersey). Crew members often come from the hinterlands of port communities; thus, the Atlantic City fleet has little directly to do with Atlantic City; the owners and crew members live primarily in old "baymen" towns like Absecon and Tuckerton, New Jersey. In ports such as Cape May and Wildwood, the New Jersey clam fleet is part of a much larger fishing fleet, all embedded in a seasonal tourist economy, where fishing is one of the very few year-round occupations.

Occupational health and safety issues loomed large in this fishery; vessels frequently sank and men's lives were often lost each year in New Jersey and Delmarva waters by the late 1980s. A study of mortality rates in New Jersey showed that fishing was one of the most dangerous occupations in the state, and these rates resulted almost entirely from the surf clam and ocean quahog fisheries (P. Guarnaccia, personal communication, September 14, 1998). For example, five clam vessels capsized in New Jersey waters in 1989. A study of fishermen's perspective on marine safety showed that sea clamming was widely seen as one of the most dangerous fisheries, partly because of its technology and partly because of the regulatory system, which created pressures to harvest and bring in as much as possible in a very short period of time, often in bad weather (McCay, 1992). Disasters affect the larger community, and in the Cape May region the resident fishing community responds by hosting parties to raise funds for the families of fishermen lost at sea. The larger community has responded by raising funds for a memorial to the region's fishermen lost at sea.

Problems and Issues That Led to Consideration of an Individual Transferable Quota (ITQ) Program

The moratorium on new clam vessels (through the MAFMC) was widely considered a success in preventing overharvest of surf clams and fostering development of the ocean quahog fishery, but it was a cumbersome regulatory system that was costly to monitor and enforce. It was characterized by numerous regulatory changes (seven amendments to the FMP between 1978 and 1987). It was complicated by the fact that after 1980, the New England Fishery Management Council took responsibility for managing the smaller fishery in the New England area (Nantucket Shoals; for a short while also Georges Bank).

Many provisions of the FMP and its implementation were seen by industry and NMFS alike as burdensome, inflexible, and in need of change. A prime example is the use of restricted fishing time to ensure relatively even distribution of the harvest over the year, to benefit the processors. Until 1987, the NMFS Northeast regional director specified the number and length of allowable trips per week or other period (up to two weeks). The vessel owner chose the day or days he or she wished to fish, notified the regional director, and then had to "use or lose" the days. In the winter, one could obtain a makeup day, but if this day also was missed, the opportunity was lost. When combined with the inability to consolidate allowable fishing time from one boat onto another, the system was obviously problematic. Moreover, the way this system was managed led to a large "ghost fleet" of mostly unused fishing capacity. Participation requirements were minimal, and owners of old and marginal vessels had incentives to retain their permits because such permits added to the value of the vessels.

Cheating (by fishing in closed areas, fishing a longer time than allowed, and taking undersized clams) was alleged to have been rampant. Much of the impetus for major changes in the management system came from concern about such administrative and enforcement difficulties.

Excess harvesting capacity was another major problem—indeed, in economic theory, the major problem. It was generated first in the open-access period, because the competition for dwindling stocks of clams provided an incentive for harvesters to use larger boats and more gear. Later, the moratorium and its grandfathering provisions allowed more boats than ever before into the restricted access fishery. Overcapitalization was intensified by (1) growth in size of the very abundant 1976 (New Jersey) and 1977 (Delmarva) year classes of surf clams, (2) technological changes such as more and larger dredges and hydraulic hoses, (3) the classic race to harvest the largest share of the TAC, and (4) increased skill and experience. These factors led to drastic increases in CPUE and equally drastic declines in allowable fishing time. Consequently, vessels were moored for much of the time unless their owners also participated in the ocean quahog fishery (which demands larger vessels with greater capacity) or the New Jersey or New York inshore fisheries (both of which are managed with limited access programs and have trip limits and other restrictions). Health and safety issues were also used to justify the development of ITQs.

Another issue identified in attempts to garner support for ITQs was that of obtaining financing from banks and other institutions, which are notoriously reluctant to support fishing ventures. The argument was that obtaining capital would be much easier if one had secure rights to a share of the total allowable catch.

Objectives of the ITQ Program

The SCOQ FMP was "preadapted" for ITQs in the sense that its objectives, from 1981, included economic efficiency and deregulation. These objectives were appropriate for the federal administration of its time and were endorsed by the Office of Management and Budget. The strong emphasis on economic efficiency was due to the participation of a neoclassical economist in the management process. Features of the 1977 SCOQ FMP, as amended in 1987 included the following:

1. "...[C]onserve and rebuild Atlantic surf clam and ocean quahog resources

by stabilizing annual harvest rates throughout the management unit in a way that minimizes short-term economic dislocations";

2. "Simplify...the regulatory requirement of clam and quahog management to minimize the government and private cost of administering and complying";

3. "...[P]rovide the opportunity for the industry to operate efficiently, consistent with the conservation of clam and quahog resources, which will bring harvesting capacity in balance with processing and biological capacity and allow industry participants to achieve economic efficiency including efficient utilization of capital resources by the industry"; and

4. "A management regime and regulatory framework which is flexible and adaptive to unanticipated short-term events or circumstances and consistent with overall plan objectives and long-term industry planning and investment needs" (MAFMC, 1988, p. 1; 1996, p. 3).

ITQ Program Development Process and the Transition to ITQs

The 1977 moratorium was intended to be a stopgap, emergency measure to be replaced by something else in a relatively short time. It lasted 12 years. However, among the alternatives being considered from the beginning was some system that would allocate quota to individual vessels: "... introduction of a per vessel allocation and some restriction on entry of new vessels (this might be a stock certificate program or an annual allocation per vessel)" (MAFMC Scientific and Statistical Committee, 1980; cited in Strand et al., 1981, p. 116). This theme appeared and reappeared throughout debates in the 1980s about how to reform management of the surf clam fishery (the ocean quahog fishery was not seen as problematic). As overcapitalization became more evident and, to some extent, costly for the participants, pressure mounted to change the system. It was intensified by frequent admonitions from NMFS to replace the moratorium with a more rational system.

By the mid-1980s, the major issue was whether and how to allow "consolidation" of fishing time among the vessels of the fleet. This incremental approach to the problem was advocated by the larger fleet owners but resisted by owner-operators and small fleet owners, concerned about the competitive advantage of the larger owners. It was also resisted by some of the big firms, concerned about rising competition from consolidation of rights to fish from the so-called ghost fleet (Marvin, 1992). Entrepreneurs accumulated the marginal, non-fishing, and sometimes sunk vessels with the hopes that they might be able consolidate their permits.

Around the same time the theme of "vessel allocation" reappeared: the notion of giving part of the quota to each vessel to minimize the costly and dangerous race for the quarterly quotas and the incentive to overload boats during the few hours they were allowed to fish. Vessel allocation was stymied by conflicts over how to make the allocations, given large differences in interest and power in the industry. An abiding concern among industry participants was that either consolidation or vessel allocation might further the monopsony (or oligopsony)[1] power of vertically integrated processors, which could lead to price collusion, forcing smaller processors and independent harvesters out of business (see Strand et al., 1981; McCay and Creed, 1990).

By 1988, the council, led by a Plan Development Team and the advice of the council's Surf Clam and Ocean Quahog Committee, was prepared to proceed beyond individual vessel allocations to ITQs, which were separable from the vessels and fully marketable. However, as of July 1988, there were still provisions in the draft FMP amendments reflecting concerns about the effects of rapid consolidation on the industry, including a "phase-in period" of three years, during which permits and allocations could be combined at no more than the rate of two for one, for each of the three years (MAFMC, 1988). However, these provisions completely disappeared in the amendment that was finally adopted by the council in October 1989 and approved by the National Oceanic and Atmospheric Administration (NOAA) in March 1990 (MAFMC, 1990).

The ITQ Program[2]

Management Units. The management unit included all surf clams and ocean quahogs in the Atlantic EEZ. This fit original Magnuson-Stevens Act policy and reversed the situation that had emerged after 1980, when management was divided between the Mid-Atlantic Fishery Management Council, concerned about overharvesting, and the New England Fishery Management Council, attempting to foster development. It also came to pose a major problem, because it included ocean quahogs being fished in federal waters by a small-scale fishery in Maine, for which there had been no logbooks and hence no historical records to use for allocation.

Initial Allocation. The initial allocation of quota share was divided among owners of all permitted vessels that harvested surf clams or ocean quahogs between January 1, 1979, and December 31, 1988. Replacement vessels were credited with the catch of vessels they replaced. These were all commercial fishing vessels, mostly working the waters of the Mid-Atlantic region.

Different formulas were used for allocations of surf clams in the Mid-Atlantic region versus ocean quahogs in both regions and surf clams in New England. For Mid-Atlantic surf clams, allocation was based on a vessel's average historical catch between 1979 and 1988. The last four years were counted twice, and the worst two years were excluded. The resulting figures were summed and divided

---

[1] A market situation in which each of a few buyers exerts a disproportionate influence on the market. (Merriam-Webster, Inc, 1998. *The WWWebster Dictionary* [Online] [Available: http://www.m-w.com/cgi-bin/dictionary] September 1, 1998).

[2] This program was first approved by the MAFMC in July 1988 and by NOAA in March 1990; it was implemented in October 1990 (MAFMC, 1990).

by the total catch of all harvesters for the period. Eighty percent of a vessel's allocation came from this ratio. A second ratio was computed on the basis of vessel capacity (length x width x depth), called a "cost factor," and this accounted for 20% of the vessel's initial allocation.

For ocean quahogs and New England surf clams, allocation was determined from the average historical catch for the years actually fished between 1979 and 1988, excluding the lowest-catch year. The average New England surf clam catch was then included in the total surf clam catch to calculate individual vessel ratios in the newly defined larger region, which incorporated both the Mid-Atlantic and New England stocks.

Nature of the ITQ. The ITQ has two components: (1) the "quota share," expressed in percentages of the TAC, which can be transferred permanently, and (2) the "allocation permit," which are in the form of tags to be attached to the large steel cages used to hold the clams after they are harvested. They can be transferred only within a calendar year. Annual individual quotas are calculated by multiplying the individual quota share by the TAC or allowable harvest in bushels. Bushel allocations are then divided by 32 to yield the number of cages allotted, for which cage tags are issued. Cage tags may be sold to other individuals but are valid for only one calendar year.

Accumulation and Transfer of Quota Shares. The minimum holding of SCOQ ITQs is five cages (160 bushels); there is no maximum holding and no limit to accumulation, except as might be determined by application of U.S. antitrust law. Anyone qualified to own a fishing vessel under U.S. law is entitled to purchase ITQs, except entities with majority foreign ownership. There are no limits on transfer of quota share. Cage tags are transferred only within a given year and cannot be transferred between October 15 and December 31 of each year. All transfers must be approved by the NMFS northeast regional director.

Monitoring and Enforcement. Monitoring the harvest of clams under the ITQ program is facilitated by the cage-tagging requirement and by mandatory reporting to NMFS by vessel owners and dealers of clams landed and purchased. Allocation permit numbers must be reported on both vessel logbook reports and dealer-processor reports. Dealers and processors must have annual permits. The cage tags are monitored closely. However, no reporting is required from truckers and other carriers.

Enforcement relies heavily on shoreside surveillance, the cage tag system, and cross-checking logbooks between vessels and processors. During seasons when state fisheries are open, at-sea and air surveillance is also required to reduce the possibility that vessels with state permits or cage tags may stray into federal waters. Allocation permits and dealer/processor permits may be suspended, revoked, or modified for violations of the FMP.

Setting of Quotas and Other Biological Parameters. The FMP is a "framework" plan that establishes the allowable range of harvest, but each year the MAFMC must recommend specific quotas, with input through various fishery management council venues, such as hearings, a public comment period, and an Industry Advisory Panel.

Unique in this fishery is the fact that the annual quotas, within constraints set for biological and long-term industry reasons, can be set "at a level that would meet the estimated annual demand" (MAFMC, 1997, p. EA-1). This policy would, in theory, meld the economic interests and incentives of ITQs with more general conservationist objectives. The policy, adopted in 1992, reflects a longer history of arguments by some segments of the industry for reducing the quota below the level warranted by stock assessments, especially for ocean quahogs. Some might see the arguments as expressions of the effects of ITQ incentives for conservation, but they existed prior to ITQs.

Administration and Compensation. No resource rents are collected from SCOQ ITQ fisheries; allocation permit fees are collected to help cover administrative costs, including the production and distribution of cage tags.

Evaluation and Adaptation. Evaluation and adaptation take place through the amendment process of the MAFMC, as well as reviews from within NOAA and studies done by outsiders. Major changes since 1990 have focused on meeting the overfishing requirements of the Magnuson-Stevens Act and dealing with the problem of the Maine mahogany clam fishery (Amendment 10). After the defeat of several lawsuits filed by industry groups challenging features of the plan, the general approach of industry appears to be acceptance and desire for consistency and predictability, as opposed to frequent change. Most industry attention is now devoted to the quota-setting process and outcomes.

Outcomes of the ITQ Program

Biological and Ecological Outcomes for the Fishery. TACs have not been exceeded during the ITQ period. MAFMC policy is to set the quota within the OY range ". . . at a level that will allow fishing to continue at that level for at least 10 years. Within the above constraints, the quota is set at a level that will meet estimated annual demand" (DOC, 1996, p. 12). For surf clams, the OY range equals 1,850,000 to 3,400,000 bushels; for ocean quahogs, the OY range equals 4,000,000 to 6,000,000 bushels (DOC, 1996).

The minimum size limit of 4.75 inches has been suspended, because the large size of most populations and incentives to search for and concentrate on aggregations of large clams mean that small clams will be avoided. These incentives come from buyers, who want large-size, high-yield clams and are strengthened by the end of competitive racing for clams due to ITQs.

One recent development is the discovery that SCOQ resources may be much more abundant and resilient than previously thought. The seemingly anomalous results of a 1994 NMFS survey of surf clams drew critical attention to NMFS survey methods and stock assessment process. In 1997, in cooperation with the industry, NMFS carried out experiments on dredge efficiency, the results of which were combined with new surveys to revise estimated total biomass. For surf clams, the results show that the stock is at "medium" level of biomass and "probably underexploited overall," although the most heavily fished area, northern New Jersey, is unlikely to result in increased catches. In addition, it is now recognized that recruitment is occurring at least annually, rather than decadally. The view on ocean quahogs, which had previously been determined to meet the "overfishing" definition, also changed; it too is now seen as at a medium to high level of biomass and to be underexploited, at the scale of the management unit, although local aggregations may be close to overexploitation (NEFSC, 1995).

Little is known about bycatch in these clam fisheries. The effects of dredging on benthic communities and habitat for other creatures are also unknown. The ITQ program is alleged to encourage targeting and selection of clam populations that meet industry demand—that is, high-yield, relatively large clams, in fairly pure aggregations. To some extent, pricing favors this strategy. Effects of such strategies on the ecology and biology of clams are unknown. Targeting of larger clams discourages harvesting densely populated beds of slow-growing clams, such as beds off Chincoteague, Virginia.

There was a shift northward in landing of surf clams and ocean quahogs during 1988-1996, partly in response to declining CPUE in waters off southern states (e.g., Virginia) as well as in heavily fished areas off southern New Jersey. The processing sector also has begun to move to southern New England ports, giving further impetus to a harvesting move north.

There has been a decline in discards under the IFQ program, typically of small clams (NEFSC, 1995). Between 1981 and 1989 there were minimum sizes, as well as area closures, to protect small clams that had not yet reached recruitment size, and both discarding and illegal harvests were substantial. Incentives for discarding were decreased when the council lifted minimum size limits because of data showing relatively low proportions of undersized clams (NEFSC, 1995), although processors continued to ask for large clams of high meat yield. ITQs may have provided some of the incentives for giving more effort to searching the locations with large clams and high meat yield, although this has not been documented.

Economic and Social Outcomes for the Industry. Appraisals of the SCOQ fishery have shown that since the introduction of ITQs in late 1990, economic efficiency in clam harvesting has increased and excess harvesting capacity has declined (McCay and Creed, 1994; Wang, 1995; Adelaja et al., 1998). Illustrative data are provided in Table G.1.

TABLE G.1 Changes in Fishing Effort, Ownership, and Catches for EEZ
Surf Clams and Ocean Quahogs, 1988 and 1994

	Year	
	1988	1994
Vessels fishing for SC	133	48
Vessels fishing for OQ	62	35
Owners of SC vessels[a]	56	28
Owners of OQ vessels[a]	25	17
Hours fished/vessel, SC	404	1,400
Hours fished/vessel, OQ	537	1,249
Average bushels/trip, SC	992	1,149
Average bushels/trip, OQ	1,458	1,491
Average trips/vessel, SC	23	52
Average trips/vessel, OQ	49	88

[a] Ownership is based on interviews to determine "true ownership," recognized in the industry as such, as distinct from official ownership in NMFS files, which is often in the name of vessel-specific corporations, leading to possible errors in reporting and judgment. Note that these data pertain to vessels actually fishing according to logbook information; owners may continue to own inactive vessels and/or quota shares. Also note that some owners (ca. 30%) have both surf clam and ocean quahog vessels and that some of the vessels are used in both fisheries.

SOURCE: Adelaja et al. (1998).

The smallest firms, in terms of either the number of vessels owned or the amount of initial ITQ allocation, were most likely to sell out in the period from 1990 to 1992. However, small firms were also resilient; two-thirds of the smallest holders kept their ITQs and about 18% actively participated in the market for ITQs by buying and selling quota, as did the majority of large firms. The medium-sized firms (i.e., holding 1-6% of the initial quota) in the surf clam fishery were most likely to purchase more ITQs; only 2 of 17 sold out. None of the largest firms (>6% initial quota) had sold out by 1992 (McCay and Creed, 1994), although some did later.

A substantial number of firms stopped fishing but held onto and leased out their quota shares. As of 1992, roughly one-third (32%) of those who held surf clam ITQs did not fish for surf clams, presumably leasing out their quota; the figure was even higher for ocean quahogs (46%). To some extent the high level of leasing or temporary transfers of cage tags was due to uncertainty about the future and about the market for ITQs on the part of people who were planning to leave the fishery. However, from interviews it was learned that many had come to recognize the nature of this new asset and its ability to generate income through leasing (McCay and Creed, 1994).

The chance that a firm would leave the clam fishery (not necessarily the ITQ program) was greatest at the beginning, decreased during the first two years, and increased again, to reach an equilibrium in the fourth year (Weisman, 1997). Being an "independent" or non-vertically integrated firm that owned only one or a few vessels had no significant effect on the chances of surviving in the fishery by the end of 1993 (Weisman, 1997).

The size of the initial allocation for ocean quahogs was directly proportional to the chance of remaining in the fishery until the end of 1993; there was no such effect for surf clams. However, for firms with both surf clam and ocean quahog initial allocations, smaller firms were more likely to leave the fishery than larger firms (Weisman, 1997).

ITQs worked in the surf clam fishery to accentuate the effects of other variables on how many clams were caught; the rapid reduction in the number of vessels used encouraged organizational changes that allowed more efficient use of production inputs (Adelaja et al., 1998). The effects were less noticeable in the ocean quahog fishery, which did not have the degree of overcapitalization present in the surf clam fishery. The major effect of ITQs in the ocean quahog fishery was the initial shake-out; those remaining after an initial round of ITQ allocation transfers had greater catch and market share than initially.

Between 1990 and 1994, clam prices were not statistically significant determinants of total catch (Menzo et al., 1997). However, the catches of firms of different sizes, as measured by average monthly landings, did respond differently to changes in price, suggesting industrial reorganization. These results fit the theory that large firms are relatively buffered against price changes, whereas small- and medium-size firms are either more vulnerable to changes in price or more flexible in responding to them.

Between 1988 and 1994, market share, an indication of firm size, had no relationship to price received for catch, suggesting the lack of monopoly in the seller's market (Adelaja et al., 1998). However, owners who leased ITQs from others for a large portion of their landings and who had large shares of the landings seemed to have some advantage in terms of the price they received for their clams (Menzo, 1996).

For surf clams, the dominance of the top four harvesting firms in terms of landed clams never exceeded 56% and hence did not meet technical definitions of "oligopoly" (>60%); their dominance varied substantially but was changed little by the onset of ITQs. However, companies with the highest market shares remain constant over the period, and the average prices received were among the lowest. The situation for ocean quahogs was similar, but with a slight decline in dominance by the top four firms in the ITQ period. Three of the operators are consistently at the top, the same companies as those at the top of surf clam landings; again, the average prices they received were among the lowest (Menzo, 1996). One explanation of this effect is that these firms are vertically integrated, making the price received at the dock less consequential for the firm's owner than for the captain and crew on the boat.

In terms of ITQ holdings, concentration of ownership increased for ocean quahogs; the largest firm in 1992 held 35.3% of the ocean quahog quota share. For surf clams, concentration of ownership did not change significantly; in 1992, the largest firm held 22.6% of the surf clam quota share. The surf clam fishery tends to have a more bimodal distribution of large versus small operators, whereas the structure of the ocean quahog fishery tends to be more evenly distributed, with a middle class of shareholders as well as large operators (McCay and Creed, 1994).

After ITQs were implemented, a few buyers-processors gained dominance (Wang, 1995; Weisman, 1997). Empirical modeling shows the general importance of buyer-seller relationships in relation to survivability in the fishery (Weisman, 1997). Reliance on a single buyer in 1990 increased the likelihood of exiting the fishery by the end of 1993; on the other hand, selling most of one's catch to the top six buyers decreased the likelihood of exiting the fishery. Distortions may also exist in the market for ITQs themselves; empirical research has not been done on this question.

Lorenz curves were constructed for 1988, 1990, and 1994 surf clam landings, by owner, showing a high degree of skewedness, expressing inequality in the distribution of wealth in terms of landings (Gini Concentration Ratio > .55), but this actually decreased with ITQs as many of the smaller firms stopped fishing (Menzo, 1996). For those who remain active in the SCOQ fisheries, the distribution of landings has become more equal, not less. Of course, there are other sources of inequality, including ownership of ITQs, which was not covered in the Menzo study, which looked solely at landings.

Economic and Social Outcomes for Fishery-Dependent Communities. Employment in the clam industry has declined, leading to downward shifts in the bargaining power of crew members and captains, symbolized and to some degree exacerbated by changes in the share system of returns to owners and crew members (McCay et al., 1990; McCay and Creed, 1994). Some owners tried to mitigate these impacts by keeping boats fishing even when not really needed, but as Table G.1 shows, the reduction of boats, and hence crew, was very rapid and very radical, even though crew reductions had already taken place during the latter years of the moratorium era (McCay and Creed, 1994).

A common practice, from the outset of this system, was for the owners of vessels to deduct the cost of leasing quota, as an operating expense, from the amount that would be shared among captain and crew members. The exvessel price paid to the crew by the vessel owner also might be reduced by leasing. For example, an owner might receive $8.00 per bushel for surf clams from a processor, but only pay "the boat" $4.00 per bushel because the cost of leasing allocation from the processor is deducted. This might be done even if the boat owner actually owns the allocation, or the owner might transfer the allocation to the processor to create a legitimate paper trail for tax purposes (Ross, 1992).

Improved safety was a major selling point for the SCOQ ITQ system. In the early IFQ period, 1990-1992, many people in the industry voiced the opinion that this was a sellout, largely because of the sinking of two clam vessels, the *John Marvin* and the *Valerie E*, in a fast-building storm in the late winter of 1991, following the loss of another boat the year before. In interviews, people said that ITQs did not help because the processors still demanded that vessels fish when the product was needed, regardless of weather conditions (Beal, 1992; McCay and Creed, 1994). Despite sharp reduction in the number of vessels in the fleet, particularly the older boats, the incidence of loss of vessels and lives at sea in the 1990s is comparable to what it was in the 1980s, when an average of one boat a year was lost. In January 1999 five clam boats and eleven lives were lost in separate events; another vessel was lost in 1997. Accordingly between 1990, when ITQs went into effect, and February 1999 nine clam boats and at least fourteen lives have been lost in this fishery. Clearly, sea clamming remains a dangerous occupation. The role of ITQs in either mitigating or enhancing its dangers is not known.

Little research has been done on the effects of ITQs—or other changes in the SCOQ industry—on local communities. However, it is clear that the appearance and disappearance of fishing vessels, and particularly processors, can have a major impact on some communities. The major source of impacts on communities is likely to be the processing sector, which has become dominated by a few large firms since ITQs began. The effect of ITQs on processor organization and concentration has not been shown in empirical economic studies; however, the vertically integrated processors had an advantage in the competition for clams over processors that did not own boats prior to ITQs. The vertically integrated processors obtained "free" quota shares during the initial allocation, whereas the others had to either purchase shares or bargain with vessel owners to supply them with clams, increasing the costs of their operations.

Administrative Outcome. Data Management. According to a 1992 NMFS evaluation, agency officials charged with administering the ITQ program found it impossible, given the nature of reporting, to determine the identity of the owning "persons" (Goodale and Raizin, 1992); many allocations were reported as owned by corporations or vessels, and the device of using addresses to identify true owners was deemed inadequate to the needs of the law. As noted below, enforcement officials were also concerned that they could not obtain real-time data on who owned how much ITQ for their purposes.

Enforcement. According to an internal NMFS review, enforcement was very problematic at the beginning of this ITQ program (McCarthy, 1992). The Mid-Atlantic region had the fewest NMFS enforcement personnel in the Northeast, and the ITQ Amendment to the FMP was allegedly designed without adequate input from enforcement officers, such that standard provisions were left out (i.e., the illegality of giving false statements to authorized officers).

Moreover, implementation was rocky; enforcement officers were not given real-time information regarding who had which cage tags or ITQs. These and other problems were mitigated somewhat by the heavy reliance on cage tags for monitoring, as well as the ability to cross-check logbooks of harvesters and processors (but not truckers, who were not required to keep records).

It has also proved difficult, if not impossible, to enforce a provision in the preamble to the final rule that the government would periodically monitor the number of quota shares owned by each person and advise the Department of Justice if any one had an "excessive share" (MacDonald, 1992). This provision was intended to justify having no limits on accumulation in the plan. As of early 1992, attorneys were unsure about how, if at all, this could be applied to the SCOQ fishery, raising questions such as whether or not the SCOQ market was a "market" within the meaning of the Sherman Anti-Trust Act (MacDonald, 1992). The excessive share provision has no definition, and courts have thus far not been concerned unless concentrations approach monopoly levels, which appears not to be the case in the SCOQ fishery (Milliken, 1994; *Sea Watch International v. Mosbacher*).[3]

Current Perceived Issues. The major issues related to the existing IFQ program include the following:

- Security of the program, given the attempts in Congress to forbid the creation of new ITQ programs and to impose sunset provisions on existing ones (Creed and McCay, 1996);
- Lack of adequate (1) stock assessments and population biology studies and (2) economic studies of supply and demand to be used with confidence in the annual quota-setting process;
- Enforcement in fisheries that include both state and federal waters;
- Concentration of shares and market power resulting from a lack of definition in the Magnuson-Stevens Act of excessive shares and small likelihood that the Sherman Act would be used to prosecute holders of excessive share (MacDonald, 1992; Milliken, 1994); and
- Need for a lien registry and other ways to strengthen the ability of IFQs to function as collateral, without transforming them into property rights.

---

[3] 762 F. Supp. 370 (D.D.C. 1991).

Alaskan Halibut and Sablefish IFQ Case Study

Prior Regulatory Conditions in the Fishery

Commercial fisheries for Pacific halibut (*Hippoglossus stenolepis*) and sablefish (*Anoplopoma fimbria*) have occurred off the Pacific Northwest, British Columbia, and Alaska for more than a hundred years. Carrothers (1941) estimates that British Columbia natives consumed more than 272 metric tons of halibut per year in the late 1880s. Development of large-scale commercial fisheries for halibut was stimulated by the completion of transcontinental railroads in the late 1880s. Carrothers (1941) reports that coastwide commercial landings of halibut exceeded 808 metric tons in 1889, 3,126 metric tons in 1899, and 9,866 metric tons in 1909. With the depletion of nearshore fishing grounds, Canada and the United States negotiated the Halibut Treaty of 1923 and established the International Fisheries Commission (later renamed the International Pacific Halibut Commission, IPHC) to investigate the halibut resource and recommend conservation measures. With the passage of the Fishery Conservation and Management Act (FCMA) of 1976 and similar legislation in Canada to establish 200-mile fishery conservation zones, and renewal of the halibut convention in 1979, the North Pacific Halibut Act of 1982 delegated limited entry and allocation decisions to the Pacific Fishery Management Council (PFMC) and North Pacific Fisheries Management Council (NPFMC). Canadian halibut fishermen were excluded from U.S. waters (and vice versa) in 1978. Recent catches of halibut and sablefish are depicted in Figure G.1.

From its inception in the 1950s through the early 1980s, the sablefish fishery off Alaska was dominated by foreign fishing operations (Figure G.2).

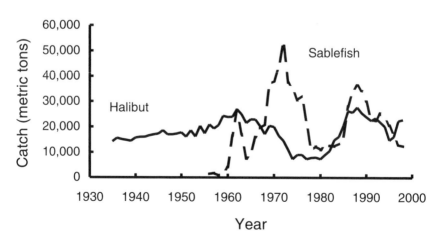

FIGURE G.1   Commercial catches of halibut and sablefish.

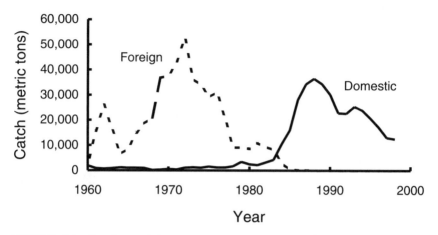

FIGURE G.2 Foreign and domestic commercial catches of sablefish.

Access. In U.S. waters, access to the halibut and sablefish fisheries was unrestricted prior to the passage of the FCMA. Following the act's implementation, various moratoriums were proposed but none were approved, so access remained open until the implementation of IFQs in 1995.

Limits on Catches. Annual limits on commercial catches of halibut are set for each IPHC regulatory subarea. Although the area boundaries have changed slightly over time, particularly in the Bering Sea and Aleutian Islands (Area 4), the 1996 regulatory areas are relatively representative (Figure G.3).

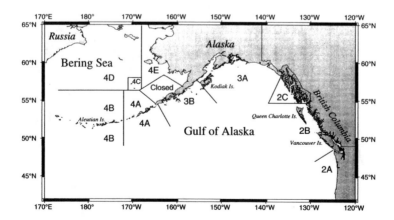

FIGURE G.3 IPHC regulatory areas (1996).

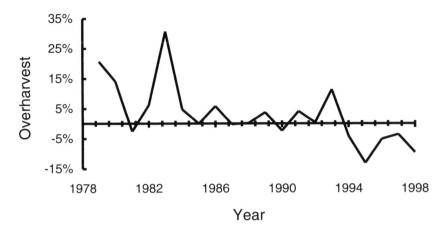

FIGURE G.4   Percentage overharvest of Pacific halibut in the directed commercial fisheries in the U.S. EEZ (IPHC regulatory areas 2C, 3A, 3B, and 4).

Individual vessel trip limits were imposed in various areas between 1988 and 1994. Trip limits have typically been applied late in the season when the remaining allowable catch was less than the unfettered fishery was expected to harvest in a single fishing period. Trip limits have been graduated by vessel class. Even with trip limits, the commercial fishery exceeded the coastwide catch limit by an average of 812 metric tons (4.9%) between 1977 and 1994 (Figure G.4).

Annual limits on catches of sablefish are set for four areas in the Gulf of Alaska (East Yakutat and Southeast Outside, West Yakutat, Central Gulf of Alaska, Western Gulf of Alaska), the Aleutian Islands, and the Bering Sea (Figure G.5). Limits on halibut and sablefish bycatch are established for combinations of target fishery and management area.

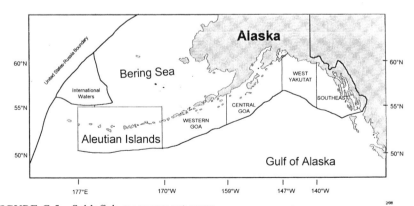

FIGURE G.5   Sablefish management areas.

Limits on Gear and Seasons. Catches of halibut and sablefish have historically been controlled through a combination of area, season, and gear restrictions. Most vessels that are engaged in these fisheries are catcher vessels that do little processing of the catch at sea. However, there are a few catcher-processor vessels (freezer-longliners) in the halibut fishery and a larger number in the sablefish fishery. The directed fishery for halibut uses longline gear. The directed fishery for sablefish uses longline, pot, and trawl gear. In the eastern Gulf of Alaska, 95% of the sablefish TAC is reserved for longline operations. Elsewhere in the Gulf of Alaska, longline fishermen are allocated 80% of the TAC. The use of pot gear for sablefish is prohibited in the Gulf of Alaska, but permitted in the Aleutian Islands and the Bering Sea. The Bering Sea TAC is split 50:50 between fixed gear (longline and pots) and trawls. Seventy-five percent of the Aleutian Islands TAC is reserved for fixed gear.

Reporting Requirements. Halibut buyers in Alaska are required to record landings on fish tickets (official landing receipts) from the Alaska Department of Fish and Game (ADF&G), which are either mailed directly to the IPHC or delivered to ADF&G offices and forwarded to the IPHC. Washington and Oregon fishery departments and the Canadian Department of Fisheries and Oceans also forward halibut landings data to the IPHC. The IPHC has also collected logbook data on an occasional basis to supplement information on the CPUE, productive fishing locations, gear configuration, and the mortality of undersized fish that are discarded.

Prior to 1986, ADF&G fish tickets were the sole source of landings data for the sablefish fishery off Alaska. Because at-sea processors were not subject to ADF&G reporting requirements, beginning in 1986, they were required to file "hail weight" reports with NMFS. These reports eventually evolved into the current Weekly Processor Reports. With expansion of the observer program in 1990, observer estimates of landings became available for some larger vessels (30% of vessels greater than 60 feet in overall length). In addition, logbook reporting requirements were strengthened to facilitate on-site verification of catches.

Prior Biological and Ecological Conditions in the Fishery

Pacific halibut and sablefish are both long-lived bottom-dwelling species. Halibut are the largest commercial species of the North Pacific, averaging 18 kg each, but occasionally exceeding 180 kg. Halibut are primarily found at 15-200 m depths on sand, gravel, or cobble substrates. Sablefish are considerably smaller (<5 kg) and occur at somewhat greater depths (100-1,500 m). Halibut are distributed from California to the Sea of Japan and into the Bering Sea (IPHC, 1987; Trumble et al., 1993). Sablefish extend this range to include waters off Baja California. Each species is considered a single stock throughout its range. Note, however, that whereas halibut are jointly managed by the United States and

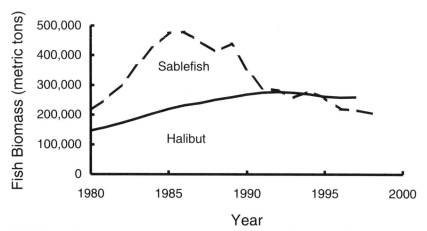

FIGURE G.6 Recent trends in coastwide estimates of halibut and sablefish biomass.

Canada, sablefish are not, and neither species is co-managed with Russia, Korea, Japan, or other principal harvesters.

Although the estimated 1997 coastwide biomass of Pacific halibut (260,423 metric tons) is 26% above the most recent 25-year average, it has declined somewhat in recent years, and based on moderate recruitment and reduced growth rates, is expected to continue to decline in the near future (IPHC, 1997). In addition, the average weight-at-age has declined 50% over the past decade.

The sablefish exploitable biomass was estimated to be 265,000 metric tons in 1996 (NPFMC, 1997a). The stock has been declining since 1986 and is 30% below the recent average. The biomass of sablefish is expected to continue to decline due to poor recruitment since 1982. The overfishing limit (OFL) for 1998 is expected to be less than 35,950 metric tons. It is anticipated that the allowable biological catch (ABC) for 1998 will be less than 17,200 metric tons. The TAC must be less than the ABC, to provide a buffer. The size of this buffer is based on the stock status and the quality of information available. See Figure G.6 for halibut and sablefish biomass trends.

Prior Economic and Social Conditions in the Fishery

Analyses of the markets before IFQ implementation are limited for halibut and nonexistent for sablefish. Crutchfield and Zellner (1962), Lin et al. (1988), Homans (1993), and Criddle (1994) describe the bioeconomics of pre-IFQ halibut fisheries using rudimentary models of the exvessel market structure. Although Herrmann (1996) provides a more realistic model of market structure, he deals exclusively with the Canadian fishery following the 1991 adoption of IVQs (individual vessel quotas), but prior to the 1995 adoption of IFQs in Alaska.

None of these models accounts for demand for halibut while simultaneously accounting for Canadian, U.S., and Russian supplies, and export markets. Figure G.7 presents a time series of real (1992) exvessel prices for catches of halibut and sablefish and Figure G.8 shows exvessel revenues.

In addition to being the focus of a directed commercial fishery, Pacific halibut is caught as bycatch in a variety of other commercial fisheries, treaty Indian fisheries, personal-use fisheries, and sport fisheries (Figure G.9). Halibut bycatch mortality has averaged 18% (6,405 metric tons) of the total halibut catch in recent years (1984-1996). Sport fishing has grown from 3% (857 metric tons) of the 1984 total catch to 11% (3,514 metric tons) of the 1996 total halibut catch. The treaty and personal-use halibut fisheries are small by comparison. The treaty Indian fisheries of the Pacific Northwest were allocated about 80 metric tons and the Metlakatla Indian Community was allocated 23.6 metric tons in the Annette Island Reserve Fishery in 1995.

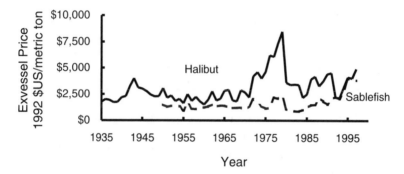

FIGURE G.7   Real gross exvessel price of halibut and sablefish in 1992 dollars.

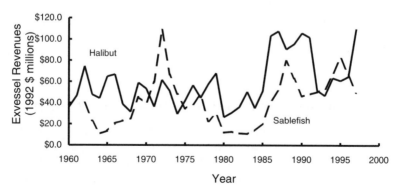

FIGURE G.8   Real gross exvessel revenues from halibut and sablefish in 1992 dollars.

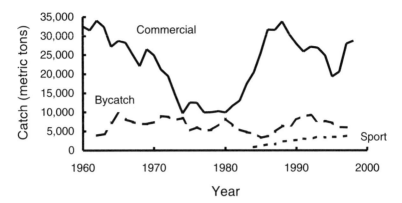

FIGURE G.9   Commercial longline catches, bycatches, and sportfishing catches
of halibut in the U.S. EEZ off Alaska.

Bycatches of halibut must be discarded if taken with other than hook-and-
line gear or if taken when the directed fishery is closed. Similar restrictions apply
to sablefish, although pots are a permitted gear in the Bering Sea.

Participants in the halibut fishery were heterogeneous. Although many vessels were specifically rigged for efficient longline operation, other vessels were
jury-rigged for halibut fishing during the short open seasons. For example,
salmon gillnetters could spool-off their nets and load longline gear on their gillnet
drums for the short halibut open seasons. Many halibut fishermen were engaged
in other (non-fishing) primary occupations and took leave to participate in the
short seasons. Figure G.10 represents the percent average (1982-1995) real
exvessel value of commercial catches off Alaska. Halibut and sablefish have
accounted for 5% and 4%, respectively, of the $1.3 billion average exvessel value
of Alaskan commercial catches.

Although halibut and sablefish together accounted for less than 10% of the
average exvessel value of Alaskan fisheries, they are regionally significant. The
1991 distribution of halibut catches by the residency of the permit holder is
represented in Figure G.11 (NPFMC, 1994a,b).

Problems and Issues That Led to Consideration of an IFQ Program

The problems and issues that led to consideration of an ITQ program for
halibut and sablefish were allocation conflicts; gear conflicts; ghost fishing due to
lost gear; bycatch loss in other fisheries; discard mortality for halibut, sablefish,
and other retainable species in the halibut and sablefish fishery; excess harvesting
capacity; product wholesomeness as reflected in real prices; safety; economic
stability in the fishery and communities; and rural coastal community develop-

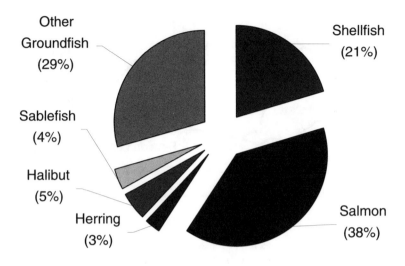

FIGURE G.10   Percent real exvessel value of commercial catches off Alaska (1982-1995).

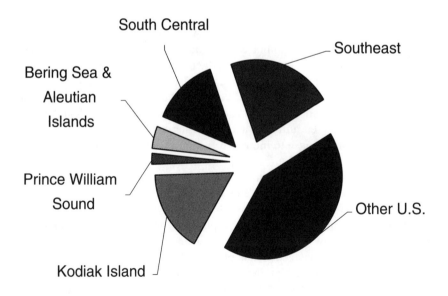

FIGURE G.11   Distribution of halibut catches off Alaska by residence of license holder (1991).

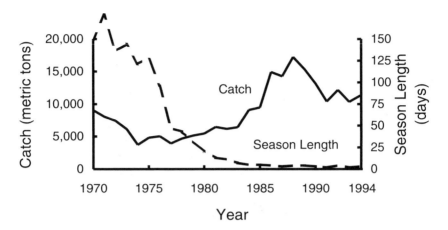

FIGURE G.12   Changes in catch and season length in the Area 3A (central Gulf of Alaska) halibut fishery from 1977 to 1994, before the introduction of IFQs.

ment of a small-boat fishery (NPFMC, 1991a,b,c). Evidence of some of these problems can been seen in time series of the number of participants, season length, fishing effort, and CPUE. The number of participants in the halibut and sablefish fisheries reached a maximum of 3,883 in 1990 for halibut and 706 in 1988 for sablefish (Pautzke and Oliver, 1997).

The central Gulf of Alaska (IPHC Area 3A) has accounted for 37-51% of the U.S.-Canadian commercial halibut catches since 1977. During this time, and despite a tripling of catch, the season length collapsed from 47 days to 2-3 days (Figure G.12). Using season length to manage fisheries becomes harder as effort increases and season length shrinks. If vessels had not been placed on trip limits after the first one-day halibut season opening in recent years, season length would have had to collapse even further to avoid overharvesting. A similar contraction of season length in response to increased fishing effort can be seen in the West Yakutat sablefish fishery (Figure G.13).

Gear conflicts can arise within or between gear types. Under the short derby seasons, conflicts between halibut and sablefish longline operations and other gear types were, by default, infrequent. Because trawling is very restricted in the Gulf of Alaska, conflict between gear types may be minor even under longer seasons. Conflict between users of similar gear can develop when some areas and times are more advantageous than others. The regulated open-access fisheries were characterized by a high incidence of lost and unrecovered fishing gear. The IPHC estimated that 1,860 "skates" (roughly 1% of the gear fished) was lost in 1990 and that the lost gear accounted for about 900 metric tons of halibut mortality (3% of the commercial catch).

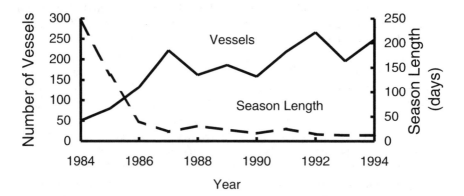

FIGURE G.13   Participation and season length in the West Yakutat sablefish fishery from 1984 to 1994, before the introduction of IFQs.

Short seasons have been cited as a contributing factor to the accident rate in the pre-IFQ halibut and sablefish fisheries. In the pre-IFQ fishery, the decision to sit out bad weather often amounted to a decision to sit out the fishing season. Another concern that led to the consideration of IFQs was the perception that exvessel prices for halibut were below what they could be if product deliveries were more distributed throughout the year.

Objectives of the IFQ Program

The Environmental Assessment/Regulatory Impact Review identifies ten problems that the ITQ program was intended to address (NPFMC, 1991a):

1. Allocation conflicts;
2. Gear conflicts;
3. Deadloss due to lost gear;
4. Bycatch loss of halibut and sablefish in other fisheries;
5. Discard mortality of halibut, sablefish, and other retainable species in the halibut and sablefish fisheries;
6. Excess harvesting capacity;
7. Product wholesomeness as reflected in prices;
8. Safety;
9. Economic stability in the fishery and communities; and
10. Rural coastal community development of a small-boat fishery.

IFQ Program Development Process and the Transition to IFQs

Following implementation of the FCMA, the NPFMC appointed a Plan Team to develop recommendations for management of the halibut fishery. The team's draft FMP proposed a limited entry program with a moratorium on new entry set at December 31, 1977. The council approved the draft FMP with the moratorium date revised to December 31, 1978. The draft FMP was shelved in late 1978 when the U.S.-Canada halibut convention was renewed. The NPFMC next approved a one-year moratorium on entry for 1982 with a cutoff date of December 31, 1981, but because the action was conditional on passage of an amended North Pacific Halibut Act and because the amended act was not passed until after the start of the 1982 fishing season, no action was taken. In early 1983, the NPFMC approved a three-year moratorium to begin on June 15, 1983. However, the NOAA administrator disapproved the NPFMC action and suggested instead that the NPFMC investigate a permanent limited entry system. The NPFMC began evaluating license limitation for the sablefish fishery in 1985 and IFQs in 1988. The NPFMC also revisited halibut license limitation and began consideration of individual fishing quotas in 1988. In January 1990, the NPFMC selected IFQs as the preferred management option for sablefish. In December 1990, the NPFMC linked further consideration of halibut license limitation and IFQs to ongoing analysis of similar measures in the sablefish fishery. In December 1991, the NPFMC approved IFQ programs for halibut and sablefish. The final rule creating halibut and sablefish IFQs was published in the *Federal Register* on November 9, 1993, for implementation in 1995.

The IFQ Programs

Management Units. The halibut IFQ program applies to all commercial hook-and-line harvests of halibut in state and federal waters off Alaska. The program does not apply to subsistence, treaty, or sport fisheries or to bycatch with trawl or pot gear. The sablefish IFQ program is limited to longline and pot gear fisheries in federal waters off Alaska and does not apply to sablefish harvested in state waters or in the trawl fisheries. Although most of the sablefish harvests are from federal waters, fishing for sablefish also takes place in state waters along the Aleutian Islands, in Prince William Sound, and in the vicinity of Chatham Strait in Southeast Alaska.

Nature of the IFQ. In the terminology adopted by the NPFMC, an individual's initial quota share (QS) allocation was set to equal the sum of his or her catches during selected qualifying years, less an adjustment for shares allocated to the community development quota (CDQ) program. The IFQ is the individual's annual allocation and is determined by dividing each individual's QS by the sum of all the QS in a region, the "QS pool," and multiplying the result by the annual fixed-gear portion of the total allowable catch. The allocation of QS was specific to area, operation mode, and vessel size category, with restrictions on transfer between vessel size classes, operation mode, and area. In addition, shares less than 20,000 pounds were "blocked" such that they could not be subdivided on transfer.

In general, owners are required to be on board when their IFQ is being fished. Exceptions are that initial QS recipients are allowed to hire "masters" to fish halibut QS in Areas 3 and 4, and that corporations and partnerships may hire masters in Southeast Alaska. Similar provisions apply to sablefish QS. With some short-term exceptions, for quota shares acquired through inheritance or divorce settlements, second-generation IFQ owners must be on board during fishing operations.

Initial Allocation. Halibut quota shares were allocated to the 5,484 vessel owners and leaseholders who had verifiable commercial landings of halibut during the eligibility years: 1988, 1989, or 1990. Allocations were based on the best five years' landings during qualifying years (1984-1990). Area-specific shares were allocated based on the geographic distribution of landings during the years used to determine quota share.

Sablefish quota shares were allocated to the 1,094 vessel owners and lease-holders who had verifiable commercial landings of sablefish during same eligibility years (1988-1990) but considered the best five-of-six qualifying years between 1985 and 1990. In determining the allocation rule, the NPFMC weighed the equity merits of a broad initial distribution based on liberal eligibility criteria against a narrower initial distribution that would provide recipients with larger initial allocations. The council's decision to allocate QS to 5,484 halibut fishermen and 1,094 sablefish fishermen represented 141% and 155% increases, respectively, over the maximum numbers of participants in any single qualifying year (3,883 for halibut and 706 for sablefish).

In December 1993, the NMFS Restricted Access Management (RAM) Division mailed Requests for Applications to all persons who, based on fish ticket and landings data, appeared to be eligible to receive QS in the initial issuance. A six-month application period (January 17, 1994, through July 15, 1994) was published in the *Federal Register.* A second mailing of requests was sent to all persons who had not completed and returned their Requests for Applications by mid-June. In addition, the RAM Division ran print and broadcast public service announcements and held 27 workshops to advertise and answer questions about the application process. Each person who submitted a Request for Application by July 15, 1994, was sent an application that detailed the official record of his or her qualifying catches, vessel size, and other relevant information. Applicants were requested to review the information and submit evidence to support corrections. The evidence was reviewed through an appeals process that could maintain or amend the official record. The issuance of QS began in November 1994 and was largely completed by January 1995.

A portion of the Bering Sea halibut and sablefish TACs was set aside for a CDQ program (see NRC, 1999a). To compensate commercial fishermen who had established catch history in the Bering Sea, a portion of the QS in the Gulf of Alaska (about 3.5%) was given to Bering Sea fishermen.

Accumulation and Transfer of Quota Shares. Rules on the accumulation and transfer of halibut and sablefish IFQs are constantly evolving. In general, there are limits on accumulation and transferability. No person (individual, company, corporation) may own more than 0.5% of the total halibut QS in combined Areas 2C, 3A, and 3B; more than 0.5% of the total halibut QS in Areas 4A-E; or more than 1% of the total QS for Area 2C. No person may control more than 1% of the total Bering Sea-Aleutian Islands and Gulf of Alaska sablefish QS or more than 1% of the total sablefish QS east of 140°W (East Yakutat and Southeast Alaska, see Figure G.5). Individuals whose initial allocation exceeded the ownership limits were grandfathered-in, but prohibited from acquiring additional QS.

Transferability is also restricted across vessel size categories. Four vessel categories were defined for halibut: (1) catcher vessels less than 35 feet in length overall; (2) catcher vessels 35 to 60 feet in length overall; (3) catcher vessels more than 60 feet in length overall; and (4) catcher-processor vessels. Three categories were defined for sablefish: (1) catcher vessels less than 60 feet in length overall; (2) catcher vessels 60 feet in length overall or larger; and, (3) catcher-processor vessels. The initial allocation of QS was based on the catch record within each vessel class. Transfer of catcher vessel QS between vessel classes was initially prohibited. However, recent program amendments permit small vessels to fish QS that was initially allocated to large vessels.

Catcher vessel QS is transferable only to "qualified" buyers of quota. Buyers must be initial recipients of catcher vessel QS, or they must be able to demonstrate 150 days of accumulated commercial fishing experience. Catcher-processor vessel QS is transferable to any person. Leasing of QS (sale of IFQ) is restricted for catcher vessels but allowable for catcher-processor vessels. Initial QS recipients were permitted to lease up to 10% of their QS during 1995, 1996, and 1997. An amendment to extend leasing provisions is under consideration. Trawlers cannot buy halibut or sablefish QS for directed fishing or bycatch. All QS transfers must be approved by the NMFS RAM Division.

Setting of Quotas and Other Biological Parameters. The setting of quotas continues to be based on the process that was in place before the adoption of IFQs. The IPHC (for halibut) and the NPFMC (for sablefish) are responsible for determining the ABC and OFL. The NPFMC is responsible for setting the TAC for commercial fisheries such that the sum of the commercial, sport, subsistence, treaty catches, and bycatch mortality is less than or equal to the OFL. The treatment of catches within state waters has been inconsistent. In some instances, catches in state waters have been subtracted from the ABC, consistent with the concept of a single stock. In other instances, consistent with a separate stock hypothesis, the state water catches have been ignored in the determination of the federal TAC. Once the TAC has been determined, determination of the IFQ is straightforward for halibut. In the case of sablefish, approximately 10% of the TAC is set aside for the trawl fishery, and the IFQ is based on the residual.

Monitoring and Enforcement. Monitoring is accomplished through a combination of real-time accounting and posttransaction auditing. Deliveries can only be made to registered buyers following a minimum six-hour advance notice to NMFS. The real-time accounting is accomplished through IFQ Landings Cards and "transactions terminals." IFQ Landings Cards function like a debit card. When a landing is made, a fisherman swipes the IFQ Landings Card through the transactions terminal and enters catch information, and the halibut and sablefish landings are deducted from his or her IFQ balance. In addition to the IFQ Landings Cards, halibut fishermen are required to submit Commercial Fish Tickets (catch reports) to the IPHC. Posttransaction auditing compares the records submitted by registered buyers with the fisherman's landings records to identify inconsistencies.

Because it can be difficult to exhaust an individual's IFQ exactly, the halibut and sablefish IFQ program has a provision for over- and underharvests. In the case of an overharvest, up to 10% of the fisherman's IFQ remaining at the time of the landing will be subtracted from the following year's IFQ. Underharvests up to 10% of the fisherman's IFQ are carried over to the subsequent year's IFQ.

Advance notification of intent to land provides an opportunity for NMFS and other enforcement personnel to observe landings, if desired. During routine boarding of halibut and sablefish vessels, the Coast Guard compares the poundage of fish on board with the balance on the fisherman's IFQ Landings Card. In addition, some of the larger vessels carry NMFS observers who are responsible for estimating the catch and discard of target and non-target species.

Administration and Compensation. The NMFS Alaska Region RAM Division was created to oversee the initial allocation of QS, approve QS transfers and leases, and monitor compliance with program requirements. There were no special taxes or fees to cover the costs of the IFQ program during 1995-1997. In keeping with new Magnuson-Stevens Act requirements, a cost recovery program is now being developed. The act provides the Secretary of Commerce with authority to levy fees up to 3% of the exvessel value of landings to cover the direct costs of IFQ management.

Evaluation and Adaptation. The first amendments to the halibut and sablefish IFQ program had been submitted to the Secretary of Commerce before the program was implemented in 1995. Virtually every meeting of the NPFMC since January 1995 has addressed one or more amendments or refinements to the program. Pautzke and Oliver (1997) briefly describe the modifications:

- IFQs less than 20,000 pounds were issued as "blocks" with increased restrictions on transferability and accumulation.
- Changes in QS associated with an effort to equalize the impact of a CDQ setaside were exempted from "block" and vessel category transfer restrictions.
- IFQ in IPHC Area 4 was allowed to shift between subareas.
- Vessels were allowed to fish in multiple management areas on a single trip if they carried an onboard observer.
- Sablefish catcher-processor vessels were allowed to fish catcher vessel QS as long as there was no processed product on board while the catcher vessel QS was being fished.
- Large-boat QS could be bought and fished on small boats.
- The sweep-up provisions of the "block" restrictions were changed.
- Weight adjustments were standardized for slime and ice on landed fish.
- The use of pot-longlines was allowed in the Bering Sea.
- The Aleutian Islands sablefish season was extended to 12 months for vessels that hold enough halibut QS to cover anticipated bycatches.
- Heirs were allowed to lease QS for up to three years.
- Ownership requirements for using skippers to fish the owner's QS were modified.
- Halibut QS ownership limits in the Bering Sea and Aleutian Islands regions were increased.

Outcomes of the IFQ Program

Biologic and Economic Outcomes for the Fishery. Gilroy et al. (1996) provide a preliminary description of the initial conservation effects of the halibut and sablefish IFQ programs. The IPHC estimates that halibut fishing mortality from lost and abandoned gear decreased from 554.1 metric tons in 1994 to 125.9 metric tons in 1995. The discard of halibut bycatch in the sablefish fishery is estimated to have dropped from 860 metric tons in 1994 to 150 metric tons in 1995. However, Gilroy et al. (1996) caution that the uncertainty of bycatch discard mortality estimates has not been determined under conditions of the IFQ fishery, and it is unclear whether the estimated reduction is statistically significant. There is no clear difference in sablefish bycatch before and after IFQs were introduced. The discard of other groundfish in the Bering Sea and Aleutian Islands sablefish fishery was higher in 1995 than in the previous four years, but there was no discernible difference in the Gulf of Alaska. There is no evidence of significant underreporting of catches of halibut or sablefish. The frequency of overharvests was significantly reduced by IFQs (Table G.2).

TABLE G.2 Frequency and Magnitude of Halibut
Overharvests

Year	Frequency of Overharvest	Average Overharvest
1977-1994	64%	6%
1995-1996	0%	−8%

Gilroy et al. (1996) found no evidence that fishermen have tried to increase the halibut or sablefish TAC. The spatial and temporal distribution of halibut catches has changed; differences in sablefish catches have not been evaluated. The biological and ecological consequences of these changes have not been evaluated. CPUE data from the commercial fishery are used in the halibut stock assessment but not the sablefish assessment. Although it is uncertain how CPUE has changed in the IFQ fishery, results from seasonal and area weighted analyses of CPUE in the Canadian IVQ fishery do not differ significantly from those in the pre-IVQ fishery (Sullivan and Rebert, 1998).

Highgrading of halibut and sablefish is prohibited under the halibut and sablefish IFQ programs. Although there is anecdotal evidence for highgrading, there is no evidence of highgrading in the halibut catch size-composition data in Alaska or Canada, nor have any instances of highgrading been documented or prosecuted. Preliminary comparison of the size distribution of sablefish in the commercial landings and catches in the NMFS sablefish longline survey do not suggest widespread highgrading.

Economic and Social Outcomes for the Fishery. It is not possible to quantify the net economic impact of the Alaskan IFQ programs because the outcomes have not yet been well studied. Although season length has increased to 245 days for both species and landings are broadly distributed through the season, it is uncertain how costs and revenues have been affected. Figure G.7 hints that the IFQ program has had a positive effect on the exvessel price of sablefish. However, Figure G.8 indicates a concomitant decrease in exvessel revenues. Without a comprehensive model of exvessel price formation that accounts for changes in landings, inventories, net exports, and exchange rate fluctuations, it is uncertain whether the observed price increase is due to the change in management regime or merely to continued declines in the sablefish TAC and hence supply. The effect of the IFQ program on halibut exvessel price is even more ambiguous. Although Hermann (1996) estimates that the Canadian IVQ program for halibut increased exvessel revenues by an average of Can$5.8 million per year, there is no comparable analysis of post-IFQ prices in the United States. The average real exvessel price for halibut under the IFQ regime is below the pre-ITQ average

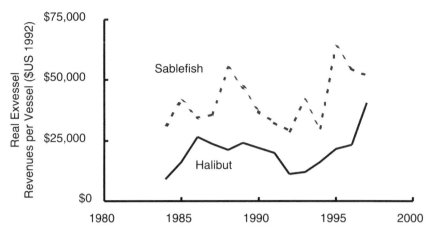

FIGURE G.14 Average per vessel real exvessel revenues for commercial catches of halibut and sablefish in the U.S. EEZ off Alaska (in 1992 dollars).

(1974-1994) (see Figure G.7). However, in contrast with sablefish, halibut exvessel revenues have increased since the introduction of IFQs (Figure G.8). Extension of the simple model developed in Criddle (1994) suggests that the real exvessel price of halibut increased by $0.56 per pound following IFQ implementation, but the 95% confidence interval on that estimate spans a range of price changes from a $0.05 decrease to a $1.16 increase. Herrmann (1996) estimates that the Canadian IVQ program for halibut increased exvessel revenues by an average of Can$5.8 million per year. The effect of these changes on the average real gross earnings of fishermen is represented in Figure G.14. A statistical analysis of the mean difference between the IFQ and pre-IFQ fisheries suggests that after accounting for changes in total landings, average per vessel real exvessel revenues have increased by $22,990 (the 95% confidence interval is $13,096 to $32,884) for sablefish and $18,658 (the 95% confidence interval is $11,813 to $25,504) for halibut. However, although this result also supports the expectation that IFQs lead to price increases, the observed change in average per vessel exvessel revenues is likely to be due, at least in part, to increases in the average number of quota shares per fisherman.

The economic effects of the IFQ program on revenues and costs for processors, consumers, and communities are even less well understood. The market power held by some processors and communities arose from the need of harvesters to deliver to locations that were near the areas fished, a result in part of the regulated open-access fishery race. There is anecdotal evidence that an increasing number of halibut fishermen are bypassing traditional processors and marketing directly to wholesalers and retailers. However, the magnitude and economic impact of this switch have not been documented. The top five halibut ports (about 50% of the TAC) have remained, with occasional rank reordering, Kodiak, Homer, Seward, Dutch Harbor, and Sitka (see Tables H.19 and H.21). Because the primary sablefish market is Japan (more than 70% of the 1995 sablefish catch was exported to Japan [Kinoshita et al., 1996]), there is less opportunity for fishermen to market sablefish directly. The list of top sablefish ports has been somewhat more variable, but the top four (about 60% of the TAC) have generally been Seward, Sitka, Kodiak, and Dutch Harbor, both before and after IFQs.

The QS transfer market has been active. More than 3,800 permanent transfers have taken place in the halibut program, and more than 1,100 permanent transfers have occurred in the sablefish program. These transfers have led to some consolidation (Tables G.3 and G.4)

The number of QS holders declined by 24% in halibut and 18% in sablefish between the initial QS issuance in January 1995 and August 29, 1997. However,

TABLE G.3   Percentage Change in Number of Halibut Quota Shareholders Between Initial Issuance and August 29, 1997

	Area				
Quota Share (pounds)	2A	3A	3B	4	Total
<3,001	−35%	−28%	−44%	−30%	−33%
3,001-10,000	−23%	−25%	−36%	−21%	−25%
10,001-25,000	4%	−3%	−2%	4%	0%
>25,000	50%	1%	12%	4%	7%
Total	−26%	−22%	−31%	−20%	−24%

SOURCE: NMFS (1997a).

TABLE G.4   Percentage Change in the Number of Sablefish Quota Shareholders Between the Initial Issuance and August 29, 1997

	Region						
Quota Share (pounds)	SE	WY	CG	WG	AI	BS	Total
<5,001	−33%	−31%	−29%	−21%	−16%	−13%	−27%
5,001-10,000	−27%	4%	−30%	−9%	21%	8%	−16%
10,001-25,000	0%	−10%	−8%	−3%	−5%	−5%	−4%
>25,000	6%	3%	−9%	3%	6%	11%	0%
Total	−20%	−20%	−22%	−13%	−7%	−8%	−18%

NOTE: AI = Aleutian Islands; BS = Bering Sea; CG = Central Gulf of Alaska; SE = Southeast Alaska; WG = Western Gulf of Alaska; WY = West Yakutat.
SOURCE: NMFS (1997a).

TABLE G.5   Average Real Exvessel Revenues for
Select QS Holdings (in 1992 dollars)

Weight (pounds)	Halibut	Sablefish
1,000	$1,240	$1,893
10,000	$12,403	$18,925
25,000	$31,007	$47,313

the number of current (as of August 29, 1997) QS holders (4,947 for halibut and 1,453 for sablefish) still exceeds the annual maximum number of participants in the pre-IFQ fishery (3,883 for halibut and 706 for sablefish). In both fisheries, the bulk of consolidation has taken place in the smaller QS holdings. Although QS holdings of less than 10,000 pounds are probably too small to serve as a primary income source for fishermen (Table G.5), they may provide an important supplement to other income sources.

There is anecdotal evidence that fishermen have reduced crew size and that QS holders are crewing for each other. However, because there are few data on pre-IFQ crewing practices, it is difficult to determine the magnitude of changes, let alone the opportunity cost of crew members who are no longer engaged in the halibut or sablefish fisheries.

Economic and Social Outcomes for Fishery-Dependent Communities. Information on the economic and social outcomes of the halibut and sablefish QS program is largely anecdotal. The regional economic impacts of fishing were not formally modeled before program implementation and have not been formally modeled after implementation. The CFEC (1996a,b) and Knapp (1997a,b) characterize changes in the regional distribution of QS ownership. Continued low prices for salmon have made halibut and sablefish catches increasingly important to regional fishing economies. The regional economic impacts of reductions in crew size are unknown because information on crew participation in the pre-IFQ and IFQ halibut and sablefish fisheries is unknown as is information on crew demographics, residency, and opportunity costs.

Administrative Outcomes and Enforcement. Currently, the increased costs of managing and enforcing the program are not being recovered from QS holders. However, as noted above, a cost recovery program is being developed that will assess fees up to 3% of the exvessel value. With the average nominal exvessel value of recent landings (1995-1997) on the order of $160 million, the fee program can be expected to generate about $5.1 million annually (see Table H.5). This compares favorably with the $3.8 million annual budget of the RAM Division.

Current Perceived Issues. Some dissatisfaction continues over the initial alloca-tion. This dissatisfaction is related to the delay between the qualifying years and the implementation date, and to the exclusion of crew members and processors from the initial allocation. The delays in implementation resulted in the exclu-sion of some fishermen who were active in the years immediately preceding implementation but were not active during the qualifying years (CFEC, 1997). Similarly, there was dissatisfaction with the award of QS to persons who were active during the qualifying years but inactive in the years immediately preceding implementation. Crew members and processors are discontented that the initial allocation, in addition to rewarding vessel owners, also changed output and factor market power in favor of QS holders.

There are ongoing concerns about the adequacy of enforcement and commu-nity impacts. With implementation, have come a heightened awareness of sub-sistence and sport catches and an effort to define limits on these competing fisheries. This competition has led to concerns about localized depletion and preemption of productive sportfishing grounds by commercial fishermen. Ex-pansion of the fishery for sablefish in Alaska State waters and possible creation of a Gulf of Alaska CDQ program are also of concern.

Florida Spiny Lobster Fishery[4]

General Description

The fishery for spiny lobster (*Panulirus argus*) is conducted primarily in the Florida Keys. It is principally a trap fishery, with additional small commercial dive and substantial recreational dive components. Most lobsters are harvested relatively close to shore in shallow water.

Prior Regulatory Conditions in the Fishery

Before the trap certificate system was implemented, the state required fisher-men to purchase "crawfish licenses." Catch was limited by a minimum carapace

---

[4] Unless otherwise noted, this information is summarized from the SAFMC/GFMC (1992). The program described for this fishery is based on individual transferable "trap certificates," a gear and effort-based system. There are no restrictions on the amount of catch, either for the fishery as a whole or for individuals. Input limitations are equivalent to output limitations only if there are no substitutes for the limited input. Because there is limited opportunity to substitute unconstrained inputs for lobster traps, the program appears to achieve many of the objectives that are also achieved by IFQ programs. Transferability allows fishermen to match the number of traps they use to the capacity of their vessel and their cycle of fishing and non-fishing activities. Nevertheless, there is some opportunity for fishermen to change their practices to increase the fishing power of individual traps through changes in the average soak time, changes in the spatial distribution of pots, and the choice of baits and other attractants.

measurement, but there was (and is) no overall TAC. The fishery was also subject to rules on trap size, markings, and buoys. A closed season was maintained during the lobster spawning period and area restrictions were maintained. Recreational fishermen were subject to bag limits. The state maintained a "trip ticket" system to record each landing of lobster. The fishery was (and is) managed under the federal Spiny Lobster Fishery Management Plan (jointly by the Gulf and South Atlantic Fishery Management Councils) even though the bulk of the fishery is conducted in Florida state waters.

Prior Biological and Ecological Conditions in the Fishery

Local populations of lobsters are seeded from current-borne lobster larvae derived from adult lobsters on reefs west of the Florida Keys. Thus, the local fishery does not fish on total spawning stock. More than 90% of legal-size lobsters are caught each year. Lobster populations are subject to habitat effects in Florida Bay from human activities in the Everglades and South Florida. Despite all of these unknown factors and variable conditions, lobster landings have been relatively constant for more than 20 years.

Prior Economic and Social Conditions in the Fishery

The lobster fishery was traditionally conducted primarily by a relatively small number of independent fishermen, perhaps less than 50. Of the more than 4,000 crawfish licenses issued per year in the 1980s, approximately half of these recorded commercial landings. The remaining approximately 2,000 licenses were held primarily to avoid the recreational bag limit or as a hedge against future limited entry systems. Six to eight hundred of the licensees were responsible for more than 80% of the harvest. The sociological makeup of the fishery is heterogeneous, with a significant Hispanic (primarily Cuban) component. Many fishermen originate outside Florida; the fishery is relatively easy to enter. Florida spiny lobster landings are a small component of the total U.S. lobster market; therefore, local fishermen must accept prices set by markets elsewhere. Spiny lobsters are a high-value, high-demand luxury product.

The Florida Keys region has developed rapidly and has reached a high degree of development and land and water use in some areas. The primary industry in the area revolves around leisure and tourism. Residents and tourists display a high degree of environmental awareness and activism. Monroe County (primarily the Keys) is known as an independent, "maverick" entity, with significant "bandit" (drug traffic) culture.

Problems and Issues That Led to the Consideration of a Tradable Permit System

Before the tradable permit system, the fishery was characterized by relatively constant landings (1975-1990), but the number of traps increased from 200,000 to more than 1,000,000 in the same period. Traps were baited with undersize lobster ("shorts") and the mortality of these juvenile lobsters was unknown. Wooden traps were dipped in oil prior to use to keep them from becoming fouled with marine organisms and infested with ship worms. Crowding had developed in the fishery, creating conflict among commercial users and between commercial and recreational users. Enforcement and administration of fishery regulations was difficult. The fishery was increasingly inefficient and subject to decreasing individual net profits due to the increased number of traps in use but the relatively constant total harvest.

Objectives of the Tradable Permit System

The objectives of the tradable permit system were several:

• To control or reduce effort so that the effort more closely matches the available fishery resource;
• To increase stability in the fishery and promote maximum net incomes for fishermen;
• To promote flexibility for fishermen in their fishing operations;
• To avoid conflict among fishermen and between fishermen and other marine users;
• To ensure that fishermen who have traditionally participated in the fishery be able to continue to due so, as much as possible in their traditional fishing patterns; and
• To make management of the fishery more efficient and effective.

Development Process and Transition to the Tradable Permit System

An outside consultant was solicited to facilitate the development and implementation of an alternative system to address the problems that were being experienced by the lobster fishery. Implementation of the project was encouraged by the State of Florida, the Gulf and South Atlantic Fishery Management Councils, and NMFS; industry expressed cautious interest. To stabilize the fishery before implementation of a new management regime, a moratorium was placed on issuance of new crawfish licenses. Socioeconomic research was performed prior to the consideration of alternatives, and the consultant teamed with industry and other constituents to develop and evaluate alternatives.

Objectives and alternatives for the fishery were developed over a two-year period through an independent, open workshop process run by the consultant. At the same time, the Florida Marine Fisheries Commission (FMFC) adopted draft rules (not developed through workshop process) to address various issues in the fishery, including the reduction of the number of traps, and these draft rules were the subject of considerable concern by the industry. The industry coalesced around a preferred alternative through the workshop process. Industry, recreational, and environmental groups agreed with the FMFC to approach the Florida legislature for authority to implement the new alternative system. The Florida legislature passed enabling legislation, which was implemented by the FMFC with assistance from industry (the alternative draft rules were dropped). Finally, the Gulf and South Atlantic Fishery Management Councils adopted the Florida program into the federal spiny lobster FMP.

The Tradable Permit System

The Management Units. The management units are individual traps, traded via trap certificates; one certificate enables the use of one trap. Actual trap usage is controlled thorough a tagging system.

The Initial Allocation of Trap Certificates. The initial allocation of trap certificates was based on individual landings history in the fishery. Each fisherman was allowed to select his or her highest individual landings from a three-year qualifying period, and a formula was developed to allocate trap certificates based on each individual's percentage of total landings. A limit was established on the total number of certificates any fisherman could receive under the initial allocation. An Appeals Panel, comprised of representative fishermen, was established to advise the Department of Natural Resources on hearing appeals regarding initial allocations.

Accumulation and Transfer of Trap Certificates. Certificates are marketable to anyone holding a crawfish license. The system includes an "antimonopoly" limit of 1.5% of outstanding certificates that can be held by any individual licensee. Certificates are subject to a transfer fee when sold, and the transfer must be registered. The total number of certificates outstanding may be reduced by the FMFC by up to 10% per year (with individual holdings reduced proportionately) as long as total lobster landings are not affected.

Monitoring and Enforcement. Traps corresponding to certificates must have individual tags. Enforcement is conducted both on land (prior to season opening) and at sea. There is ongoing monitoring of biology, ecology, and administrative effectiveness by the Florida Department of Natural Resources and the FMFC.

Administration and Compensation. The crawfish license fee doubled with the implementation of tradable permits, from $50 to $100. Certificate transfer fees and tag fees were established, with tag fees rising in price (up to $2 per tag) as the number of traps decreased, to maintain constant revenue. A "windfall profit" surcharge of 25% was applied to the first transfer outside a fisherman's immediate family. Revenues from fees and charges are divided among several beneficiaries. Ninety percent are devoted to dedicated funds for research, monitoring, enforcement, and education related to the spiny lobster fishery. The remaining 10% is allocated to the General Fund of the State of Florida. Provisions for the capture of profits can be implemented at the discretion of the FMFC, the governor, and his cabinet. Record keeping is administered by the state. The federal FMP requires state certificates and tags for fishing in federal waters.

Evaluation and Adaptation. Enabling legislation and administrative rules have been amended, with input from agencies and industry. The permit system is monitored by the state and by academic scientists.

Outcomes of the Tradable Permit System

General. The number of traps decreased from more than 1,000,000 prior to implementation of the program in 1992 to approximately 550,000 in 1996 due to reduced initial allocation of certificates and subsequent annual 10% reductions. Spiny lobster landings have remained stable and trap reductions are on hold for now.

Biological and Ecological Outcomes for the Fishery. Catches have remained stable, with a record high catch in 1995. Little ecological change is traceable to trap reduction, but the system is presumed to have positive economic and biologic benefits.

Economic and Social Outcomes for the Fishery. The total number of crawfish licenses has decreased from more than 4,000 to approximately 2,500, primarily due to exit of recreational fishermen from the license list and trap fishery (although they may not have been active in the first place). The cost of individual trap certificates has risen from the earlier range of $0.50-10.00 per trap to the present range of $50-70 per trap, in response to total trap reduction. The general configuration of fishing operations has remained constant.

Economic and Social Outcomes for Fishery-Dependent Communities. The cost of entry into the lobster fishery has increased due to the need to purchase trap certificates. Many recreational and marginal commercial fishermen have exited the trap fishery, although they are not precluded from participation.

Administrative Outcomes. The system was designed to be revenue positive, but has fallen somewhat short. Enforcement is perceived to be inadequate.

Current Perceived Issues. Some fishermen feel that the system was unnecessary in the first place because the issues involved were primarily social and economic, and some feel a loss of flexibility due to the cost of certificates and the burden of administrative requirements. There is concern about "hidden," localized consolidation of certificate ownership among groups of fishermen.

SUMMARY OF FOREIGN EXPERIENCE

The Icelandic Individual Transferable Quota Program

Viewed on a world map, especially one with a Mercator projection (which expands the size of high-latitude countries), Iceland does not appear small, but in terms of population it certainly is. The entire population of Iceland was just under 270,000 in 1996, or slightly more than one-half that of Alaska. It goes without saying that Iceland or the Icelandic economy can hardly be noticed in any international statistics, with one exception, fishing. In 1994, Iceland ranked as the fifteenth largest fishing nation, ahead of Spain and Mexico and just behind North Korea and Denmark. In per capita terms, Iceland is roughly comparable to its neighbors in terms of gross domestic product (GDP) and many other indicators of living standards. In 1994, the per capita GDP, measured as purchasing power parity, was 90% of the GDP of the United States, 83% of Denmark's, and 30% higher than in the United Kingdom.

The Icelandic economy is heavily dependent on its fisheries. About 73% of the value of goods exported in 1996 consisted of fish and fish products. Approximately 20 years ago, fishing accounted for as much as 90% of exports. The decline is largely due to the development (since the late 1960s) of energy-intensive metal production (aluminum and ferrosilicon), which accounted for 12% of exports in 1996. In terms of total receipts of foreign currency, the fishing industry is less important but still accounts for more than half (52% in 1995). Tourism is an increasingly important source of foreign currency (12% in 1995, as estimated by the Central Bank of Iceland), but neither tourism nor services are a net source of foreign currency income; in recent years the services account has been roughly in balance.

Like other developed economies, the Icelandic economy is characterized by a large service sector and a high degree of urbanization. In 1995, about two-thirds of all employment was in private and public services, while only 11% of the population was employed in fishing and fish processing, with these latter industries contributing about 15% of the GDP. The productivity of the fishing industry therefore appears reasonably high and is probably higher than that of other industries in Iceland. About 90% of Iceland's population lives in villages and towns with more than 200 inhabitants, and 60% lives in the capital city of Reykjavík and its suburbs. The towns and villages are located primarily on the coast and scattered almost all around the island, with fishing being a dominant

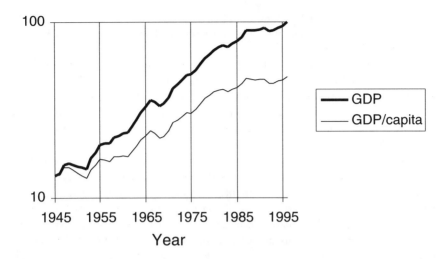

FIGURE G.15   Relative change in gross domestic product of Iceland (logarithmic scale).

industry in most of these. All towns and villages have road connections, although not necessarily good ones that are dependable in harsh winter weather. Iceland does not have any indigenous population of an ethnic origin different from the rest of the population. Immigration has been very limited.

The waters around Iceland used to be fished by both Icelanders and foreigners, with foreigners taking about one half of the catches of groundfish, of which cod is most important. Both world wars provided a temporary reduction in fishing pressure, as foreign fleets disappeared from the fishing grounds because of dangers to fishing vessels from military actions and the shift in manpower from fishing to fighting; when the wars were over fishing activities rebounded. The total catch of cod peaked in the mid-1950s, and cod became fully exploited and possibly overexploited as early as the late 1950s.

Icelanders are keenly aware of their dependence on the sea. The key to economic growth and rising standards of living was perceived to lie in ever-increasing fish catches. Accomplishing this was believed to require the elimination of foreign fishing around Iceland. The first attempt to reserve the fish stocks around Iceland for Icelanders was the passing of a law in 1948 claiming ownership of the living resources in the waters above Iceland's continental shelf. This law was inspired by the Truman Declaration of 1946 that claimed all resources on and beneath the seabed on the U.S. continental shelf as the federal property of the United States. On the basis of the 1948 law, Iceland extended its fishing limits several times in the 1950s and 1970s, sometimes before recognition of such extensions as international law and in open conflict with some of its neighbors, most notably Great Britain, which traditionally had been heavily involved in fishing around Iceland.

Economic growth in Iceland was somewhat variable but still rather impressive from World War II until the mid-1980s, about 4% per year (see Figure G.15). In 1987, the economy entered its longest period of stagnation since the Second World War; the per capita GDP was not much higher in 1996 than it had been in 1987, nine years earlier. In the last two or three years, growth has resumed again and was 6% in 1996.

The economic growth over most of the 40-year period from 1945 to 1985 was to a large extent driven by increases in fish catches and productivity in the fishing industry. Catches of cod, the most valuable groundfish, increased from about 200,000 metric tons per year in 1945-1948 to about 300,000 metric tons in 1954-1956. Around 1960, there was a herring bonanza, with catches rising from less than 100,000 metric tons per year before 1960 to 400,000-600,000 metric tons in 1962-1966. This development was brought about by the so-called power block, a device that made it possible to pull nets mechanically instead of by hand, which led to a vast increase in the size of nets (purse seines) and vessels.

Prior Economic and Social Conditions in the Fishery

Establishment of the 200-mile EEZ internationally legitimized the 1948 claim of ownership of the living resources above Iceland's continental shelf since virtually all of the continental shelf around Iceland is contained within the zone. As a result, Iceland gained virtually full control of the demersal stocks around the island, which are largely confined to the waters of the continental shelf. Icelanders rapidly replaced foreigners in the catching of cod and other demersal fish; foreign fishing around Iceland virtually came to a halt in 1976, and the Icelandic catches of cod increased from around 250,000 metric tons annually in 1971-1975 to an all-time high of 461,000 metric tons in 1981. In terms of resource conservation, however, little happened. Iceland embarked on an ambitious vessel construction program in the early 1970s and expanded rapidly into the void created by the displacement of foreign fleets. Only a few years after the establishment of the EEZ, reports began to document overcapacity of the fleet and overexploitation of Icelandic fish stocks, particularly cod. Gradually, it was recognized that it would be necessary to reduce fishing effort and the capacity of the fishing fleet in order to build up the stocks and increase the catches and profitability of the industry.

In principle, there were means available to control both fleet capacity and fishing effort in Icelandic fisheries. Investment in large fishing vessels was usually financed by loans from public investment funds. From 1977 onward, attempts were made to limit overinvestment in the fishing fleet by making it more difficult to obtain such loans, and after 1980, fishing vessels could no longer be imported without special licenses. In a number of cases, however, political pressure was applied to allow allegedly disadvantaged towns or villages to acquire fishing vessels even though the fleet was considered oversized. Measures to prevent overcapacity were not particularly effective; in the period 1977-1983 the value of the fishing fleet increased by about 17% (2.6% annually).

The first serious limitations of Icelandic fishing effort were temporary bans on fishing on particular grounds. Later, trawlers were limited in the number of fishing days per year, a number that declined over time. After the EEZ was established, the catch of cod was supposed to stay below the TAC, but the TAC was consistently exceeded despite the limitation on the number of fishing days.

By 1982, politicians and interest groups increasingly believed that more radical measures would be needed to limit effort. A vessel quota system was suggested in 1983 to deal with the ecological and economic problems of the fisheries; this system would divide the TAC among industry participants. The precise allocation of catches was debated, until it was agreed late in 1983 that each vessel was to be allocated an annual quota on the basis of its average catch over the past three years.

ITQ Program Development Process and the Transition to the ITQs

Herring and Capelin. The existing Icelandic ITQ program was preceded by developments in the pelagic fisheries. ITQs were first applied in the fishery for a local Icelandic herring stock. In the late 1960s, the Atlanto-Scandian herring stock collapsed, probably because of lower sea temperatures and excessive fishing pressure by Icelandic and Norwegian vessels allowed by the invention of the power block. Two smaller, local Icelandic herring stocks also collapsed, and one is believed to have disappeared altogether. The second herring stock was put under a moratorium in 1972, and after a partial recovery the fishery was opened again on a small scale in 1975. Vessels with a catch history were allowed to participate, but there were many more vessels than could be easily accommodated. There were regulatory attempts at limiting the number of vessels, such as requiring all herring to be salted on board, which disadvantaged the smaller and less seaworthy vessels, but ways were found around this; the regulations did not prevent the salting process from taking place when the vessels were in harbor. In 1976, vessel quotas were introduced, but each vessel received a very small allocation, due to the low TAC and the large number of vessels with a catch history. At first, the quotas were not transferable, but due to the small size of the quotas and the difficulty of fishing them profitably, transfers were allowed from 1979 on.

In 1980, vessel quotas were introduced in the capelin fishery, and in 1986 they were made transferable. This case is of some interest because it is sometimes alleged that ITQs cannot be applied to highly volatile fisheries. Yet, capelin is a short-lived species, and only one or two year classes are fished. Since the size of the year classes is highly volatile, the TAC is also very variable; there have been years when no fishing for capelin has been allowed. As in other Icelandic fisheries, capelin ITQs are determined as shares of the TAC. The Icelandic experience is not unique; in Norway, boat quotas are applied in the capelin fishery and other pelagic fisheries, but these quotas are not transferable. The Norwegian capelin fishery is even more volatile than the Icelandic one.

Groundfish. By 1982, Icelandic politicians and interest groups increasingly believed that radical measures would be needed to prevent collapse of the cod stock. Also, it was argued, a new approach was needed to reduce overcapitalization in terms of fishing vessels. An ITQ program was introduced by the Icelandic Parliament in 1983 to deal with the problems of the cod fisheries. When the ITQ program was first implemented, each fishing vessel over 10 tons was allotted a fixed proportion of future TACs for cod and five other demersal fish species. Catch quotas for each species, measured in metric tons, were allotted annually on the basis of this permanent ITQ share. Moreover, a new licensing scheme stipulated that new vessels could be introduced to the fisheries only if one or more existing vessels of equivalent size (in GRT) were eliminated in return. The ITQ program has been revised several times, but remains in force.

The ITQ program divided access to the resource among vessel owners on the basis of their fishing record during the three years preceding implementation of the program. Initially, ITQ shares could only be bought or sold undivided along with the fishing vessel to which they were originally allotted, although they could be leased relatively freely; that is, ITQ shares were not fully divisible or independently tradable. The ITQ program was initially put in place for only one year and was seen by many as a temporary emergency measure, to be abolished when the stocks recovered. It was, however, successively prolonged for two or three years at a time, and in 1990 a program of quotas of indefinite duration was emplaced. Quotas did not, therefore, constitute true private property rights. Nevertheless, the program introduced in 1984 was an individual *transferable* quota program, albeit one with restrictions on transferability. In 1990, several radical alterations were made to the existing ITQ program.

With the Fisheries Law passed by Parliament in 1990, the program was reinforced and extended into the distant future. First, the program was extended by allocating ITQ shares to approximately 900 smaller vessels (6-10 GRT) that had been fishing without restrictions. As a result, the number of ITQ holders increased by 156% (from 451 in 1990 to 1,155 in 1991). Second, the ITQ program was extended to include all major fisheries. Finally, and arguably most significantly, the ITQ program was made indefinite in duration, and ITQs became fully divisible and independently transferable, making them more akin to permanent property rights.

The program prior to 1990 had two major loopholes, the effort quota option and a general exemption for vessels smaller than 10 GRT. The effort quota option was included in the program partly to accommodate fishermen who felt they had been shortchanged by the initial quota allocation and partly to comply with demands for making allowance for differences in fishing expertise. Under the effort quota option, a vessel was allocated a certain number of fishing days, subject to an upper limit on how much cod could be caught. By taking this option, the catch record of a vessel could be improved, and its allocation under the quota option could be increased the next time quotas were allocated. Because the effort quota vessels usually succeeded in improving their records, the quota allocations for the vessels that consistently stuck to the quota option eroded over time. The effort quota was particularly popular in 1986 and 1987; in these years, less than 40% of the total catch was taken by vessels with catch quota allocations.

Because they were exempt from the quota program, the number of vessels of less than 10 GRT increased from 1,128 in 1984 to 2,023 in 1990, and their share of the total catch of cod increased from 3.3% in 1982 to 13.1% in 1991-1992. These vessels are typically owner operated with only one person on board. Over the years there have been periodic attempts to limit the fishing of these vessels. In 1986, the number of fishing days was limited. In 1988, vessels of 6-10 GRT were incorporated into the quota program, the number of such vessels was frozen, and a new vessel could be acquired only if another vessel was scrapped. Vessels of less than 6 GRT are an ongoing contentious issue, and attempts are still being made to incorporate their activities into the quota program.

The ITQ Program

According to the new fishing law in 1990, most fish stocks around Iceland were incorporated into the quota management program. For groundfish, the main exemption is that vessels less than 6 GRT are subject to limitations in the number of fishing days and an overall limit on how much they can catch, and only one-half of the catch taken by vessels fishing with longlines in the winter months is counted against the quota. Quota shares can be leased or permanently sold. Quota allocations are of an indefinite duration and could be revoked by the Icelandic Parliament at any time, but the prices of permanent quota shares suggest that this is not considered a very high risk; in the summer of 1997, permanent quota shares for cod were trading at about eight times the cost of renting quota shares for a year. Leasing of quota shares cannot be repeated indefinitely; in order to retain their quota share allocations, quota shareholders must fish at least half of their quotas every second year.

There are some restrictions regarding transferability and ownership of quota shares. In order to be eligible for holding quota shares, a person or company must have access to a vessel to which the quota shares are allocated. In most cases the person or the company owns the vessel, but cases have been reported in which a quota shareholder allocates quota shares to someone else's vessel. These cases are exceptional, however, and considered legally tenuous. Such leasing also goes against the spirit of this regulation, which is meant to prevent absentee ownership. If a quota is to be leased or sold to a vessel operating from a different place, the consent of the municipal government and the local fishermen's union must be obtained. This restriction does not seem to have had much effect; consent to quota leasing and selling appears usually to be given virtually automatically. Trading of quotas appears to be brisk; in the "fishing year" 1993-1994 the trading of cod and saithe quotas amounted to 44% and 96%, respectively, of the total catch. Note, however, that the same quota can be traded more than once.

For groundfish, there is a certain flexibility built into the program. Twenty percent of a year's quota can be shifted to the subsequent year without a penalty, but the overage is subtracted from the quota allocation in the following year. This is less injurious to conservation of stocks than it might appear; the exploitable stock of groundfish consists of ten year classes or more, which smoothes the pattern of catches over time despite large variations in the size of year classes.

Objectives of the ITQ Program

During the policymaking process when ITQs were set for the cod fisheries, ITQs were credited with several positive characteristics. It was argued that under an ITQ program,

- the size of fish stocks would be stabilized, both because it would be easier to ensure that the total catch stayed below the TAC and because harvesters would show greater responsibility in their treatment of the resource;
- fishing would become more efficient and overcapacity would be reduced;
- the quality of landed fisheries products would improve and, therefore, their economic value would increase;
- the management program would become simpler, less "political," and therefore, more efficient; and
- fishing would be safer, resulting in fewer accidents and injuries at sea.

Outcomes of the ITQ Program

Biological and Ecological Outcomes for the Fishery. Since the collapse of Icelandic herring stocks in the late 1960s, management of the herring stock has been very successful. The instantaneous fishing mortality rate for fish age 4-14 years has been kept moderate (it has varied between 0.15 and 0.35 per year and has been about 0.2 per year in recent years, which is probably near optimal for a relatively slow-growing stock consisting of more than ten year classes), and the stock has been built up gradually. Catches have also increased gradually, from less than 20,000 metric tons in 1975 to about 140,000 metric tons in the 1994-1995 season, but they fell in the 1996-1997 season to about 100,000 metric tons.

Whether or not ITQs have contributed to this recovery is difficult to determine. The primary tool for conservation is the TAC. To the extent the ITQs have kept the total catch below the TAC, they have helped promote conservation.

Management of the Icelandic cod stock, which is also under full Icelandic control, has been much less successful than management of herring, despite the fact that it is also part of the IFQ program and much more important for the Icelandic economy. The cod stock reached an all-time low in 1992 but has recovered somewhat since then. The primary reason for the population decline is probably an excessive TAC; the TAC set by the Icelandic government has consistently exceeded the recommendations of the Icelandic Marine Research Institute. Moreover, catches have surpassed the excessive TAC; in 1984-1996 the excess of catches over TAC was about 12% annually, and the excess of the TAC over the amount recommended has been of a similar magnitude. ITQs cannot be blamed for the depletion of the cod stock; on the contrary, the excess of catches over the TAC is due to exemptions from the quota program, for example, fishing by vessels less than 6 GRT and the hook-and-line fishery in winter. If anything, ITQs should have mitigated the situation by capping the catch of the vessels fully under quota. The effort controls that preceded ITQs were ineffective in keeping the catch below the TAC.

The government has consistently exceeded the recommendations of the Marine Research Institute because of the importance of the cod stock for Iceland's economy and an unwillingness to accept large short-term losses to achieve longer-term gains. Obviously, such trade-offs cannot be made without reference to economic and social conditions, so the recommendations made by the Marine Research Institute, which do not take such factors into account, are not sacrosanct. Nevertheless, it would seem that the Icelandic government has been unduly careless in its trade-offs between the present and the future; in 1992, the cod stock reached an all-time low, although the situation appears to be improving. In 1995, the TAC was set for the first time on the basis of a "TAC Rule," proposed in a bioeconomic study of the fishery. According to this rule, the TAC should be either 25% of the fishable stock or 155,000 metric tons, whichever is greatest. Except for the minimum of 155,000 metric tons, this appears to be a prudently conservative rule for a long-lived and slow-growing species such as cod. The historical minimum and maximum of the annual catch from the Icelandic cod stock are 169,000 (1995) and 546,000 (1954) metric tons per year, respectively.

One of the arguments for the development of ITQs emphasized that the privatization inherent in quota programs would encourage stewardship, as the new "owner" of the resource (or fishing rights) realized that he or she would benefit directly from caring for the resource. Discarding small and immature fish during fishing operations and highgrading the catch seem, however, to continue to be a serious problems in the Icelandic fishery and these problems may have escalated with ITQs. Since quotas are fixed and excessive catch is a violation of the law and subject to prosecution, a quota shareholder tends to land only the portion of the catch that generates the highest income.

It is not uncommon for vessels that have finished their cod ITQs to accidentally catch a few tons of cod while fishing haddock or another demersal species. If they land the cod, they must acquire an equivalent amount of cod ITQs to cover their catch to prevent loss of their fishing licenses. The price of ITQs leased for this purpose tends to fluctuate considerably in relation to supply and demand. According to many fishermen, this results in considerable amounts of dead fish being thrown back into the sea, especially toward the end of the fishing year when ITQs are scarce and the lease price is inordinately high. ITQs may, therefore, contribute to the waste of living resources, resulting in the erosion of ecological responsibility. It is difficult to estimate the scale of such practices, but it may be noted that the Icelandic Parliament expressed grave concerns and passed strict laws on the "treatment" of fishing catches in June 1996.

Economic and Social Outcomes for the Fishery. ITQs in the herring fishery have led to a substantial increase in economic efficiency. The number of vessels participating in the herring fishery has fallen drastically. In 1996, there were 29 vessels participating in the herring fishery, a decrease from the peak participation year of 1980, when there were more than 200 vessels. At the same time, the total catch has increased, from 53,000 metric tons in 1980 to almost 140,000 metric tons in the 1994-1995 season. It is noteworthy that the number of vessels having quota allocation is considerably higher than the number that actually participated; in 1996, 44 vessels with quota allocations did not participate in the fishery, and 6 participating vessels that had no quota allocation rented their quota from others. Fishing on this stock is seasonal, with a duration of a few months (October to February); all vessels fishing for herring are engaged in other fisheries for the remainder of the year.

The ITQ program appears to have improved the profitability of Icelandic fishing firms considerably. The price that fishing firms are prepared to pay for renting cod quota is a possible measure of this profitability. This price has risen from the equivalent of $US0.05-0.09 per kilogram in 1984 to $US0.90-1.00 per kilogram in 1994, and quotations from the summer of 1997 showed prices up to $US1.25 dollars per kilogram, which is more than one-half of the normal exvessel price. The increase in quota price is much greater than the rate of inflation, so the real price has undoubtedly risen substantially. It must be noted, however, that these figures reflect not only the increased profitability of fishing operations, but also the increasing scarcity of cod. It must also be kept in mind that the prices that fishing firms are prepared to pay for small amounts of quota do not necessarily reflect their long-term profitability; a vessel owner may be willing to pay an amount equal to the difference between the exvessel price of fish and the marginal operating cost, but the profit margin after taking into account capital costs may be much lower.

TABLE G.6   Profits as a Percentage of Gross Revenue

Year	Profit	Year	Profit
1980	−5.8	1988	−5.0
1981	−5.7	1989	−2.1
1982	−8.6	1990	1.4
1983	−9.1	1991	−0.3
1984	−9.2	1992	3.2
1985	−4.9	1993	4.0
1986	1.3	1994	5.1
1987	0	1995	4.0

Figures compiled by the Icelandic National Economic Institute show a rising profitability of the fishing industry in recent years (Table G.6). These figures have been compiled from annual accounts of harvesting and processing firms in the fishing industry (note that the largest firms are vertically integrated, so it probably makes most sense to consider the entire sector rather than just the harvesting sector, even if the quotas are allocated to vessels, thus primarily affecting the harvesting sector). The method of calculation corrects for price changes to correct for earlier years with rampant inflation that could distort profitability figures.

Analyses of productivity in the fishing industry carried out by Ásgeir Daníelsson (1997) at the National Economic Institute indicate a very substantial growth in productivity from the mid-1980s until the present. The total productivity of capital and labor in the fishing industry showed extreme sensitivity to changes in the size of the fish stocks in the period 1973-1985. This effect is expected; it is usually cheaper to catch a ton of fish from a plentiful stock than a depleted one. Since 1985, productivity has increased without a similar increase in the stocks. Although this is no proof that ITQs have increased productivity, it is certainly consistent with such an effect. Total productivity of capital and labor in the fishing industry increased by 67% over 1973-1990, despite the fact that the fish stocks were less plentiful in 1990 than in 1973. In the economy as a whole, the total productivity of capital and labor increased by only 20% over the same period.

As previously mentioned, the catch capacity of the Icelandic groundfish fleet had grown well beyond what was needed to catch available stocks by the 1970s. One of the main arguments for ITQs is that they should prevent overcapacity in the fleet or reduce it whenever it has developed. The number of decked vessels began to decline in 1990 when it had reached a peak of about 1,000 and had fallen to 800 by 1996. The size of the fleet in terms of GRT remained relatively steady, however, from 1990, the year ITQs were made indefinite in duration, but increased from 120,000 GRT in 1994 to 130,000 GRT in 1996. Thus, there has been a development toward fewer and larger vessels.

Whether or not ITQs have reduced the excessive capacity of the fleet in ITQ fisheries is still an open question. The size of the entire Icelandic fishing fleet in terms of GRT has increased slightly since 1990, the year when quotas became long term and could be expected to have an impact on fleet size. Some of the increase in capacity may be justified because of increased distant water fishing (the Barents Sea, Flemish Cap), which requires large vessels that can make long trips. Furthermore, reduction in capacity is a process that will (and should) take some time. The way this process works is that vessel owners will not invest in redundant vessels or those that are too large when the time comes to get rid of the old vessel. With poor second-hand markets in used vessels, fleet reduction may take a long time, because it is profitable to continue using an old vessel as long as it recoups its operating costs, even if it will never be replaced by a new vessel.

It may also be noted that the trend toward increased fishing in distant waters has been encouraged by ITQs, because the owners of the largest vessels have leased their groundfish quotas in Icelandic waters to other fishermen and dispatched the vessels to distant waters rather than chasing the limited amount of fish available in the Icelandic EEZ and making little or no contribution to the overall value of the fishery. Such an extension of Icelandic fishing outside Icelandic waters, however, creates classic common-pool problems internationally. Vessel owners are racing for fish on disputed fishing grounds, inviting conflicts with foreign governments.

The Icelandic government initiated a buyback program in 1994, aimed at removing vessels from the fisheries. The existence of this program indicates that the expectations of the ITQ program and the market approach to management for eliminating or reducing overcapacity have not been fulfilled.

Effects on Equity. One way to examine the changing social distribution of quotas is to arrange quota holders into discrete groups, based on the size of their quota share, and examine the number of quota holders in relation to the size of their quota shares. Gísli Pálsson and Agnar Helgason at the University of Iceland have provided such an analysis for the demersal fishery (Pálsson and Helgason, 1995). To simplify, they distinguish between "giant" quota holders (the group with the largest quota shares holding more than 1% of the total quota each), "large" quota holders (holding 0.3-1%), "small" quota holders (with 0.1-0.3%), and "dwarves" (with less than 0.1%). Figure G.16 shows distributional changes of quota shares among vessel owners for an 11-year period (excluding, for the sake of comparison, 6-10 GRT vessels that were incorporated into the system in 1991). There is a steady decrease in the total number of quota holders. A gradual increase in the number of giants is concurrent with a decrease in the numbers of the other three groups.

A more telling way of elucidating distributional changes is to compare the aggregate permanent shares of the groups defined above. Figure G.17 indicates

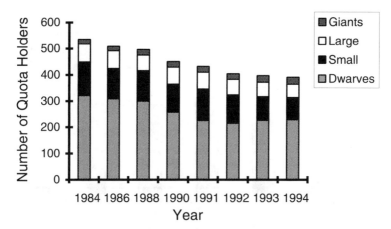

FIGURE G.16   Number of quota holders in Iceland (1984-1994).

changes in the relative distribution of quotas for the same 11-year period. Evidently, while the giants have grown in number through the years, they have been accumulating quotas to a disproportionate degree. At the same time, the shares of other groups have diminished in relation to their reduced numbers.

These data indicate a sizable increase in the level of inequality in the distribution of quotas from 1984. Many vessel owners have dropped out of the program, and a large majority of these were the smallest operators. At the same time, quotas are becoming concentrated in the hands of fewer vessel owners and companies. Only the giants average quota share shows a substantial increase, going from 1.64% to 1.91%.

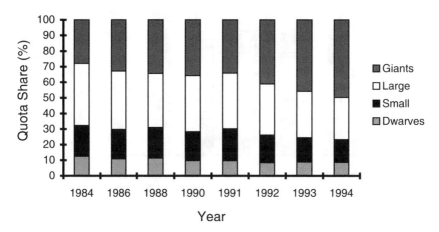

FIGURE G.17   Quota distribution in Iceland (1984-1994).

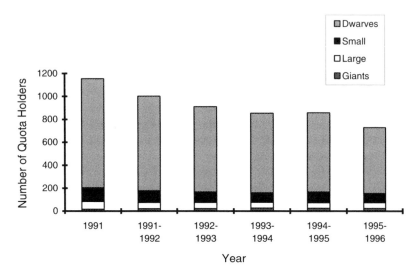

FIGURE G.18   Number of quota holders in Iceland (1991-1996).

The 1990 fishery law resulted in the inclusion of 704 new small-scale ITQ holders not included in the figures presented.   Figures G.18 and G.19 show changes in the number of quota holders and the distribution of holdings since quotas became fully transferable.   This time, the 6-10 GRT vessels that were incorporated into the system in 1991 are included.

During this period, the number of quota holders was reduced from 1,155 to 729.   Only the group of giants has grown in terms of number.   At the same time,

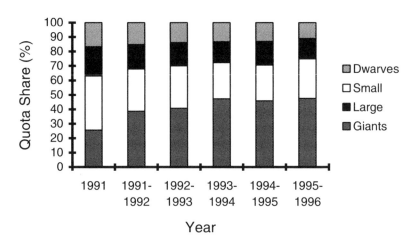

FIGURE G.19   Distribution of quotas in Iceland (1991-1996).

the quota share of the giants increased from 25.6% to 47.5%. The share of the dwarves, in contrast, decreased from 16.8% to 11.05%. Together giant and large quota holders own 75% of the total quota in the demersal fisheries. Currently, 24 giants own almost half of the total quota. The share of the largest quota holder is about 6%.

If the TAC is reduced, ITQ shares effectively become devalued; that is, all operators in the industry suffer reductions in the amount of fish they are permitted to catch, even though their actual ITQ share remains constant. This aspect of the ITQ program is highly relevant for the distributional developments. After bleak estimates of the fish stocks in Icelandic waters by marine biologists, the Ministry of Fisheries has repeatedly reduced the TACs (particularly for cod). As a result, many small companies have found themselves increasingly left with insufficient catch quotas to keep their vessels active throughout the fishing year. To give some indication of the extent of these devaluations, a vessel owner who controlled an ITQ share in cod of 0.1% (the upper limit of a dwarf) was entitled to approximately 254 metric tons of cod in 1987, 200 metric tons in 1991, but only 106 metric tons in 1994.

Some, if not all, of the giant companies are owned by a large number of shareholders. One could argue, therefore, that the concentration of quotas masks a more egalitarian distribution of access and ownership. However, it is possible that some individuals own shares in several different companies; thus, the distribution of ownership of quotas may be even more unequal than the raw figures on distribution indicate. Also, the distribution of holdings within the largest companies may be very uneven, with few individuals controlling the majority of the shares. Finally, even though shareholders turned out to be more numerous than before, in actual practice a small group of managers has immense power.

Two things must be noted, however, at this juncture. First, some concentration of quota holdings is inevitable and desirable. The purpose of the ITQ program is to restrict access and reduce overfishing and overcapacity. This is bound to mean that not everyone who desires will be able to participate in the fishery, and the reduction of overcapacity means that some participants will have to leave the industry. Second, concentration of quotas in large firms is probably an inevitable consequence of increasing the efficiency of the industry. There are most likely economies of both scale and scope in fishing and fish processing, and the fact that the quotas have been bought and sold freely in an open market indicates that they have gravitated to the most cost-effective firms. Even the largest Icelandic fishing firms are still small when compared with foreign companies. Although it is true that the large Icelandic fishing firms can be quite powerful in their local labor markets, the Icelandic economy cannot afford to miss the advantage of such economies of scale. More than 95% of all fish caught by Icelanders is exported, and more than one-half of all the foreign currency earnings of Iceland result from selling fish and fish products to markets dominated by large wholesalers and retail chains that have a wide range of alternative suppliers. Only the largest firms in Iceland have attempted to export on their own, and the remainder are partners in a few large (on an Icelandic scale) exporting firms that sell Icelandic products abroad.

Effects on Remuneration and Relative Power. Vessel owners have been permitted to lease their ITQs from the onset of the program. ITQ leasing was originally proposed by administrators as a way for vessel owners to fine-tune their operation to meet short-term needs arising from unexpected "devaluations" of ITQ shares; fluctuations in local, regional, and national markets; and bycatch problems (for example, by trading haddock ITQs for cod ITQs). At first, ITQ leasing did not seem to be a particularly common practice, and it was probably undertaken mainly on a small scale by operators who needed extra ITQs after a particularly successful fishing season. The lessors in most of these cases were operators actively engaged in using their own ITQs. Over time, however, some ITQ shareholders came to realize that considerable profits could be earned through leasing ITQs on a larger scale, particularly with many fishing operations suffering from the devaluation of ITQ shares resulting from repeated reductions in the TAC for cod after 1988.

Recently, new and more formalized modes of ITQ leasing have begun to emerge. These transactions involve long-term contracts between large ITQ holders and smaller operators, in which the former provide the latter with ITQs in return for the catch and a proportion of the proceeds. One such arrangement, usually referred to as "fishing for others," is becoming increasingly widespread within the industry. Invariably, in such arrangements, the supplier of the ITQs is a large vertically integrated company that controls two or more trawlers and a processing plant. The smaller operator's vessel fishes the ITQs and delivers the catch to the supplier's processing plant in return for a payment that usually amounts to about 50-60% of the market value of the catch. There are limits, however, to how far this activity can develop. Fish processing firms must own vessels in order to own ITQs, and they are required to have fished at least one-half of their own quotas on their own vessels over the previous two years.

Strictly speaking, then, there is no lease price paid up front for the ITQs. However, the small-scale operator is effectively paying a lease price of up to one-half the value of the catch. Understandably, the lessee vessel owners cannot make the same level of profits when fishing for others as they can when fishing their own ITQs. Their outlay is identical in both cases, but when fishing for others their income is cut by 40-50%. As a result, they try to compensate for their losses by reducing the shares paid to the crew members. Fishermen receive a share of the value of the catch adjusted to the price of oil on the international market. Before the fishermen's shares are calculated, however, the vessel owner is permitted to deduct maintenance costs from the proceeds of the catch. Increasingly, the lessee vessel owners have resorted to reckoning the fishermen's shares from the amount left *after* the lease price has been subtracted from the value of the catch. The result is that fishermen working for lessee companies may suffer up to 50% wage reductions. This is not, however, permitted according to the wage contracts between vessel owners and the fishermen's union.

The typical lessee operator is either an owner of a relatively small vessel that has finished its own annual supply of ITQs or the owner of a vessel that has virtually no ITQs of its own and is operated solely on leased ITQs. Through ITQ leasing, vessel owners with small ITQ holdings manage to prolong their fishing operations throughout the year. Moreover, by lowering the shares of their crews, they are just about able to make such practices economically feasible. For the suppliers of ITQs, however, leasing represents a rather lucrative business. By leasing its ITQ shares, a company can free itself from the expenses of actually catching the fish, while still procuring up to half the market value of the resulting catch. Moreover, it keeps the company's processing facilities well supplied. During recent years, dwarves and small ITQ shareholders have been the typical lessees. This reflects, on the one hand, the distributional changes described earlier and, on the other, the emergence of the relations of tenancy associated with fishing for others. In a number of cases the lessors are integrated fishing firms that have leased part of the quotas of their trawlers and dispatched the latter to distant waters such as the Barents Sea.

Evidently, then, the Icelandic fishing industry is undergoing an extensive re-structuring process, in which large vertically integrated companies have strengthened their position while smaller operators are being marginalized or forced out of business. Some of the small operators seem to be persevering by entering into contracts to fish for larger ITQ holders.

Effects on Property Rights. ITQs remain, according to the first clause of the 1990 fisheries management legislation, the "public property of the nation." During debates on the 1990 fisheries laws, some members of the Icelandic Parliament raised doubts about the "legality" of the ITQ program, arguing that proposed privileges of access might imply permanent, private ownership that contradicted some of the basic tenets of Icelandic law regarding public access to resources. Lawyers concluded that the kind of ITQ program under discussion in Parliament was in full agreement with the law and that ITQs represented temporary privileges, not permanent private property. The laws that eventually were passed reinforced such a conclusion by stating categorically that the aim of the authorities was *not* to establish private ownership.

The issue of ownership, however, is still contested. The Icelandic tax authorities have decided, one may note, that ITQs are to be reported as "property" on tax forms and that the selling of ITQs involves a form of "income." Some evidence indicates that in legal practice, quota shares are gradually acquiring the characteristics of full-blown private property, despite legal clauses to the contrary. Owners of quotas may write them off for tax purposes over five years. In practice, quota shares are passed on as inheritance from one generation to another. Normally, however, the quota shareholder remains the same (a fishing firm), although a new generation is taking over. A case was contested in courts in which a woman divorcing her husband, the owner of a firm with a sizable quota holding, demanded her share of the estate. The Icelandic Supreme Court ruled on December 3, 1998 in favor of the woman, which may be seen as one further step to the formal recognition of quota shares as private property. Thus, the use rights of fish resources are becoming increasingly entrenched as private property while the resources themselves (i.e., the fish stocks) are proclaimed as being publicly owned. The implications of such a contradictory situation are unclear. Could the Icelandic Parliament, for example, change the ITQ program fundamentally, without compensating quota shareholders for rights they would perceive as having lost? The Icelandic constitution protects the holders of property rights if they can prove their ownership. The issue of protection of resource property rights is being debated by the Icelandic Parliament.

There has been a long discussion over whether quotas can be used as collateral for obtaining loans. Without this possibility, it is considerably more difficult to obtain a loan to buy a fishing vessel, because a vessel without ITQ is worth much less than a vessel with ITQ. The law is unclear on this point, stating on the one hand that quotas cannot be used as collateral but on the other that vessel owners cannot sell their quotas without the consent of whoever has a lien on the vessel. Again, if economic and legal practice recognizes quotas as collateral, it will be a further step in the recognition of quotas as private property, undermining the significance and effect of the statement in the current law on public ownership.

Recently, Örlygsson (1997) analyzed the legal status of fishing quotas. Among his conclusions are the following:

• Quota shares are not to be regarded as the private property of quota holders.

• Quota shares may, however, achieve the characteristics of private property as time passes. Quota shareholders may gain increased legal protection of their shares under clauses concerning the "right to work."

• Quotas are not in any meaningful sense the property of the "nation." The legal clause on fish stocks being the property of the nation expresses the intent of the lawmakers that the resource be managed for the benefit of the public (the equivalent of the U.S. doctrine of the public trust), but the nation does not constitute an owner. On the other hand, with changes to the Icelandic constitution the *government* may become an owner of the resource.

• The existing program of quota management is in accordance with Icelandic law.

• The government may legally withdraw quota shares and cease to issue quotas without compensation to existing quota shareholders.

• A resource fee would be allowed under the current law. Community quotas and ceilings on the size of quota holdings would also be in accordance with the law.

The 1990 fisheries law is still controversial, however; on December 3, 1998, the Icelandic Supreme Court unanimously concluded that the clause in existing fisheries laws (Art. 5, 38/1990) which privileges those who derive their fishing rights from ownership of vessels during a specific period (during which their "fishing history" was established) is unconstitutional. This privilege, the Court concludes, violates both the Constitutional rule against discrimination (Art. 65) and the rule about the "right to work" (Art. 75). The Court reasoned that while temporary measures of this kind may have been both necessary and constitutional in the beginning, to prevent the collapse of fish stocks, the indefinite legalization of the discrimination that follows from Art. 5 38/1990 is not justified. That Article, in principle, the Court went on, prevents the majority of the public from enjoying the right to work, and the relative share in the common property represented by the fish stocks, to which they are entitled. The implications of the Court's decision will, no doubt, be far-reaching.

Effects on Communities. Some companies that have encountered economic difficulties have sold their quota to companies located elsewhere. Also, when TACs are decreased, some quota holders sell out because their share is not viable anymore. Whatever the reason for movement of quota out of communities, it affects the entire community (Pàlsson and Helgason, 1995). This has caused employment problems and eroded the tax base of certain municipalities, while companies in other municipalities have increased their quota holdings. The pattern of changes in the regional distribution of quota, however, is a complex one. The main accumulators of quota are companies in the larger towns of the northern part of Iceland. Small communities, with fewer than 500 inhabitants, have lost a much larger share of their quota than larger communities. In some cases, rural municipalities have tried to reverse the process of decline by buying or leasing quota or investing in local fishing firms.

Loss of quota in the smallest communities is particularly painful. For one thing, often there are no alternative jobs. Also, smaller communities are characterized by small vessels and household units. Frequently, an entire family is engaged in the operation of a small vessel; the housewife is likely to take care of financial accounts as well as baiting lines and clearing nets. Once the opportunity for fishing has been sold (sometimes because the quota share is too small to provide available fishing operation), the family as well as the fisherman becomes unemployed. At the same time, the family's house is likely to decrease in value because other residents are leaving the area, limiting the chance of establishing a household in another place.

Effects on Safety. Studies of fishing in Iceland and several other contexts—including the United States (notably Alaska), Canada, New Zealand, and Great Britain—have found an excess of work-related deaths and injuries in fisheries. Analyses at the University of Iceland show that between 1966 and 1986, 132 fishermen had fatal accidents at sea (108 died by drowning) (Rafnsson and Gunnarssdóttir, 1992). The number of person-years for the same period, a mea-

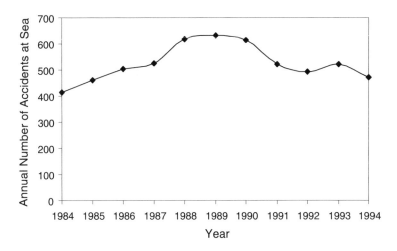

FIGURE G.20   Reported accidents at sea. Quotas were introduced in various fisheries between 1976 and 1983. SOURCE: G. Pálsson, unpublished analysis.

sure of the number of fishermen at risk, was 147,649, which suggests a mortality of 89.4 per 100,000 person-years. The mortality rate for all accidents did not change appreciably during this period. As mentioned earlier, one of the objectives of the ITQ program was to make fishing safer, resulting in fewer deaths and injuries at sea. It is difficult to evaluate the impact of the quota program in isolation because many other developments have taken place at the same time (the structure of the fleet has changed, as well as the number of fishermen at risk; there are new regulations on safety precautions), and no systematic study has been conducted. Interviews with the people responsible for recording and analyzing accidents at sea do not, however, indicate significant changes in terms of safety and accidents. Data provided from the National Insurance Institute show that the frequency of accidents at sea (including non-quota fisheries in international waters) increased from the onset of the ITQ program to 1994 (see Figure G.20). Additionally, cod are plentiful in the winter, so fishing effort is concentrated then, despite the bad weather. There may also be pressure under the quota program for absentee owners to disregard crew safety.

Current Perceived Issues.   Current discontent with the ITQ program can be summarized in several points:

1. Many people oppose the privatization of fishing rights entailed by ITQ management because fishing rights are still very much intertwined with the symbolic notions of national sovereignty and equity. The "Cod Wars" with Britain in

the 1970s, it is often argued, established a common, national fishing space within a 200-mile limit, not a privatized territory for a few vessel owners with quota.

2. The initial allocation of quota to vessel owners is often criticized. Crew members point out that prior to the program, fishing was typically regarded as a "co-venture" of vessel owners and crew. Now, they say, vessel owners have become millionaires, while crew members, some of whom have a long fishing history, are disenfranchised. In some contexts, for example, in the Alaska halibut and sablefish fisheries, the allocation of crew shares to individual fishermen would face practical difficulties, due to inadequate records on the fishing history of crew. In Iceland, such difficulties were negligible. Records on crew are just as good as those on vessel ownership. The fact that crew have been left out in the initial allocation in most cases, in Iceland as elsewhere, seems to reflect a common bias toward capital ownership in the theorizing about and applications of ITQs.

3. In the public debate, the idea that the fishing industry should pay for the privilege of holding harvesting quotas has a long history, and proposals to this effect have also been framed in the Icelandic Parliament, without obtaining sufficient support. At the present time, industry pays very little in the way of user fees; a fee of up to 0.4% of the catch value is collected to defray the costs of ITQ regulations. The fishing industry is, not surprisingly, adamantly opposed to any collection of fees beyond what would be needed to cover the cost of fisheries management.

4. Many Icelanders are wary of the rapid concentration of ITQs in the hands of large vertically integrated companies. A committee appointed by the Ministry of Fisheries recommended that a ceiling for any single quota holder be fixed by law. Parliament decided in 1998 to set the limit at 10% for cod and haddock and 20% for other species.

5. There is much resistance to profit-oriented exchange of fishing rights. Vessel owners who engage in such transactions are labeled "quota profiteers." One recent survey established that 60% of vessel owners believed that the buying and selling of ITQs was morally wrong.

6. There is much concern with the emergence of the relations of dependency associated with fishing for others. Often, heavily loaded feudal metaphors are used to describe this state of affairs. In public discussion, the large firms that have been accumulating quota shares are habitually referred to as "quota kings" or "lords of the sea." The lessor quota kings are likened to medieval landlords, and conversely, small-scale lessees become "tenants" or "serfs." In January 1994, fishermen went on a national strike, protesting against the ITQ program, especially the effects of the so-called tenancy system. The leading slogan they employed was "No More Profiteering!" To many fishermen, this was a battle aimed at eliminating the ITQ program. As it turned out, the strike resulted in a two-week standstill in the fishing industry. Ultimately, the strike was terminated by temporary laws that forced fishermen back to work. Not content with this turn of events, fishermen went on strike again in May 1995 and once again in Febru-

ary 1997. A committee established by the government to resolve the conflict suggested changes to the ITQ program in line with some of the fishermen's demands. For one thing, all transactions in quota shares should become transparent and pass through a quota market. Also, fishing and processing should be more clearly separated through the establishment of an office for registering the market prices of fish. Laws that incorporated these changes terminated the strike.

7. One of the arguments for quota programs is that they obliterate the bureaucratic "jungle" of temporary regulations typical for traditional methods of dividing access, making it possible to avoid endless revisions and local debates. It is often argued, however, that the complexity of bureaucratic practices and regulations has not been significantly reduced under quota programs.

8. Finally, there is much concern with the threat of municipal bankruptcy in fishing villages that have lost most or all of their quota, with massive unemployment and dissolution of communities. There are demands for effective limitations on quota transfers between regions and communities, to avoid extreme uncertainty in employment. Such limitations are applied in Norwegian fisheries, for example.

In summary, many Icelanders seem to have a sense of having been cheated by the designers of fisheries policy, drawing attention to the failures of the democratic political process. The critical decision on ITQs in 1983 was implemented in haste without sufficient public political debate. Neither then nor later has the electorate been presented with clear alternatives for fisheries management because most of the political parties have been divided on the issues involved. Originally, the ITQ program was presented as a short-term "experiment." Given, however, the relative irreversibility of social transformations of this kind, the ITQ program was hardly the innocent experiment that policymakers tended to speak of. Moreover, the program was presented as a fairly limited and technical exercise. There were no serious indications or warnings of the large-scale structural transformations that later took place. In fact, some of the proponents of the program indicate that a pure market-based program was introduced in moderate doses to avoid public rejection at an early stage.

It is one thing to abolish an ITQ program and quite another to change it. Although there is much public discontent with the program, there is no consensus concerning a potential alternative. Many Icelanders, however, insist that certain changes have to be made to the existing fisheries legislation.

New Zealand's Individual Transferable Quota System

Prior Regulatory Conditions in the Fishery

The legislated management of New Zealand's fisheries began with the Fisheries Act 1908, which remained in force until 1983. The act provided the statutory authority for regulatory policies aimed at the biological conservation and protection of fishery resources. Prior to the declaration of the 200-mile New Zealand EEZ in 1978, marine fisheries were small and confined to an inshore domestic industry, fishing mostly in depths of less than 200 meters. New Zealand extended its jurisdiction initially to 3 miles, then to 12 miles. Outside the territorial sea, the fisheries were exploited by foreign fishing vessels, primarily from Japan, Korea, and the Soviet Union.

Before the introduction of ITQs in 1986, a number of fundamental changes were made in the way that fisheries were managed. From 1938 until 1963, the inshore fishery was managed using a restrictive licensing system involving extensive gear and area controls that required vessels to fish from specific ports. In 1963, the inshore fishery was completely deregulated. During the period of open entry that followed, the federal government encouraged investment in the fishing industry through investment incentives, capital grants, allowances, and tax breaks. The domestic industry expanded rapidly during this period, laying the foundation for the development of the deepwater sector following the declaration of the EEZ. The government's economic objectives remained unfocused, and its policies encouraged overcapitalization of the fishing industry.

In 1978, a moratorium was introduced on the issuance of additional permits to fish for rock lobsters and scallops. This was followed in 1982 by a moratorium on the issuance of new permits to fish for finfish. The moratoriums limited entry into the fisheries but did not limit fishing power, which continued to increase. In 1979, a number of separately managed limited entry fisheries were established for rock lobsters. Licenses were nontransferable, and entry to and exit from the fisheries were managed by a government licensing authority. This system of limited entry failed to control the increase in effort and investment in these fisheries.

When the EEZ was declared in 1978, fisheries inside and outside the 12-mile territorial sea were initially managed separately. For the zone outside 12 miles, a policy of limited domestic expansion, joint venture arrangements, and licensing of foreign fleets was applied. The moratorium continued to operate inside 12 miles.

Subsequently, the Fisheries Act 1983 was passed. This new act consolidated previous fisheries legislation and introduced the concept of fishery management plans. The act, and by extension the management plans, recognized the goal of maximizing the economic returns from fisheries as well as biological objectives. The act did not, however, integrate the economic goals with the goals of biological conservation. The suite of regulatory controls that were used to address the biological goals was largely retained. Most remain in force in 1999.

Also in 1983, the government issued a Deepwater Fisheries Policy that introduced a system of enterprise allocations for the deepwater trawl fisheries based on company individual quotas. This system created the basis for the introduction of the ITQ program into the inshore fisheries.

In 1986, the government passed an amendment to the Fisheries Act 1983 that allowed for the introduction of the ITQ program in the inshore fishery and for its broader application to the deepwater fishery.

Prior Biological and Ecological Conditions in the Fishery

Prior to the introduction of ITQs in 1986, there was a widespread perception within government and industry that the inshore fisheries were biologically overfished.[5] However, because there had been only limited stock assessment research before this time, this perception was supported by little quantitative information.

Initial TACs for most of the inshore finfish stocks were based on average reported landings during periods when the catches were considered to be sustainable. This was a largely qualitative rather than quantitative assessment. For a number of the prime inshore species, the initial TACs were set at levels up to 75% below the catches reported immediately prior to the introduction of ITQs.

Prior Economic and Social Conditions in the Fishery

Prior to the introduction of ITQs in 1986, there was a widespread perception within government and industry that the inshore fisheries were also economically overfished.[6] Again, there was limited economic information to support this perception. The only published information available was a statement that the harvesting sector was overcapitalized by about $NZ28 million, based on insured value (Anon., 1984).

Until recently, New Zealand's economy depended mostly on the primary production industries of agriculture, horticulture, forestry, and fishing. In recent years, there has been a rapid growth in tourism in rural areas. Because of this mixture of industries there are no communities that depend solely, or even primarily, on fishing. One notable exception is the Chatham Islands where farming and fishing, and increasingly tourism, are the mainstays of the islands' economy.

Problems and Issues That Led to the Consideration of an ITQ Program

The problems and issues that led to the introduction of the ITQ program were based on the perception that New Zealand's fishery resources were suffering from biological and economic overfishing. The industry was overcapitalized, crippled by "excessive government management intervention" (Crothers, 1988), and subject to rapidly declining economic performance. Recreational fishermen were also concerned about the decline of the amateur fishery.

---

[5] If total harvest from a stock can be increased by a reduction in the amount of fishing, the stock is biologically overfished.

[6] If total profits from fishing stock can be increased by a reduction in the amount of fishing, the stock is economically overfished.

Objectives of the ITQ Program

During the development of the proposed ITQ program, the government issued a consultation document titled *Inshore Finfish Fisheries—Proposed Policy for Future Management* (Anon., 1984). This document clearly stated the objectives and aims of the proposed ITQ program:

- To achieve the long-term, continuing, maximum economic benefits from the resources; and
- To preserve a satisfactory recreational fishery.

A proposed management regime was developed and used as the basis for discussion. Within this management regime, ITQs were seen as the best mechanism for maintaining the balance between the harvesting sector and the fish stocks, delivering government restructuring assistance, and maintaining profit and equity within the industry.

The government proposed a management policy with the following characteristics:

- Future management of the inshore fishery would be by ITQs.
- Restructuring assistance would be provided by the government under a competitive tendering scheme to those who voluntarily reduced catches of key species.
- Resource rentals would be introduced with the ITQ property right.
- Future adjustment of TACs would be by purchase or sale of quota by the government by competitive tender.
- The program would be introduced first into the finfisheries and subsequently into the rock lobster and shellfisheries.

The government decided that it would not consider any assistance for the inshore fishery unless it was assured of the following:

- There was a high level of support, cooperation, and involvement from industry in the proposed policy.
- Benefits gained through catch reductions and restructuring assistance were permanent.

The aims of the proposed management policy using ITQs as the main management mechanism were as follows:

- To rebuild fish stocks to their former levels:
- To ensure that catches would be limited to levels that could be sustained over the long term;
- To ensure that these catches would be harvested efficiently with the maximum benefits to fishermen and the nation;

- To allocate catch entitlement equitably based on fishermen's commitment to the industry;
- To manage the fishery so that fishermen would retain maximum security of access to fish and flexibility of harvesting;
- To integrate the ITQ programs of the inshore and deepwater fisheries;
- To develop a management framework that could be administered regionally in each fisheries management area;
- To assist the harvesting sector financially to restructure its operations to achieve the above aims; and
- To enhance the recreational fishery.

ITQ Program Development Process and the Transition to ITQs

The important features of the ITQ development process and the transition to the ITQ program have been described in detail by Clark and Duncan (1986), Clark et al. (1988), Crothers (1988), Dewees (1989), Boyd and Dewees (1992), and Davies (1992).

The following is a summary of the important steps leading up to implementation of the ITQ program.

1. A long period of consultation occurred between 1983 and 1985 in which possible solutions to address the biological and economic overfishing issues facing the industry were explored by government and industry. Two broad types of solutions were considered: (1) regulatory intervention based on input controls and (2) intervention to establish long-term economic management principles, followed by the reductions of government interference to allow market forces to operate within biologically sustainable levels. After consultation, ITQs were chosen as the preferred management option, with industry support.

2. During 1982, a moratorium on new entrants into the inshore fishery was implemented. During 1983-1984, regulations prohibited the participation of part-time fishermen. A participating fisherman had to earn a minimum of 80% of his or her income from fishing or $NZ10,000 per year (or both) from fishing. Fishermen not meeting this criterion were excluded from fishing by having their permits removed.

3. In 1982, an enterprise allocation scheme for seven important species in the deepwater and offshore trawl fisheries was introduced.

4. In 1986, the Fisheries Amendment Act 1986 was passed, making the introduction of ITQs possible.

5. TACs were established for the inshore and deepwater finfish species that were included in the program.

6. TACs were allocated among fishermen based on their catch history over a period of qualifying years.

7. The government provided adjustment assistance to the fishing industry in the form of a buyback of quota entitlements in certain fisheries.

8. A computerized reporting system was implemented in 1986, including monthly reports from fishermen and fish buyers, catch logs for vessels, and reports of all quota transfers.

9. The ITQ program was implemented on October 1, 1986, and the tendering process was completed by the end of 1986.

The ITQ Program

The important features of the early years of the quota management system (QMS) have been described in detail by Clark and Duncan (1986), Clark et al. (1988), Crothers (1988), Dewees (1989), Boyd and Dewees (1992), and Davies (1992).

ITQ Management Units. As of October 1, 1997, there were 30 species or species groups in the QMS. The fishery for each species in the QMS is divided into a number of different fishery management units, officially designated as *Fishstocks.* The number of Fishstocks ranges from 2 to 10 for any given species, with a total of 179 different Fishstocks in the QMS. There are 10 different quota management areas (QMAs) in the QMS, and each Fishstock is composed of one or more QMAs. The government plans to introduce all remaining commercially harvested species into the QMS, which will increase the number of Fishstocks by more than 100 from the present number.

Initial Allocation of ITQ. The initial allocation of ITQs was made free of charge. ITQs were allocated in perpetuity and authorized the holders to take specified quantities of each species annually in each quota area (as opposed to a percentage share of an annually adjusted TAC).

Except for the species included in the enterprise allocation system introduced into the deepwater and offshore fisheries in 1983 that is described below, initial allocation was made on the basis of catch history, modified by the results of a buyback scheme and administrative reductions made to match effort more closely to the available resource. Fishermen who held permits in May 1985 were advised in mid-1985 of their individual catch by species for the three years ending in September 1984. They were allowed to choose two of these three years, the average of which would form their ITQ. They had the right to object to these catch histories before regional objections committees on the basis of statistical error, changed fishing patterns, or distortions of their normal catch record due to vessel breakdowns or bad health. Of 1,800 individuals notified of their catch histories, objections were lodged by 1,400. After the objections were heard, fishermen received notification of their catch histories, some of which had been amended, and provisional allocations of quotas shares were made. These provisional allocations could be amended by the government administratively, reducing provisional quota shares among all fishermen on a prorata basis, or by fishermen offering quota shares back to the government as described below. A Quota Appeal Authority was established to hear objections to the provisional quota share allocations. More than 1,100 appeals were lodged with the authority.

In March 1982, an enterprise allocation system was introduced for seven important species in the deepwater and offshore trawl fisheries. Initial allocations to this sector were made on the basis of investment in catching, onshore capital, and onshore throughput. These allocations were converted to ITQs in 1986. Although catch history was used as the first mechanism for existing fishermen, in the case of some species, for example hoki and orange roughy, the Crown also held quota shares (the difference between the ITQs allocated by catch history and the TAC) and these were allocated by tender. At least one large company gained its foothold in the deepwater fishery by being a successful tenderer. The allocation of some of the ITQs by tender had an effect on the perception of the strength of the property rights involved. This was used as an argument by industry that those rights were stronger than they would have been if ITQs were simply a recognition of the fishing history of an individual.

The most important aspect in the initial implementation of the ITQ program was the adjustment assistance offered in the form of a buyback of the fixed quota entitlements. The mismatch of the fleet capacity to available catch and the need to achieve significant reductions in the catches of many inshore fish stocks were major problems for the introduction of ITQs. Reductions in catch of as much as 83% for the inshore Fishstocks were required to ensure biological sustainability. Because the reductions in catch levels would not have been spread evenly across the species, the historic catch mix of individual fishermen would have been upset, leading to economic and bycatch problems.

The government provided adjustment assistance to the fishing industry by purchasing all or a portion of participants' quota shares. The sum of ITQs based on catch histories was often too high and exceeded the TACs for many inshore species. The government offered to buy back enough of the provisional allocations based on catch histories so that the sum of the remaining ITQs did not exceed the TACs. The use of voluntary reductions through a buyback scheme allowed individuals to decide whether to remain in the fishery at their historical catch level, sell out, or restructure their operations by selling only part of their provisional quota shares. Two tender rounds took place. The first was a competitive tender in which fishermen made bids to leave the fishery or reduce their effort. This round succeeded in establishing the price levels at which fishermen were seeking to retire quota. However, the full reduction to match the sum of the ITQs to the TACs was not achieved, so a fixed-price offer 20% below that determined by the competitive tender was made to those remaining. This still did not achieve the target reductions for a number of species, and administrative cuts were prorated across the remaining quota shareholders, who were provided with a guarantee that the cuts will be restored when stocks recover and TACs are increased. Government paid out $NZ45 million to buy back 15,800 out of 21,500 metric tons in catch reductions sought.[7]

Accumulation and Transfer of ITQs. Maximum and minimum holdings of ITQ have been set. No person or company can hold more than 35% of the total of ITQs (for all areas combined) for each of the seven deepwater and offshore species originally allocated under the enterprise allocation scheme, or more than 20% of the total ITQ for any single Fishstock area for any other species. These limits apply to both owned and leased quota. These upper limits were introduced to address concerns over monopolistic aggregation of quota. A minimum quota holding of 5 metric tons was specified for finfish species and 1 metric ton for shellfish. The minimum limits were introduced to address concerns about excessive splitting of quota share holdings, resulting in too many vessels operating on fishing grounds, and to reduce the administrative costs of servicing and policing many small quota holders.

ITQs may not be held by persons not ordinarily resident in New Zealand or by companies with overseas control. ITQs may not be allocated to or held by owners of licensed foreign fishing vessels. Government has the sole right to lease ITQs to foreign fishing vessels.

Except for the restrictions described above, ITQs are freely transferable on the open market. A national fish quota exchange operated by the New Zealand Fishing Industry Board became operational on January 10, 1987. The exchange ceased operation after a few years because of the low volume of trades. Most quota has exchanged hands through direct negotiations between quota shareholders or through the small number of private quota brokers that have become established.

Monitoring and Enforcement. The New Zealand ITQ monitoring and enforcement system is based on documented product flow control that establishes and tracks a fish "paper trail." Fishermen must sell only to licensed fish receivers. These actions are documented, and details are submitted to the Ministry of Fisheries. All persons selling, transporting, or storing fish must keep business records establishing that the product has been purchased from a licensed fish receiver. Enforcement is largely land based. Fishery officers enforce product flow, and fishery auditors examine business accounts and records to monitor quota compliance. Cost-effective enforcement is enhanced by the use of sophisticated electronic monitoring and surveillance information and analytical systems.

Quota Monitoring and Reporting System. The system through which quotas are reported and monitored is based on three documents that can be cross-

---

[7] The government had previously obtained income from the sale of IFQs. In retrospect, there is some question about whether the industry, rather than the government, should have borne this risk.

checked—the Catch Landing Log, the Quota Management Report, and the Licensed Fish Receivers Return.

1. *Catch Landing Log*—The Catch Landing Log provides an on-site record. It must be completed by the skipper of the fishing vessel when catch is landed. The log does not have to be submitted at regular intervals, but must be available on demand to any fishery officer or examiner. The log may be used to verify both of the other reports. It provides information on fishing activity and sale of the fish.

2. *Quota Management Report*—This is the basic document for monitoring catch against quota. It must be completed by the quota shareholder and submitted to a Ministry of Fisheries registration office every month, or at shorter intervals if specified. It details (by area) the quantity of fish caught for each species for which quota shares are owned or leased.

3. *Licensed Fish Receivers Return*—The Licensed Fish Receivers Return must be submitted to a Ministry of Fisheries registration office monthly, or more often if specified, by all persons licensed to receive fish from commercial fishermen. It contains the quota shareholder's name and fisherman identification number, and the species and weights on landing for all fish received. The report is designed to monitor commercial fish receiving operations beyond the landing point. It makes receivers as responsible as fishermen and quota shareholders in monitoring the use of the resource.

Other Reporting Systems. The Ministry of Fisheries obtains information from three other systems that can be compared with the information submitted through the Quota Monitoring and Reporting System: Catch and Effort Returns, Observer Programme, and Vessel Monitoring System.

1. *Catch and Effort Returns*—The Ministry of Fisheries operates a compulsory catch and effort return system for 100% of New Zealand's fishing fleet. All fishermen are required to provide details of their fishing operations, including area(s) fished, species caught, amount of effort expended, and other relevant information. Depending on the size of the vessel and the gear type used, information is required either on a trip-by-trip basis or for each day fished. Returns must be supplied monthly.

2. *Observer Programme*—The Ministry of Fisheries operates an at-sea observer program primarily in the deepwater and offshore fisheries that employ larger vessels. The Observer Programme averages 4,000 at-sea days per year and attains coverage of up to 25% of the total at-sea days by the fleet in some fisheries. Observers keep detailed logbooks that capture the same type of information as the Catch and Effort Returns and Catch Landing Logs and can be cross-checked with these other data sources. Observers also monitor at-sea transshipments of fish between vessels.

3. *Vessel Monitoring System*—The Ministry of Fisheries operates a satellite system that monitors the position of fishing vessels. The system is currently required on all vessels greater than 43 m in length and on smaller vessels in selected fisheries. Consideration is currently being given to extending the system's coverage to smaller vessels in other fisheries.

Treatment of Offenses. Offenses against the ITQ program are treated not as traditional fishing violations but as commercial fraud. Penalties include significant fines, forfeiture of fish, vessel, and quota shares and are part of an effective deterrent.

• *Quota busting*—Reliable estimates of illegal catch are notoriously difficult to obtain and New Zealand is no exception. Quota busting is known to occur in some fisheries, especially those for high-value species such as rock lobster, paua, snapper, and orange roughy. The illegal catch of rock lobsters in 1993 was estimated at 715 metric tons, about 25% of the total New Zealand TAC (Annala, 1994). In previous years, the estimates of illegal catch from individual Fishstocks were as high as 68%. The accuracy of these estimates is not known.

Boyd and Dewees (1992) concluded that quota busting has been substantially suppressed. A few recent well-publicized prosecutions have resulted in heavy penalties, including loss of quota shares, vessels, and plant and equipment. Industry is taking a more active role in helping to reduce illegal fishing, especially in the rock lobster and paua fisheries. An industry-initiated management plan for the east coast North Island rock lobster fishery, which had the highest estimated level of illegal catch, has apparently reduced the level of illegal catch substantially. The fishery is now closed during summer months, the traditional period of greatest illegal activity, and all pots must be removed from the water during the closure period to assist enforcement.

• *Discarding and highgrading*—The discarding or "dumping" of species in the QMS is illegal, except in very limited circumstances. Discarding has been experienced since the introduction of the QMS, but because it occurs at sea, it is difficult to prove. In the multispecies inshore trawl fisheries, fishermen have been known to dump quantities of non-target QMS species rather than use one of the legal mechanisms for dealing with bycatch. In the deepwater trawl fisheries, vessels carrying observers have reported larger quantities of non-target QMS species than vessels fishing the same area that do not carry observers, indicating that discarding probably occurs on vessels without observers. Highgrading has occurred in both the inshore and the deepwater fisheries when a premium price is paid for fish of a certain size or quality and when small fish are discarded because of their unsuitability for processing.

Administration and Compensation. The New Zealand ITQ program is administered primarily by the Ministry of Fisheries. The one major exception is quota trading, which is carried out directly between quota shareholders or through private brokers. The Ministry of Fisheries is currently consulting with fisheries stakeholders on the transfer of responsibility to the commercial industry for administering the ITQ program.

Some of the major administrative issues encountered during the first 10 years of the New Zealand ITQ program are described below.

Bycatch Problems in Multispecies Fisheries. Most species in the QMS are caught in multispecies trawl fisheries. Bycatch problems have been experienced (mainly in the inshore fisheries) because the TACs set initially in 1986 were not set in proportion to pre-QMS landing levels and because of natural variations in stock size. TACs for the overexploited inshore species were set at levels from 25 to 75% of the pre-QMS levels, depending on the biological status and management objectives for each Fishstock. TACs for under- and fully exploited species taken in the same mixed fisheries were set at levels equal to or greater than their pre-QMS levels. This resulted in an imbalance in the catch mix relative to the available quota. Bycatch problems have resulted in both TAC overruns and underruns.

There are no constraints in the QMS that require fishermen to stop fishing in multispecies fisheries when the quota (either ITQ or TAC) of a particular species has been filled if the quota of other associated species has not been caught. The preferred method for dealing with bycatch problems is for fishermen to change their methods of operation to match their catches to their quota allocation. Where this approach is not successful, a charge of overfishing can be avoided if fishermen obtain quota shares to balance catch within a short period of time (before the fifteenth of the next month) using one of the following mechanisms:

- Purchasing or leasing ITQ from other holders;
- Fishing on behalf of another ITQ holder by either fishing against the other's quota or declaring catch against the other's quota;
- Catching up to 10% in excess of their ITQ for a given species for a given year or carrying over up to 10% of their ITQ to the following year; and
- Leasing to the Crown for the remainder of the fishing year an equivalent value of unfished ITQ of another species that has been approved by the Ministry of Fisheries for the QMA (the bycatch trade-off scheme). The scheme operates only for selected inshore species in certain areas and is not permitted in deepwater fisheries.

If a fisherman cannot obtain quota to cover bycatch using one of the mechanisms described above, there are two additional mechanisms that can be used to avoid an overfishing charge:

- Surrendering the exvessel price value of overcaught fish to the Crown; or
- Paying a "deemed value" for overcaught fish to the Crown. Fishermen have until the fifteenth of the month following the month of the overcatch to obtain quota to cover the overcatch, and if they are successful the deemed value

will be returned to them. The deemed value is set at a level that discourages dumping of fish caught without quota but does not encourage targeting without quota.

Annala et al. (1991) presented information on the extent of TAC overruns during the first two years of the QMS. In 1986-1987, 15 TACs were exceeded out of the total of 169 Fishstocks. The major reasons for these overruns were the 10% ITQ overrun provision in ten Fishstocks and surrenders and the bycatch trade-off scheme in five Fishstocks. Seven of the fifteen overruns were due to overcatching non-target species in multispecies trawl fisheries.

In 1987-1988, a total of 33 of the 169 TACs were exceeded. The 10% ITQ overrun provision accounted for 9 of the overruns, and surrenders and the bycatch trade-off scheme for 24 overruns. Seventeen of the 33 overruns were due to overcatching non-target species in multispecies trawl fisheries. The increased use of the bycatch trade-off scheme in 1987-1988 was due to a combination of an increase in the number of species and areas in the program and fishermen becoming more familiar with its operation.

The occurrence of TAC overruns has decreased since 1987-1988. Reported landings for the 1993-1994 fishing year indicated that only 22 of the 179 TACs were overcaught. Of these 22 Fishstocks, 6 were overcaught by 2% or less, the same number as were overcaught by this amount in 1987-1988. The 10% overrun provision was the main cause of 12 overruns and surrenders-deemed values-bycatch trades for the other 10.

The reduction in the number of TAC overruns has resulted from changes in the methods of operation of fishermen as they have gained more experience with the QMS. Fishermen have adjusted their quota holdings to reflect their expected catch mix more accurately. Some fishermen have stopped targeting certain species when their ITQ for associated bycatch species is filled if they know they will continue to catch the bycatch species. Industry has actively encouraged the reduction of TAC overruns for bycatch species by introducing codes of practice in some fisheries. An example is the reduction in the bycatch of hake, ling, and silver warehou in the west coast hoki (*Macruronus novaezelandiae*) fishery. Where these steps have not been possible or successful, the "overfishing" provisions described earlier have been used. Fishery managers have worked in consultation with the fishing industry to fine-tune the use of the overfishing provisions on an annual basis to reduce the amount of overcatch.

Complicated Nature of the Quota Management System. The complicated nature of the QMS, especially related to the overfishing provisions described earlier, has required very complex computer systems to track catch against quota. The inclusion of provisions such as 10% overruns and underruns, "fishing-on-behalf" arrangements, and the deemed value and bycatch trade-off systems have added complexities that have often strained computer systems to near breaking. Catch-against-quota balances have often been late and sometimes inaccurate.

This has reduced the quota shareholders' faith in the system to produce accurate information. These problems are being addressed in the QMS simplification project described below.

Evaluation and Adaptation. Evaluation. One of the glaring gaps in the New Zealand ITQ program is the lack of any systematic, quantitative evaluation of the benefits and costs of the program either by government agencies or by the fishing industry. There is not much in the way of objective, quantitative information available, but there is a great deal in the way of perceptions.

Adaptation. A number of adaptations have been made in the first 10 years of the New Zealand ITQ program. The important ones are described below.

1. *Bycatch problems in multispecies fisheries*—The adaptations made to the program to address bycatch problems in multispecies fisheries have been described in the previous section.

2. *Settlement of Maori fisheries claims*—In 1987, a High Court injunction was obtained by the Maori that prevented the QMS from being expanded to incorporate all commercially significant species. This led to a great deal of uncertainty within the industry, both with regard to existing Maori property rights and future implementation of the QMS.

In September 1992, in-depth negotiations began between the government and representatives of Maori fisheries interests. The Maori negotiators had a confirmed mandate from the great majority of Maori tribes to finalize a settlement of claims on certain conditions. The negotiations resulted in the signing of a Memorandum of Understanding that set the basis of the benefits and conditions for final negotiations.

Intense negotiations were carried out on the basis of the Memorandum of Understanding and resulted in the final settlement. The settlement, consisting of the Deed of Settlement and the Treaty of Waitangi (Fisheries Claims) Settlement Act 1992, provided for a full and final discharge of the government's obligations under the Treaty of Waitangi. The settlement provided the following:

- $NZ150 million for the purchase of 50% of Sealord Products, New Zealand's largest fishing company, which had 25% of total allocated fish quota and achieved annual sales of about $NZ250 million;
- The transfer to the Maori of 20% of the quota for all new species entering the QMS;
- Regulations to recognize and provide for the customary food gathering and the special relationship between the Maori and those places that are of customary food-gathering importance to the extent that such food gathering is not conducted in a commercial manner.

The settlement provided for the transfer of NZ$500 million in assets to the

Maori, giving them almost 40% of the New Zealand commercial fishery, with the potential to acquire a larger proportion of the resource with the funds provided by the settlement. The settlement made the Maori the single largest participant in the industry.

The settlement also protected the livelihoods of existing quota shareholders by bringing security to the commercial fishing industry. The settlement prevented the Maori from advancing further commercial fisheries claims through the courts and ensured that the management of fisheries was not compromised by the Maori acting contrary to sustainable management practices. The purchase of Sealord Products had little effect on other quota holders since it was simply a transfer of ownership of a fishing company and had little impact on the distribution of quota shares within the industry.

3. *Change to proportional ITQs*—When the QMS was introduced in 1986, ITQs were denominated as a fixed tonnage. TACs were to be increased or decreased by the government entering the marketplace and either selling or buying quota. When the QMS was being developed, it was proposed to create a "revolving fund" for such transactions. Resource rentals and revenues from the sale of quota would have gone into the fund, which would be used to buy back quota as necessary. However, the fund was never created.

In the late 1980s, the government was faced with substantial costs to reduce the TACs for, primarily, two orange roughy Fishstocks. The government announced its intention to change the QMS from a program based on fixed-tonnage ITQs to one in which ITQs were denominated as a proportion of the TAC. ITQs would be increased or decreased in proportion to the changes in the TAC.

The government and the fishing industry entered into negotiations over the change from fixed to proportional ITQs. The outcome of the negotiations was an agreement widely known as the "Accord." The main results of the Accord are as follows:

- ITQs were changed from a fixed to a proportional basis on October 1, 1990.
- Resource rentals for all species in the QMS were frozen for five years from October 1, 1989, except for increases in line with movements in the Consumer Price Index.
- During the period from October 1, 1989, to September 30, 1994, compensations for TAC reductions were paid out of the resource rental pool. The compensation period for hoki could be extended beyond this period if TAC reductions exceeded certain limits.
- The price of compensation was to be agreed between the government and the fishing industry, and failing agreement, was to be arbitrated.
- A TAC reduction of 4,000 metric tons for orange roughy on the Chatham Rise was agreed to, and other TAC reductions were to be discussed with industry.
- A body with 50:50 government-industry representation was created to

advise the Minister of Fisheries on TAC changes. This body has developed into the TAC Advisory Council.

• Quota shareholders who suffered administrative cuts without compensation would receive first preference when TACs are increased.

• Resource rentals were paid on quota shares held until the end of the compensation period.

• The bycatch trade-off system continued, subject to annual review.

• TAC reductions would be accomplished by the Crown first canceling quota owned by it, subject to its various obligations.

The terms of the Accord provided the basis for the move to proportional ITQs. Since October 1, 1990, there have been substantial changes in some TACs, for example, those for orange roughy, snapper, and rock lobster discussed earlier.

The terms of the Accord related to compensation for TAC reductions and payment of resource rentals expired on September 30, 1994. Resource rentals ceased on that date and were replaced by the introduction of cost recovery on October 1, 1994. From October 1, 1994, all of the avoidable costs of managing, researching, and enforcing commercial fisheries are to be paid by the fishing industry. For the 1994-1995 fishing year, about $NZ37 million was recovered from the fishing industry from the total government expenditure of about $NZ46 million in these three areas.

4. *Strategies for adjusting TACs in situations with limited information*—For many Fishstocks there is limited information available on stock size and population dynamics that can be used to estimate yields to provide a basis for varying TACs. Resource constraints make it unlikely that substantially more information will become available in the near future. For some of these Fishstocks there have been substantial increases in catch in recent years, and the existing TACs either are constraining the catch or are being overcaught. The fishing industry has suggested that these increased catches have not occurred through an increase in effort but are the result of increases in stock abundance.

The fishing industry has proposed increases in TACs for some Fishstocks, using a process that has become known as "adaptive management." The process is not strictly a form of adaptive management as defined by Hilborn and Walters (1992), where an experimental approach is taken in which TACs are set at various levels from total closure to deliberate overexploitation in an effort to obtain better estimates of stock size, productivity, and sustainable yields. The New Zealand approach has been to increase the TACs for certain Fishstocks where anecdotal information suggests that increased catch levels are likely to be sustainable and then to monitor the effects.

The Ministry of Fisheries has recognized that there could be considerable benefits from increasing some of the TACs, as proposed by the fishing industry. These potential benefits include the following:

- Economic benefits to the industry through increased catches;
- Development of precedents for cooperative management of fisheries with the industry, especially insofar as industry is willing to assist with increased data collection and analysis; and
- Changing catch levels in a controlled fashion, which may provide better information that can be used as a basis for future management decisions.

The following criteria and guidelines under which adaptive management changes could be considered were agreed:

- Catch level increases could be considered for Fishstocks for which abundance appears to have increased, for example, in situations of stable or increasing landings or increasing CPUE where effort has not increased.
- Contingency plans agreed upon by the Ministry of Fisheries and the fishing industry should be finalized before TACs are changed.
- Contingency plans should specify the data that will be collected and the responses to the results of the data collection and subsequent analysis. For example, they should specify what level of change in an indicator variable would indicate that stock size is increasing or decreasing.
- There should be an agreed period over which the TAC changes will take place. There is general agreement that five years is a useful period over which to assess the effects of such changes. However, the effects of TAC changes are assessed each year during the annual stock assessment and TAC-setting process.
- There should be agreement on the nature and extent of cooperative data gathering and research projects and how resources are to be provided for these projects.
- The choice of which Fishstocks to include in the scheme should be based largely on what effective monitoring programs can be established. Inclusion of a large number of Fishstocks is seen as too ambitious, and the scheme should be developed incrementally.

The adaptive management scheme was first implemented for the 1991-1992 fishing year. During the first four years of operation of the scheme, the fishing industry requested TAC increases for 39 Fishstocks (there have been no requests for TAC decreases under the scheme). Increases have been granted for 19 Fishstocks.

The effects of these increases are being monitored in a number of ways. The Ministry of Fisheries operates a mandatory catch and effort logbook system for all fisheries, and data are being collected and analyzed from this system. In addition, data are being collected for some of the adaptive management Fishstocks during the course of Ministry of Fisheries research programs (e.g., trawl surveys) for these and other species. The fishing industry has introduced a voluntary logbook program for some of the adaptive management Fishstocks to collect more detailed and specific catch and effort and biological information. The fishing industry is also providing resource-specific research projects for some of the Fishstocks and has employed two scientific staff members to work coopera- tively with Ministry of Fisheries staff specifically on the adaptive management Fishstocks.

During the 1997 stock assessment and management consultative meetings, criteria were developed by the Ministry of Fisheries, in conjunction with stake- holders, to determine the suitability of the TAC increases that had been granted under the adaptive management program. Criteria included the determination of whether or not the TAC increases were sustainable, based on the available infor- mation, and whether effective monitoring programs had been put in place by the industry. The criteria were applied to each of the adaptive management Fishstocks using a decision tree approach. As a result of this evaluation, the TACs for nine Fishstocks were reduced to or toward their lower pre-adaptive management pro- gram levels.

Outcomes of the ITQ Program

Biological and Ecological Outcomes for the Fishery. Following are the major biological and ecological outcomes of New Zealand's ITQ program.

Improved Biological Status of the Resource. Before the QMS was intro- duced in 1986, there had been only limited directed stock assessment research in New Zealand. Most fisheries research had been directed at gathering basic bio- logical information on a few commercially important species. Moreover, be- cause of the lack of a mandate to carry out stock assessment research, abundance estimates were not available for most species. For many species in the QMS, data on age, growth, mortality, fecundity, abundance, and other important factors are still not available.

The QMS was introduced in 1986 because of the perception that many, if not most, of the inshore fish stocks were suffering from high levels of biologic and economic overfishing. TACs for the overexploited inshore species were set at levels from 25% to 75% of the pre-QMS catch levels, depending on the biologi- cal status and management objectives for each Fishstock. On the other hand, the offshore, deepwater species were relatively newly exploited, and most were likely to still be in the "fishing-down" phase.

New Zealand legislation has constrained the ability to vary TACs. The criterion that TACs are set and altered to allow the stock to move toward a level of biomass B that will support the maximum sustainable yield (B_{MSY}) has been interpreted very strictly. Estimates of maximum constant yield and current an- nual yield have been used as reference points when varying TACs and have not necessarily translated directly into TACs (see Annala, 1993, for definitions). TACs cannot be changed unless it can be demonstrated by the stock assessment process that the stock is moving toward a size that will support the MSY, even when other data suggest that the TAC is at an inappropriate level.

How has the QMS performed with regard to improving the biological status of Fishstocks? Sissenwine and Mace (1992) concluded that there was little evidence of improvement in the condition of fisheries resources since 1986, but because stock assessment information was limited, it was difficult to know. They stated, "The general conclusion is that TACs are not closely tied to the best available assessments of the fisheries resources, nor are catches strongly controlled by the TACs. Some valuable stocks have probably declined in abundance. To date, the track record of ITQ management with respect to conservation is not good."

The current situation based on the 1997 stock assessments (Annala and Sullivan, 1997) is much more positive than the picture painted by Sissenwine and Mace. Of the 179 Fishstocks in the QMS as of October 1, 1997, 30 were created for administrative purposes around an offshore island group that is only lightly fished for a few species. Of the remaining 149 Fishstocks, only 11 (7.4%) were estimated to be below B_{MSY}. Sixteen (10.7%) Fishstocks were estimated to be above and 27 (18.1%) at or near B_{MSY}. The status of the remaining 95 (63.8%) Fishstocks relative to B_{MSY} was not known. Unfortunately, most of the inshore Fishstocks that experienced large reductions in catch levels in 1986 are included in the latter category, and it has not been possible to monitor the rate of stock rebuilding, if any. However, a series of inshore trawl surveys initiated in the late 1980s and early 1990s will provide future estimates of abundance for some of these species so that sustainable yields can be estimated and the rate of rebuilding determined.

Most of the major Fishstocks that are below B_{MSY} are now being rebuilt. The TAC for the largest orange roughy fishery on the Chatham Rise has been reduced from about 38,000 metric tons for the 1988-1989 fishing year to 7,200 metric tons for 1997-1998. The TAC for the Challenger orange roughy fishery was reduced from 12,000 metric tons in 1988-1989 to 1,900 metric tons in 1990-1991 and has remained at this level since. The TAC for the largest snapper fishery in QMA 1 was reduced from 6,000 metric tons in 1991-1992 to 4,900 metric tons in 1992-1993 and 4,500 metric tons for 1997-1998. In addition, a management plan is being developed to rebuild the stock to B_{MSY}. The combined TAC for the eight North and South Island rock lobster Fishstocks was reduced from 3,275 metric tons in 1990-1991 to 2,383 metric tons in 1993-1994. A 10-year management plan has been developed to address various biological issues to improve the probability of rebuilding stocks.

The need for some large TAC reductions, especially in the orange roughy fisheries, prompted the change from fixed tonnage ITQs to ITQs as a proportion of the TAC in 1990. The change to proportional ITQs has removed the financial burden from the government to buy back quota and has made the reduction of TACs easier.

Open and Transparent Stock Assessment and TAC-Setting Process. One of the strengths of the New Zealand QMS is the completely open and transparent stock assessment and TAC-setting process. The process is open to all users of the resource and all groups with interests in the fisheries, including the Maori, the commercial industry, recreational fishermen, and environmental-conservation groups. User groups can be represented by consultants, and they have been employed particularly by the commercial industry. All stock assessment data collected by the Ministry of Fisheries are made available (at cost) to all participants in the process. The usual caveats regarding commercial sensitivity apply to the release of catch and effort data collected from the industry. The data are provided only in an aggregated form so that individual fishermen and/or companies cannot be identified.

The foundation of the stock assessment process lies in the Fishery Assessment Working Groups. The working groups analyze the available fishery and research data and prepare draft reports giving the details of the stock assessments and status of the stocks according to agreed terms of reference for all 179 Fishstocks in the QMS.

Fishstocks for which the stock assessments indicate a substantial change in the yield estimates or status of the stocks are referred to the Fishery Assessment Plenary. The plenary session is open to all participants in the process and reviews the data and analyses produced by the working groups. The stock assessment results from the plenary are used as a basis for preparation by the Ministry of Fisheries of an initial position paper providing advice to the Minister of Fisheries as to which Fishstocks may be considered for changes to TACs and other management measures for the following fishing year. Other information (e.g., socioeconomic, environmental) is included in the discussions at this stage. This advice paper is also made available to the users and forms the basis for discussion at a series of consultative meetings between the Ministry of Fisheries and these groups.

After the consultative meetings between the Ministry of Fisheries and the user groups, the Minister of Fisheries holds a series of meetings with users to obtain their views on any proposed management changes. The final authority to decide on TAC changes lies with the Minister of Fisheries.

Economic and Social Outcomes for the Fishery. The following major economic and social outcomes of New Zealand's ITQ program have also been identified.

Secure Access to the Resource. The allocation of quota shares in perpetuity has guaranteed security of access to the resource. When rock lobsters were introduced into the QMS in 1990, court action taken by Maori was settled when the government agreed to rock lobster ITQ being issued for a 25-year term. The settlement of Maori fishing rights issues with the passage of the Treaty of Waitangi (Fisheries Claims) Settlement Act in 1992 resulted in the lifting of the injunction.

A Market-Oriented Industry Structured by Market Forces. Dewees (1989) found that the fishing industry responded rapidly to the guarantee of access to the resource conferred by ITQs. Many fishermen in the Auckland region had switched from maximizing quantity to maximizing quality, which he attributed largely to the easing of the "race for fish." By May 1987, 40% of the quota shareholders were changing to methods that allowed the onboard handling of individual snapper to supply the lucrative *ike jime* market in Japan. A more recent development has been the export of live fish (*ike dai*) to Japan.

Other market developments have been attributed at least partly to the security conferred by ITQs. The export of live rock lobsters has increased from 1,947 metric tons in 1990 to 2,722 metric tons in 1993. Fishermen are also spreading their fishing effort and catching more of their rock lobster quota during months with traditionally low catch rates when the price is higher. New Zealand's largest fishery is the trawl fishery for hoki. The fishery has been carried out primarily on the west coast of the South Island during the July to September spawning period. Most fish were caught by vessels that headed and gutted the fish or with onboard surimi plants. Catch rates of up to 200 metric tons per tow were not uncommon. In recent years, New Zealand companies have invested in vessels with onboard filleting lines designed specially for hoki. These vessels fish for hoki year-round away from the spawning ground and typically target smaller quantities of fish, usually in the range of 4 to 5 metric tons per tow to improve product quality. The catch of hoki caught away from the spawning ground outside the spawning season increased from 30,000 metric tons in 1990-1991 to 69,000 metric tons in 1992-1993.

Reduced Overcapitalization. Some commentators have stated that the QMS has resulted in a reduction in overcapitalization (e.g., Clark, 1993). However, there are few actual data or analyses to support this assertion. One indirect measure that could be used to evaluate the reduction in overcapitalization involves changes in quota holdings. Between 1987 and 1989, the 10 largest ITQ shareholders increased their share of the total quota from 67% to 82% (Anon., 1987; Bevin et al., 1989). However, this share had fallen back to 68% by March 1994 (Parker, 1994), so the concentration of quota may not be a good measure of the reduction in overcapitalization.

The New Zealand fishing industry has experienced strong export revenue growth since the QMS was introduced. The total primary value of the catch (which is estimated as landed catch multiplied by an estimated exvessel price) corrected for inflation increased from $NZ427 million in 1986 to $NZ456 million (1986 dollars) in 1993, while total seafood export receipts increased from $NZ657 million to $NZ896 million (1986 dollars) over the same period. In an ITQ program, the expected rates of return are capitalized into the quota value when it is traded. The strong revenue growth has led both directly and indirectly to large increases in the value of quota for some species. For example, the average sale price of pa (abalone) (*Haliotis iris*) quota increased from about $NZ50,000 per metric ton in 1991 to about $NZ190,000 (1991 dollars) per metric ton in 1994.

Greater Industry Freedom, Flexibility, and Responsibility. The easing of the race for fish that has resulted from the ITQ program has undoubtedly given fishermen the freedom and flexibility to structure their operations to maximize the value of their catch rather than the volume. Some examples given in the section above describe changes to operations that have resulted in maximization of the value of the landed catch.

There has also been increased industry responsibility and cooperation since the start of the QMS. Some examples follow:

- Quota shareholders in three of the rock lobster Fishstocks are funding two full-time scientific staff members to carry out research in their areas that will contribute to stock assessment.
- Quota shareholders in one of the paua Fishstocks have asked for and taken a voluntary 10% reduction in TAC because of their concern about the state of the Fishstock.
- Quota shareholders in the Chatham Rise orange roughy fishery have spent in excess of $NZ1 million on exploratory fishing ventures, bathymetric surveys, and trawl surveys to estimate relative abundance, which will directly impact the assessment for this fishery.
- Scallop (*Pecten novaezelandiae*) quota shareholders have formed a company to fully fund an enhancement program for New Zealand's largest scallop fishery. Included in the arrangement is the provision that they will also enhance recreational-only fishing areas.
- In 1989-1990, high levels of bycatch of hake (*Merluccius australis*), ling (*Genypterus blacodes*), and silver warehou (*Seriolella punctata*) in the west coast hoki fishery resulted in the TACs for these three species being exceeded. Quota shareholders developed a voluntary code of practice and the catches of these three species have been reduced.

Improved Industry Efficiency, Competitiveness, and Profitability. It is a widely held perception that the efficiency, competitiveness, and profitability of the industry have increased (e.g.. Clark, 1993). Once again, there are few actual data or analyses to support these claims.

An index of the seafood industry's competitiveness has been calculated for 1988-1993 (Parker, 1994). The index was computed as the difference between the seafood industry's costs of production (input index) and the revenue received for output (output index). This competitiveness index increased by about 20% from 1988 to 1993.

Since 1986, an Annual Enterprise Survey of the seafood industry has been carried out by the New Zealand Department of Statistics. The survey provides a financial picture of industry performance over the fishing year. It details costs, assets, and revenue for the catching and processing sectors, which are combined to avoid double counting.

Because it is a survey, the entire industry is not covered, different companies are sampled each year, and it is difficult to compare results between years. However, survey results since the first year of the QMS do not support the perception that industry profitability has increased. For the six years from 1986-1987 to 1991-1992 the returns on assets (after interest, rentals, and tax) for the major quota holders in the survey were 11.1%, 3.0%, 10.2%, 10.8%, 6.6%, and 11.5%, respectively.

Economic and Social Outcomes for Fishery-Dependent Communities. New Zealand has few communities that are largely dependent on fishing. The economic and social outcomes of the ITQ program for these communities have not been analyzed.

Administrative Outcomes. Major economic and social outcomes of New Zealand's ITQ program have been identified. Sissenwine and Mace (1992) concluded that the QMS had not reduced government intervention. Indeed, the advent of the QMS saw the introduction of new recordkeeping and reporting requirements, such as the quota monitoring and reporting system (described in Clark et al., 1988) and the bycatch trades system (Annala et al., 1991). In addition, most input controls, for example, minimum size restrictions, closed seasons and areas, have remained in place.

Although the government paid for the management systems, there were no incentives for individuals in the industry to minimize the system costs. Instead, it was apparently to the industry's advantage to seek an elaborate management system to make it more convenient. The government was generally sympathetic to the requests for convenience, but it either underestimated the costs of elaborating the systems or was reluctant to pay to construct proper systems. The result was poorly constructed complex systems that led to significant dissatisfaction with aspects of the QMS for many years. The lessons from this included:

1. Simple systems that work well and are relatively inexpensive have much to commend them compared to complicated systems.
2. Having the industry pay for the management systems and having a large role in their design and operation will help balance aspirations for complexity and controlling costs.

Current Perceived Issues. In 1996, a new Fisheries Act was passed by the New Zealand Parliament. The act concluded the review of fisheries legislation that had been ongoing since 1991. It provided a complete rewrite of the Fisheries Act, building on the strengths of the QMS; refined some aspects of the QMS; and added other fisheries management features.

The act has the following principal components that address many of the current issues with regard to the ITQ program.

Environmental Principles. The act provides the following general environmental principles:

- Stocks must be maintained at or above defined levels. TACs must be set at a level that will maintain stocks at or above a level or move them toward the level that will produce the maximum sustainable yield.
- The effects of fishing on associated and dependent species must be taken into account.
- The biological diversity of the aquatic environment must be conserved.

Consultation. The act formalizes the processes for consultation with sector-user groups. This replaces the current informal advisory group structure. The creation of a National Fishery Advisory Council, with representation from all sector-user groups, has been authorized.

Conflict Resolution. The act formalizes the resolution of conflicts concerning access to resources. The process first encourages various sector-user groups to resolve their differences. If the parties are unable to negotiate a solution, the Minister may appoint a commissioner to hold an inquiry and report back to the Minister. All such disputes will be resolved by the Minister.

Addition of New Species into the Quota Management System. The government intends to move all commercially harvested species into the QMS over the next three years. Twenty percent of all new quota will be allocated to Maori. For most species, quota will be allocated on the basis of catch history. There will be an appeals process for quota allocations, but the process will be stricter than previously. The process will not result in any increases to TACs, and there will be a time limit for lodging appeals.

Simplification of the Quota Management System. At present, the central rule in the QMS is that fishermen must hold quota before going fishing. The manner in which this rule is administered has resulted in the overfishing provisions described above and drives much of the complexity of the QMS.

The new Fisheries Act separates the property right (ITQ) from the catching right by introducing a system of annual catch entitlements (ACEs). For most species, fishermen will no longer be required to hold ITQ before going fishing but will be required to hold an ACE. At the beginning of each fishing year, every person who holds quota will be allocated an ACE based on the amount of quota held. ACEs are superficially similar to an annual lease of quota and are tradable rights like ITQs. When the catch exceeds the ACE, a deemed value is payable. The separation of catching rights from ITQs is expected to assist investment in fisheries by increasing the security of ITQs.

The existing 10% overrun of ITQ provision will be abolished. Consultation is occurring with stakeholders to determine which mechanisms will be retained to assist with managing bycatch issues in multispecies fisheries.

Institutional Reform. Another issue is reform of the delivery of fisheries management services. Recent reforms include the provision of services by agencies outside the Ministry of Fisheries (including fisheries research), the transfer of fisheries stock assessment research into a Crown Research Institute, and the establishment of a stand-alone Ministry of Fisheries. The role of the Ministry of Fisheries is being reduced to one of policy advice; determining the standards and specifications for and purchasing, monitoring, and auditing the contestable services; liaison and facilitating conflict and dispute resolution; and enforcement, compliance, and prosecutions. Contestable services will potentially include all the other functions currently performed by Ministry of Fisheries, including administration of quota and permit registries, catch and effort data management, satellite vessel monitoring, and the observer program. Consultation is occurring on the direct contracting of some of these services by the industry.

APPENDIX
H

Potential Economic Costs and Benefits of Individual Fishing Quotas to the Nation

The charge to the committee in the Sustainable Fisheries Act mandates an evaluation of the "potential social and economic costs and benefits to the nation" of individual fishing quotas (IFQs) (Sec. 108[f]). Other chapters of this report have described and elaborated on the specific effects of these programs. The purpose of this appendix is to attempt to quantify some of the costs and benefits that may accrue to the nation as a whole from the implementation of IFQ programs. The committee found it impossible to conduct a true cost-benefit analysis of any U.S. IFQ fishery because of insufficient data. The committee presents the following data and analyses as order-of-magnitude estimates. Recommendations are also presented regarding data needed to improve such estimates.

In evaluating the potential economic costs and benefits to the nation from IFQ programs, several factors need to be examined. First, calculating the benefits and costs of IFQ programs would have to take into consideration the changes in value to consumers (in terms of both product mix and quality). These calculations would require an extensive analysis of the particular fishery and therefore would go well beyond the information in this appendix. Further, although revenues generated from taxes or fees associated with an IFQ program may offset some of the monetary costs associated with administering these programs, they are not the principal benefits of an IFQ program. The real net benefits of an IFQ program consist of increases in revenues over costs, including management and enforcement costs. Tax revenue are transfers from individual entities to the government and should not be confused with real changes in net benefits to the nation. Their fiscal effects are important, however, because they provide a funding base for management and enforcement costs that are currently paid almost exclusively from public revenue. The factors that must be assessed in order to ascertain the overall costs and benefits from the implementation of an IFQ program include the following:

- Potential generation of revenue from fees or taxes from the IFQ program;
- Potential changes in the cost of administration for the fishery;
- Potential changes in the economic efficiency of fishing operations;
- Potential changes in employment in the fishery and associated industries; and
- Potential long-term effects on fishery-dependent communities.

In many cases, it is difficult to distinguish between costs incurred by general fishery management practices and those incurred in IFQ management. As the following analysis will shows, there are not adequate data to perform a complete cost-benefit analysis for any U.S. IFQ fishery.

POTENTIAL REVENUE FROM FEES AND TAXES

Although open-access fishery management systems may require a permit, endorsement, license, or other form of certification, fees from these systems are usually insignificant compared to the other costs required to engage in fishing (e.g., labor, gear, fuel, equipment) and usually do not generate substantial tax revenue. Exceptions, however, include among others a 2% landings tax in Alaska that provides significant revenues to local and state governments.

Several difficulties arise in assessing the potential revenues generated by IFQ management compared to other forms of fishery management. In many cases, it is difficult to obtain accurate estimates of quota price and therefore the value of the quota in the U.S. programs. The Alaskan halibut and sablefish IFQ programs have the best information available concerning variations in quota price, due in part to research conducted by the National Marine Fisheries Service (NMFS), the Alaska Commercial Fisheries Entry Commission (CFEC), and independent quota brokers, as well as the large volume of transactions.

There are several ways in which revenue can be derived from IFQ fisheries. First, the potential increases in net revenue due to the improved efficiency of the IFQ fishery can provide increased revenues to the fishery (e.g., reduced variable costs through the use of less gear and cost efficient equipment). Second, revenue can increase due to changes in harvest patterns and product forms (e.g., increased exvessel prices due to handling and closer alignment of landings with market demand). Fiscal revenue (transfers of revenues from the fishery to government), can be generated through a combination of corporate and business taxes, capital gains taxes, landings taxes, and vessel fees. The Magnuson-Stevens Act provides several mechanisms for assessing fees on the operation of IFQ fisheries. The act specifically provides for the assessment of a fee of up to 3% of the exvessel value of the fish landed under an IFQ program (Sec. 304[d][2]). Further, up to 0.5% of the value of a limited access permit or IFQ can be collected upon registration and transfer of the title of a permit, for the operation of a limited access system administration fund, or for the general administration of the fishery from which the fee is collected (Sec. 305 [h]). Non-IFQ fisheries do not specifically provide for the assessment of additional exvessel fees, so the adoption of an IFQ program could generate significant tax revenue compared with other forms of fishery management. The following sections evaluate the potential revenues from capital gains taxes, the 3% exvessel fee, and the 0.5% registration and transfer fee. These revenue projections are limited by the terms of the Magnuson-Stevens Act. Thus, these projections do not reflect what could be possible with more liberal provisions in future amendments to the Magnuson-Stevens Act nor do these projections consider changes in state or federal corporate or other business tax receipts that could arise from increases in the profitability that are attributable to the implementation of IFQs.

Potential Revenue from Capital Gains Taxes

Capital gains taxes can be assessed on the increased value of commodities such as stocks, bonds, real estate, and other "capital assets," including IFQs. Capital gains are not subject to taxation until they are "realized," generally considered to occur when the appreciated asset is sold. To conduct an assessment of the potential revenue from capital gains taxes, it is important to have relatively accurate information on the amount of quota being traded, and the price at which the quota shares were purchased (if any), and the sale price. Unfortunately, not all IFQ fisheries have reliable transfer and price information to provide the baseline assessment of the value of quota shares. The surf clam/ocean quahog (SCOQ) IFQ program does not require that price information be reported and there are no independent brokerage firms. There has been very little trading in the wreckfish IFQ program, according to information available from the South Atlantic Fishery Management Council. Thus, capital gains are not likely to be significant.

Independent brokers are active in the Alaskan halibut and sablefish programs, and can provide data on the range of quota prices. The quality of data available and the types of data collected vary among the independent broker firms. There have been significant variations in price, and precise determinations of potential capital gains revenue are not possible without access to proprietary data. Transfer price information for any new or existing IFQ programs would improve the ability to quantify the potential capital gains revenues. Data on the number of transfers, the number of pounds transferred, and the sale price per transfer in the Alaskan halibut and sablefish IFQ program are maintained by the

TABLE H.1 Total Number and Pounds of Permanent (IFQ/QS) Transfers in the Alaskan Halibut and Sablefish IFQ Programs (1995-1997)

Year	Halibut Transfers	Halibut Pounds	Sablefish Transfers	Sablefish Pounds
1995	1,198	4,515,771	351	2,565,832
1996	1,419	3,706,462	359	2,217,580
1997	1,395	4,546,431	440	1,913,780

NOTES: Transfers to surviving spouses are excluded. Permanent transfers include sweep-ups (even to the same person). Pound are the pounds actually transferred and may not necessarily equal the pounds of fish that the quota transferred will yield in a given year.
SOURCE: NMFS RAM Division.

NMFS Restricted Access Management (RAM) Division from information provided by NMFS, CFEC, and the Alaska Department of Fish and Game (Table H.1, Figures H.1-H.4).

To determine the value of quota share (QS), data on the number of transfers, the amount of quota transferred, and the price of the transfer are required. Unfortunately, the data available from both NMFS and independent quota brokers are not sufficiently detailed to provide an accurate determination of the value of the quota. Data on the ranges of quota market prices were provided to the committee by the CFEC and one of the larger IFQ brokerage firms, Access Unlimited, Inc. (Tables H.2-H.4). The actual transaction prices vary depending on the negotiating positions of the buyer and seller and seasonal fluctuations in the market. Prices offered for quota may also vary depending on whether the quota offered is blocked or unblocked, and the relative size of the quota share being offered. Typically, unblocked quota shares sell for higher prices. Additionally, prices offered for quota shares for small-vessel classes (Class D) tend to be lower than for larger-vessel classes (Class C and B), although these trends seem to have been affected by regulations allowing Class C and B quota to be used, or "fished-down," on smaller vessels (Class D).

Providing a range of the total value of the quota is necessary because it is not appropriate to assume that the average price of the quota is the mean of the minimum and maximum values reported. The information provided comes from a larger IFQ brokerage firm; however, trades also occur outside brokerage firms. The indication from testimony and information provided to the committee is that these trades may occur at a lower price than those conducted through a brokerage firm.

Several general trends can be inferred from the market price of quota share in the Alaskan halibut and sablefish IFQ programs. As might be anticipated, the price of quota was lower during the first year of the program than in subsequent

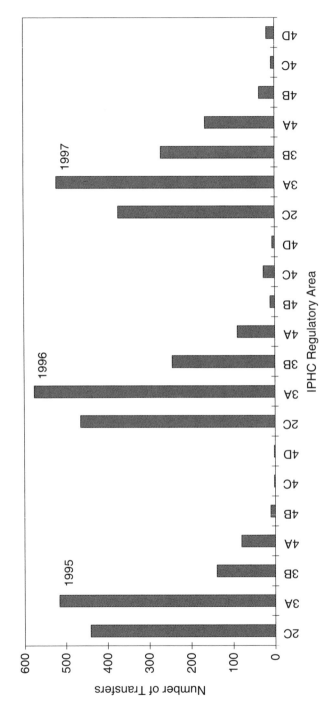

FIGURE H.1 Number of permanent (IFQ/QS) transfers per International Pacific Halibut Commission (IPHC) regulatory area in the Alaskan halibut IFQ program (1995-1997). NOTES: Transfers to surviving spouses are excluded. Permanent transfers include sweep-ups (even to the same person).
SOURCE: NMFS RAM Division.

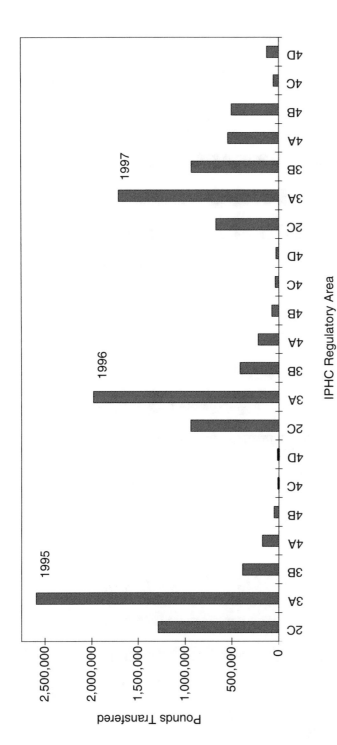

FIGURE H.2 Pounds of permanent (IFQ/QS) transfers per International Pacific Halibut Commission (IPHC) regulatory area in the Alaskan halibut IFQ program (1995-1997). NOTES: Transfers to surviving spouses are excluded. Permanent transfers include sweep-ups (even to the same person). Pounds are the pounds actually transferred and may not necessarily equal the pounds of landings that the quota transferred will yield in a given year.
SOURCE: NMFS RAM Division.

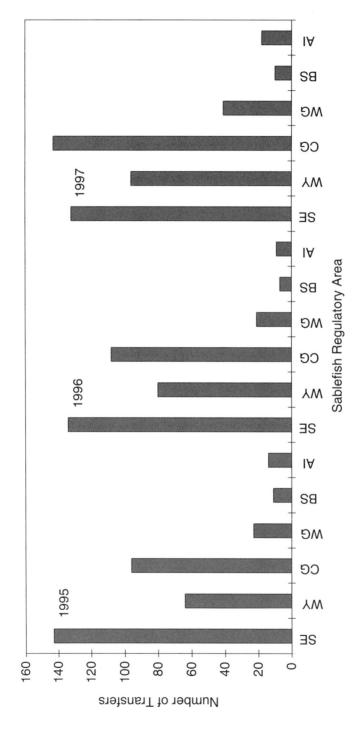

FIGURE H.3 Number of permanent (IFQ/QS) transfers per sablefish regulatory area in the Alaskan sablefish IFQ program (1995-1997). NOTES: Transfers to surviving spouses are excluded. Permanent transfers include sweep-ups (even to the same person). SOURCE: NMFS RAM Division.

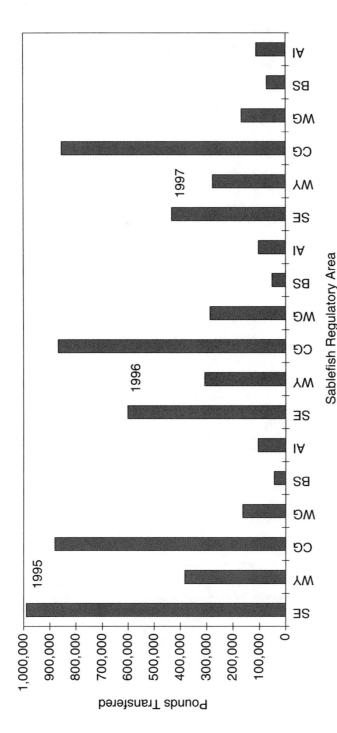

FIGURE H.4 Pounds of permanent (IFQ/QS) transfers per sablefish regulatory area in the Alaskan sablefish IFQ program (1995-1997). NOTES: Transfers to surviving spouses are excluded. Permanent transfers include sweep-ups (even to the same person). Pounds are the pounds actually transferred and may not necessarily equal the pounds of landings that the quota transferred will yield in a given year. SOURCE: NMFS RAM Division.

TABLE H.2   1995 Market Prices of Permanent (IFQ/QS)
Transfers per International Pacific Halibut Commission (IPHC)
and Sablefish Regulatory Area per Quarter in the Alaskan
Halibut and Sablefish IFQ Programs

Species	Area (vessel class)	2nd Quarter (price/pound)		3rd Quarter (price/pound)		4th Quarter (price/pound)	
		High	Low	High	Low	High	Low
Halibut	2C (D)	$7.50	$6.00	$7.90	$6.00	$7.75	$5.00
	2C (C)	$9.25	$8.50	$9.25	$8.25	$9.00	$7.75
	2C (B)	$9.25	$8.50	$9.25	$8.25	$9.00	$7.75
	3A (D)	$8.00	$6.00	$7.25	$6.00	$7.00	$5.00
	3A (C)	$8.50	$7.75	$8.50	$7.25	$8.00	$7.00
	3A (B)	$8.00	$7.25	$8.50	$7.25	$7.75	$6.75
	3B (D)	$7.75	$6.00	$8.25	$6.00	$6.75	$5.50
	3B (C)	$7.75	$6.00	$8.25	$7.25	$7.25	$6.75
	3B (B)	$8.25	$7.75				
Sablefish	SE (C)	$8.10	$5.50	$8.00	$5.50	$8.00	$5.50
	SE (B)	$8.00	$7.50	$7.50	$7.00	$7.50	$7.00
	WY (C)	$8.10	$5.50	$8.00	$5.50	$8.00	$5.50
	WY (B)	$8.00	$7.50	$7.50	$7.00	$7.50	$7.00
	CG (C)	$7.25	$5.50	$7.25	$5.50	$7.25	$5.50
	CG (B)	$7.25	$6.25	$7.25	$6.75	$7.25	$6.75
	WG (C)	$6.50	$5.50	$6.50	$5.50	$6.50	$5.50
	WG (B)	$6.50	$5.50	$6.50	$5.50	$6.50	$5.50

NOTES:

1. Refer to Appendix G for maps displaying the halibut and sablefish regulatory areas.

2. No trades were recorded in the first quarter since the program was not fully implemented until March 1995.

3. Market prices for halibut regulatory areas 4A-4D, sablefish regulatory areas AI (Aleutian Islands) and BS (Bering Sea), and vessel class A are not reported because data on market prices for quota for these vessels and regions are not sufficiently complete. Moreover, Class A transfers represent a small fraction of the total quota share transferred.

4. Market prices are derived from values reported by IFQ brokerage firms. The actual price of quota may vary from these ranges. The high and low designations are subjective definitions provided by the brokerage firms and do not correspond to the average price of quota shares transferred. Both blocked and unblocked quota are considered within the price ranges given.

5. Sablefish regulatory areas: SE (Southeast), WY (West Yakutat), CG (Central Gulf), WG (Western Gulf).

6. Halibut vessel Class B = catcher vessel greater than 60 feet in length; halibut vessel Class C = catcher vessel between 60 and 35 feet in length; halibut vessel Class D = catcher vessel less than 35 feet in length.

7. Sablefish vessel Class B = catcher vessel greater than 60 feet in length; sablefish vessel Class C = catcher vessel less than 60 feet in length.

SOURCE: Access Unlimited, Inc.

TABLE H.3   1996 Market Prices of Permanent (IFQ/QS) Transfers per International Pacific Halibut Commission (IPHC) and Sablefish Regulatory Area per Quarter in the Alaskan Halibut and Sablefish IFQ Programs

Species	Area (vessel class)	1st Quarter (price/pound)		2nd Quarter (price/pound)		3rd Quarter (price/pound)		4th Quarter (price/pound)	
		High	Low	High	Low	High	Low	High	Low
Halibut	2C (D)	$7.50	$5.00	$10.00	$6.50	$9.10	$8.25	$9.25	$8.25
	2C (C)	$8.50	$6.50	$10.75	$6.50	$10.00	$9.00	$10.00	$9.00
	2C (B)	$9.10	$8.75						
	3A (D)	$7.50	$5.00	$8.00	$6.00	$8.00	$6.75	$8.50	$7.00
	3A (C)	$8.00	$7.00	$10.00	$6.50	$10.00	$6.75	$11.00	$7.50
	3A (B)	$8.00	$6.75	$10.00	$7.00	$10.00	$6.75	$11.00	$7.50
	3B (D)	$6.75	$5.25	$7.00	$5.75				
	3B (C)	$8.00	$6.00	$7.50	$6.75	$9.50	$6.00	$12.00	$7.50
	3B (B)	$7.25	$6.50	$7.50	$6.00	$9.50	$6.00	$12.00	$7.50
Sablefish	SE (C)	$8.00	$5.00	$9.00	$6.50	$9.00	$7.25	$9.00	$7.25
	SE (B)	$9.00	$7.00	$7.75	$7.00	$9.00	$7.25	$9.00	$7.25
	WY (C)	$7.75	$5.00	$9.00	$7.25	$9.00	$7.25	$9.00	$7.25
	WY (B)	$9.00	$7.00	$7.75	$7.00	$9.00	$7.25	$9.00	$7.25
	CG (C)	$8.00	$5.50	$7.00	$5.75	$7.25	$5.75	$8.00	$5.75
	CG (B)	$8.00	$5.50	$7.25	$5.75	$7.25	$5.75	$8.00	$5.75
	WG (C)	$6.75	$5.25	$6.75	$5.25	$6.75	$5.25	$6.75	$6.50
	WG (B)	$6.75	$5.25	$6.75	$5.25	$6.75	$5.25	$6.75	$6.50

NOTES: See notes for Table H.2.
SOURCE: Access Unlimited, Inc.

TABLE H.4　1997 Market Prices of Permanent (IFQ/QS) Transfers per International Pacific Halibut Commission (IPHC) and Sablefish Regulatory Area per Quarter in the Alaskan Halibut and Sablefish IFQ Programs

Species	Area (vessel class)	1st Quarter (price/pound)		2nd Quarter (price/pound)		3rd Quarter (price/pound)		4th Quarter (price/pound)	
		High	Low	High	Low	High	Low	High	Low
Halibut	2C (D)	$9.50	$8.50	$9.50	$8.50	$12.00	$8.00	$12.50	$8.00
	2C (C)	$12.00	$9.00	$13.00	$9.00	$15.00	$10.00	$15.00	$10.00
	2C (B)	$12.00	$9.00			$15.00	$10.00	$15.00	$9.00
	3A (D)	$8.50	$7.50	$8.00	$6.50	$10.50	$6.50	$11.50	$6.50
	3A (C)	$11.50	$7.00	$12.25	$6.50	$12.00	$6.50	$13.00	$6.50
	3A (B)	$11.50	$7.00	$12.25	$7.00	$12.00	$6.50	$13.00	$6.50
	3B (D)	$8.25	$7.50	$8.00	$6.75	$10.00	$6.00	$9.50	$6.00
	3B (C)	$12.00	$7.50	$11.50	$7.50	$12.00	$6.00	$12.00	$6.00
	3B (B)	$12.00	$7.50	$12.00	$7.50	$12.00	$6.00	$12.00	$6.00
Sablefish	SE (C)	$10.00	$7.25	$10.50	$7.25	$13.50	$7.25	$15.00	$7.75
	SE (B)	$10.00	$7.25	$10.50	$7.25	$13.50	$7.25	$14.00	$7.75
	WY (C)	$9.00	$5.00	$10.00	$7.25	$11.50	$7.25	$14.00	$7.25
	WY (B)	$9.00	$7.00	$10.00	$7.25	$12.50	$7.25	$14.50	$7.25
	CG (C)	$10.00	$5.75	$10.00	$5.75	$9.00	$5.50	$12.00	$5.50
	CG (B)	$10.00	$5.75	$10.00	$5.75	$11.75	$5.75	$13.25	$5.50
	WG (C)	$6.75	$6.50	$6.75	$5.25	$8.50	$5.00	$8.50	$5.00
	WG (B)	$6.75	$6.50	$6.75	$5.25	$8.50	$5.00	$9.50	$6.50

NOTES:　See notes for Table H.2.
SOURCE:　Access Unlimited, Inc.

years due to several factors. Litigation pending by the Alliance Against IFQs, the unfamiliarity of lenders with the IFQ programs and the risks involved, the newness of the program, and the fishermen's lack of familiarity with the regulations concerning quota exchange all contributed to create market conditions that did not favor significant long-term capital risks. As might be anticipated, the confidence of lenders has increased as the price of quota has increased. As the program has continued, familiarity with it has improved, and litigation has been appealed and denied. Additional factors such as an increased exvessel price of product or shifts in the markets from frozen product to higher-priced fresh product have most likely also contributed to increased quota prices. In recent months, the economic instability of Asian financial markets has affected seafood imports and may be reflected in fluctuations in quota prices.

The prices of quota share in the future may be affected by a wide range of factors. As an example, recent drops in exvessel prices in halibut, due in part to large stock size, increased total allowable catch (TAC), and a large inventory of frozen product, could cause fluctuations in quota price. Lower exvessel prices in sablefish have also been observed, possibly due in part to the unfavorable market conditions in Japan and other Asian countries that are the principal markets for this product. Future quota prices for sablefish may continue to reflect these conditions. The implementation of the North Pacific Loan Program provided for in the Magnuson-Stevens Act (Sec. 108[g]) has recently been announced, and loans to entry-level fishermen of approximately $5,000,000 are anticipated to be available for fiscal year 1998. These loans (having favorable interest rates) may also affect the price of quota share, particularly for smaller amounts of quota. Tables H.2-H.4 summarize the quarterly market values of quota traded in the Alaskan halibut and sablefish IFQ programs and provide an indication of the range of variation in market price since the inception of the program.

It is important to note that capital gains taxes can be assessed on any sale of quota shares that results in a capital gain. In the case of initially allocated quota, the entire value of the quota sold would be subject to capital gains taxes. In the case of quota that is purchased and sold, only the net capital gain (the sale price less the original purchase price) would be subject to taxation. It is difficult to determine the precise amount of revenue generated from capital gains taxes on the sale of quota for several reasons:

1. Not all quota is sold. In the case of the Alaskan halibut and sablefish IFQ programs there have been numerous in-kind trades, particularly during the first two years of the program. These exchanges included trades for shares of quota in other regions or exchanges for other limited access permits (e.g., salmon or herring permits). These exchanges are not subject to capital gains taxation. In some cases, quota may be transferred in exchange for a percentage of the harvest revenue, gear, or other equipment, and these agreements also would not be subject to capital gains taxation. However, trades that included goods and services

may be easily transferred to cash, and thus could be subject to other forms of taxation.

2. In some cases, the sale of quota was between family members, and thus, price for the quota may have been substantially lower than its market value, and the amount of capital gains revenue collected would have been much lower than that generated from quota shares traded on the open market. However, these exchanges could be subject to other forms of taxation as gifts or inheritance.

3. The price of quota differs between the various classes, regions, and the times of sale. As an example, the prices for quota in more desirable regions for halibut and sablefish tend to be higher (e.g., halibut in Area 2C, sablefish in Area SE; Tables H.2-H.4).

4. Purchased IFQs are considered a depreciable asset. Any purchased quota shares will depreciate by 1/15th of its purchase price value per year. After 15 years or more, the potential tax revenue from capital gains taxes that could be collected from the sale of quota will be on the entire sale value of the quota, just like shares received through an initial allocation. Only a portion of sold quota shares is subject to capital gains taxes before 15 years, with the portion of the asset taxed increasing over time.

An accurate determination of the potential revenue generated by capital gains tax in the Alaskan halibut and sablefish IFQ programs would require more accurate information on the price and amounts of quota traded than is available through existing sources of data. The committee recommends mandatory reporting of information on the price of quota exchanges, in the format appropriate for determining the value of the quota sold for all individual fishing quota and limited access programs so that the potential tax revenue from these programs can be calculated accurately.

Because of the difficulty of determining the total amount of quota that could be subject to capital gains taxes, the committee was unable to determine the precise amount of tax revenue generated. Nevertheless, simple calculations suggest that at current market prices of about $7.50 per pound of halibut QS and $10.00 per pound of sablefish QS, the total asset value of QS holdings is on the order of $2-3 billion for halibut QS and $3-4 billion for sablefish QS. Consequently, there is substantial potential for capital gains revenues. Efforts to quantify the amount of capital gains tax revenue generated by the trade of IFQs should be encouraged, and future programs could be designed to provide data that would estimate the revenues generated from quota trading (e.g., using zero-revenue auctions; See Chapter 6).

Potential Revenue From a 3% Exvessel Landing Fee

A preliminary analysis of the potential future revenue from a 3% exvessel landing fee requires knowledge about both the amount of fish landed and the average exvessel price of the fish. The variability in stock conditions and harvest

TABLE H.5   Potential Revenue From the 3% Exvessel Fee for the Surf Clam, Ocean Quahog, Wreckfish, Halibut, and Sablefish IFQ Fisheries Based on Total Harvest and Average Exvessel Price

Fishery	Total Harvest (pounds)	Average Exvessel Price	Estimated Revenue
Surf clam (1996)	63,567,458	$0.60	$1,144,214
Ocean quahog (1996)	46,566,263	$0.45	$628,644
Wreckfish (1996)	442,561	$1.74	$23,102
Halibut (1997)	49,294,628	$2.12	$3,135,138
Sablefish (1997)	28,651,250	$2.38	$2,045,699

NOTES:

1. The annual average exvessel prices for SCOQ and wreckfish are based on an average of monthly data reported to NMFS in these fisheries. Landings of very small quantities of fish with exvessel prices outside the range of the average values reported for other months were not used in this estimation.

2. Preliminary price estimates for halibut and sablefish are derived from the State of Alaska Commercial Operator Annual Reports submitted by processors. They have not been verified with fish ticket information because 1997 fish tickets had not been submitted to the CFEC at the time of this analysis.

3. The weighted average annual exvessel price estimates for halibut and sablefish represent a statewide average price estimate.

4. The sablefish price is derived from the longline "Eastern Cut" sablefish reported delivery price ($3.78/pound), converted to a round weight delivery price.

5. The estimates for halibut and sablefish include only prices for landings in State of Alaska ports. Normally, deliveries outside Alaska (e.g., Seattle or other Puget Sound ports) will yield somewhat higher prices; however, prices for deliveries outside Alaska cannot be estimated until fish ticket data are available.

SOURCES: NMFS, CFEC.

size and variability in price make it difficult to predict the exact amount of revenue that will be generated by the 3% exvessel fee.

A historical analysis of the stock assessments for IFQ fisheries indicates that there is substantial interannual variability in the level of the stocks. The TAC set for a given year reflects this variation. In providing an estimate of potential revenue, a range of TACs reflecting historical data is used as a basis of assessing potential catches. Based on 1996 or 1997 harvest levels and the average exvessel price, the revenues from the 3% fee are described in Table H.5.

Potential Revenue from a 0.5% Registration and Transfer Fee[1]

The potential revenue generated from a 0.5% registration and transfer fee value will vary widely depending on the amount and value of quota traded. It is

---

[1] This is based on the dollar value of quota registered or transferred, not pounds landed, as in the case of the 3% fee analyzed in the previous section.

not possible to determine precisely the potential revenue generated from this fee unless the amount of each sale is recorded and available in a centralized data source. Because of these considerations, the committee was not able to estimate precisely the potential revenue generated from the 0.5% fee. The Alaskan halibut and sablefish fishery has the best available data for assessing the potential revenue from a 0.5% registration and transfer fee. However, an analysis of the potential revenue generated from the 0.5% fee in the Alaskan halibut and sablefish IFQ programs is problematic for several reasons:

1. Historic transactions may not reflect future quota share prices or trade volume.

2. The high level of uncertainty regarding IFQ markets and the relatively short time that this program has been in operation limit the precision of estimates based on previous exchanges for determining potential future revenue.

3. A precise evaluation of the potential revenue from past transfers would require access to proprietary data on the actual price paid by the purchasers of the quota. By assuming that a 0.5% fee would be assessed once for both the transfer and the registration of the quota, the cost could be split between the purchaser and the seller of the quota.

General Considerations for Assessing Quota Value

To determine the potential value of the quotas in an IFQ program, it is important to note that the value of the quota depends on a wide range of factors, including the stability of the market, the accessibility of capital (in turn affected by the confidence of lenders in the security of quota share as a financial instrument and any IFQ-specific loan programs), the degree of consolidation and the incentives and disincentives for consolidation, and the financial situation of the seller. It is likely that the more freely transferable a quota share is, the more likely it is to be traded, and the higher is the potential revenue generated from capital gains taxes and the 0.5% transfer fee. The revenue generated by the fees assessed under an IFQ program is greater than the revenue generated by fees generated under open-access fisheries (essentially zero). Even if a similar 3% exvessel landings fee were assessed on open-access fisheries, IFQs and other limited entry permits provide capital gains tax revenue that are not available from open-access fisheries.

POTENTIAL CHANGES IN THE COST OF ADMINISTRATION
FOR THE FISHERY

Both the literature and the experiences of other nations indicate that the implementation of IFQ programs may increase the cost of managing a given fishery. Increases in administrative costs in U.S. fisheries may be due to increases in the costs of development of regulations for a new program, additional evaluation by the NMFS General Counsel, and increases in staff time required at the regional councils and among NMFS and Coast Guard staff for administration, monitoring, and enforcement. Another cost of such systems is the cost of data collection, management, and distribution. Efforts to include a wider range of potential participants in the development and implementation of IFQ programs will require additional meetings and other activities that could add to the development costs of a program.

In particular, the enforcement of an IFQ program may be more costly than for an open-access fishery. Because IFQ programs rely on the accurate reporting of individual catch and landing data, unless these data are already collected in existing management systems, new enforcement and monitoring activities may be needed. Depending on the nature of the fishery, these costs could be significantly greater than for an existing management regime. In general, fisheries with a large number of participants with small vessels, landing at numerous ports in regions with easy access to markets for unprocessed product (e.g., New England fisheries, Gulf of Mexico shrimp) will be the most difficult to monitor and enforce. Fisheries with these characteristics would require greater increases in expenditures to provide adequate monitoring and enforcement. Moreover, the same conditions that facilitate misreporting or cheating in an IFQ fishery are likely to encourage similar behavior in other non-IFQ management regimes. It is difficult to determine the appropriate level of monitoring and enforcement for a given fishery, although regional councils and fishermen in the region are likely to be able to determine what types of monitoring and enforcement programs would have to be designed for a given fishery.

A preliminary analysis of the cost of the existing management of the Alaskan halibut and sablefish IFQ programs follows. These programs were chosen because data are more readily available and more detailed than for either the SCOQ or the wreckfish IFQ programs. Based on data provided by the NMFS RAM Division, the NMFS Enforcement Division, and U.S. Coast Guard responsible for overseeing IFQ enforcement, it is clear that there were significant new expenditures for personnel, contractual services related to the establishment and maintenance of computer technology, and the computerized transaction terminals that were used in this IFQ program. Table H.6 provides the actual and projected costs of the RAM Division, Office of Administrative Appeals (OAA), and NMFS enforcement for the Alaskan halibut and sablefish IFQ programs. It should be kept in mind that in some cases, enforcement personnel also provide services to non-IFQ fisheries so that the numbers presented in Table H.6 are maximum numbers. In the case of the RAM Division, the tasks assigned to personnel also include other limited access management programs, and some of the personnel costs indicated here would include expenditures for activities other than IFQ management.

TABLE H.6　Actual and Projected Implementation and Maintenance Costs for the Alaskan Halibut and Sablefish IFQ Programs for the RAM Division, the OAA, and NMFS Enforcement (thousands of dollars)

Expense Category	Fiscal Year					
	1993	1994	1995	1996	1997	1998
RAM personnel	$18.0	$433.4	$567.1	$621.0	$658.0	$706.5
Travel	$2.4	$41.4	$87.5	$31.5	$21.3	$43.5
Printing		$17.7	$18.4	$16.6	$14.5	$15.0
Mailing, phone, freight, rent		$96.2	$133.5	$130.0	$128.1	$131.7
Office supplies	$2.1	$50.0	$40.0	$13.0	$1.9	$1.9
Equipment						
Office		$98.6	$11.1	$15.3		$14.5
Transaction terminals			$275.0			
Grants				$65.6	$71.8	$72.0
Training				$5.5	$10.8	$11.0
Contractual services						
Computer programming	$57.1	$329.2	$322.2	$310.3	$265.8	$342.0
Appeals officers (OAA)			$87.2			
RAM subtotal	$79.6	$1,066.5	$1,542.0	$1,208.8	$1,172.2	$1,338.1
NMFS Enforcement			$3,052.0	$2,600.0	$2,600.0	$2,600.0
Totals	$79.6	$1,066.5	$4,594.0	$3,808.8	$3,772.2	$3,938.1

NOTE:　Enforcement costs are estimates from Northwest Fishery Science Center Enforcement Division.
SOURCE:　NMFS RAM Division, Northwest Fishery Science Center Enforcement Division.

The data in Table H.6 do not include the costs that may be incurred by other agencies involved with the management and enforcement of the halibut and sablefish fisheries, such as the U.S. Coast Guard (USCG) and the North Pacific Fishery Management Council (NPFMC). Although other agencies—such as the Alaska Department of Fish and Game, the Commercial Fisheries Entry Commission, and the International Pacific Halibut Commission (IPHC)—all have some involvement in the management of the IFQ fisheries, the committee assumed that expenditures by these agencies dedicated to halibut and sablefish management were not significantly higher under IFQ management than under the previous open-access system. A more precise estimate of the management costs for the halibut and sablefish fisheries would have to consider the budgets of these agencies in more detail.

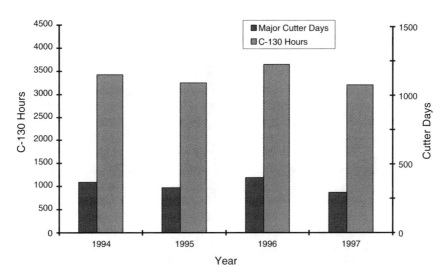

FIGURE H.5 Number of major cutter days and C-130 hours dedicated to fisheries enforcement by Coast Guard District 17 in the North Pacific region.
SOURCE: USCG, 17th District.

In the case of the Coast Guard, since the implementation of the IFQ programs, the number of search and rescue (SAR) operations has decreased dramatically (see Table 3.2). Overall, the total vessel and aircraft time dedicated to fisheries enforcement in Coast Guard District 17 activity has been relatively steady for the past few years (Figure H.5). The halibut fishery has the largest number of vessels in the North Pacific region, and it would be expected that significant efforts would have to be dedicated to enforcement and monitoring operations for this fleet as the season lengthens.

Both ship time and aircraft hours dedicated to enforcement operations in the halibut and sablefish fisheries have increased with the implementation of IFQ programs (Figures H.6-H.7). However, in 1997, the overall time spent in fisheries enforcement operations in the Alaskan halibut and sablefish IFQ fisheries decreased from the 1995-1996 period, possibly as a result of improved enforcement techniques, shifting enforcement operations to other fisheries, or other unrelated events. Based on Coast Guard data, it appears that both the total number of fisheries violations and the percentage of boardings resulting in violations have decreased with the implementation of IFQ programs (Figure H.8).

An average price per hour for aircraft and vessel operations was determined based on data from the U.S. Coast Guard and used to estimate enforcement costs prior to and following the implementation of IFQs. These costs reflect the time allocated to fisheries enforcement operations categorized by the Coast Guard for the halibut and sablefish fisheries, but do not include expenditures by the Coast

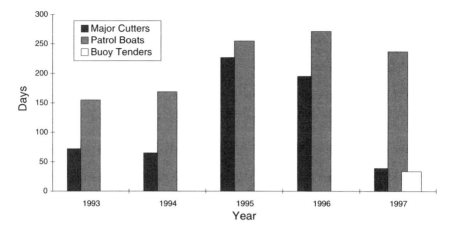

FIGURE H.6 Number of vessel (major cutters, patrol boats, buoy tenders) days dedicated to fisheries enforcement by Coast Guard District 17 in the Alaskan halibut and sablefish fisheries.
SOURCE: USCG, 17th District.

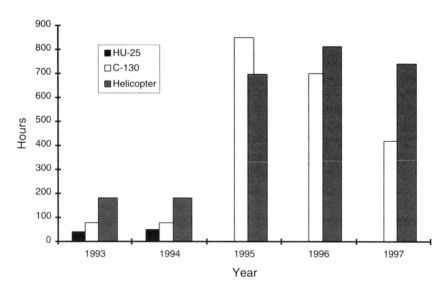

FIGURE H.7 Number of aircraft (HU-25, C-130, helicopter) hours dedicated to fisheries enforcement by Coast Guard District 17 in the Alaskan halibut and sable-fish fisheries.
SOURCE: USCG, 17th District.

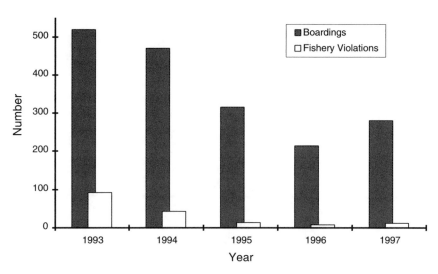

FIGURE H.8 Number of boardings and fishery violations reported by Coast Guard District 17 in the Alaskan halibut and sablefish fisheries.
SOURCE: USCG, 17th District.

Guard for dockside enforcement operations or any increases in personnel costs that may be required to support the increased level of vessel operations in these fisheries. These costs are provided in Table H.7.

From the data provided, overall vessel and aircraft enforcement costs have increased under IFQ management for operations in Coast Guard District 17. However, it appears that these costs were reduced in 1997 compared with the 1995-1996 period, possibly due to improved techniques, increased familiarity with enforcement requirements by fishermen, and/or a shift of enforcement efforts to dockside operations or other fishery operations, drug interdiction efforts, or other missions requiring the reallocation of assets. The estimated costs may not reflect the total costs of all enforcement operations; however, they should indicate the relative trends in enforcement expenditures. **Both the Coast Guard and the NPFMC should monitor future trends in enforcement costs in order to provide a long-term perspective of the potential costs of the enforcement of IFQ programs over time.**

It should be noted that the costs derived from this analysis are not necessarily applicable to other fisheries. The halibut and sablefish IFQ fisheries are conducted in a relatively remote region by more than 2,000 vessels under potentially hazardous weather conditions, in a region with relatively high operating costs. It is reasonable to assume that enforcement costs under any IFQ regime will be higher if there are a large number of small vessels operating in remote areas.

Table H.7 Enforcement Costs for the Operation of Vessels and Aircraft in the Alaskan Halibut and Sablefish IFQ Program Reported by Coast Guard District 17

| | | Fiscal Year | | | | |
Equipment	Cost/Hour	1993	1994	1995	1996	1997
Major cutters	$2,768	$4,783,104	$4,318,080	$14,971,104	$12,860,640	$2,572,128
Patrol boats	$525	$1,953,000	$2,129,400	$3,213,000	$3,427,200	$2,986,200
Buoy tenders	$1,113					$908,208
C-130	$4,244	$326,788	$326,788	$3,697,400	$2,970,800	$1,782,480
HU-25	$3,888	$155,520	$194,400			
Helicopters	$3,837	$326,788	$326,788	$3,607,400	$2,970,800	$1,782,480
Annual Total		$7,545,200	$7,295,456	$25,488,904	$22,229,440	$10,031,496

NOTE: Rates (cost/hour) are assumed constant throughout the period reviewed. Costs are based on data provided by the USCG, 17th District.

SOURCE: USCG, 17th District.

Comparison of Potential Revenue and Anticipated Expenditures

Without detailed information about the amount of capital gains revenue or changes in corporate or business tax revenues, it is not possible to precisely quantify the potential fiscal revenue from the Alaskan IFQ programs. Although the total revenue from the Alaskan halibut and sablefish IFQ fisheries cannot be precisely quantified given the existing data, the program does have the potential to generate revenue through a variety of taxes and fees. The cost of management conducted prior to implementation of the program did not have cost recovery fees specifically associated with it.

No attempt was made to determine whether the revenues generated from the SCOQ and wreckfish IFQ fisheries were sufficient to cover the additional administrative costs that have been incurred in managing these fisheries because clearly definable costs attributable to the IFQ management of these programs and accurate price data were not available. It does appear, based on testimony and our analyses, that IFQ management is likely to be more costly than traditional forms of management. However, it is likely that IFQ-managed fisheries could provide a positive net revenue flow to the nation if the capital gains revenue and changes in corporate and business tax revenues are sufficiently high and if the 3% exvessel fee and 0.5% registration and transfer fee are implemented. Again, the calculations used here assume the maximum fee provided for under the Magnuson-Stevens Act; higher or lower fees will affect the revenue generated.

It should be noted that the Magnuson-Stevens Act authorizes up to 25% of the fees collected under the mandated 3% exvessel fee in existing IFQ programs to be used to aid in financing the purchase of IFQs by fishermen who fish from small vessels or who are the first-time purchasers of quota (Sec. 304[d]). In the North Pacific region, these fees are mandated to be used to finance small-vessel (Classes B, C, and D) and entry-level fishermen (fishermen who do not own halibut and sablefish IFQ, who harvest less than 8,000 pounds in a given year, and who participate aboard a vessel) in the Alaskan halibut and sablefish IFQ programs (Sec. 108[g][2]). Additionally, fees collected under the assessment of the 0.5% transfer fee must be used for the implementation of a central registry system for limited access system permits and for administering and implementing the Magnuson-Stevens Act provisions for the fishery in which the fee is assessed (Sec. 305[h]). The specific use of the fee is determined by the Secretary of Commerce (Sec. 305[h]). These reallocations of funds could reduce the total amount of revenue that could be used for the general management of an IFQ program.

Although it is difficult to predict the costs of implementing other IFQ programs, it is possible that if more programs were implemented, costs would be reduced based on experiences with previous programs. In addition, facilities developed for monitoring existing IFQ programs might be extended to additional programs with little additional costs. For example, the computerized transaction terminals used in the Alaskan halibut and sablefish IFQ programs could be used for recording landings of additional species in other programs with little or no increase in hardware costs if those fisheries used similar ports and facilities. It is reasonable to assume that if IFQ management were attempted in regions with significant IFQ experience and infrastructure, the costs of initial implementation would be lower than for previously implemented programs. Further, it is likely that if IFQs were implemented for large fisheries with fewer vessels, a well-organized biological data gathering system, limited enforcement costs, and high exvessel revenues (e.g., Alaskan pollock), the capital gains taxes and exvessel fees assessed from these larger fisheries would be more than sufficient to cover additional administrative costs and could generate significant net tax revenue to the nation, as well as potentially improving the operational efficiency of these fisheries.

For fisheries with large numbers of vessels and complicated enforcement, IFQ-based management could be more costly than traditional forms of management if fees assessed are significantly lower than the costs of additional enforcement that may be required under IFQ management. It should be noted that other forms of limited access management may be equally costly; however, unlike IFQ programs, other limited access programs do not have specific cost recovery fees established and may consequently have higher net costs than IFQ management. If there are sufficient data to determine both the costs and the revenues generated by IFQ programs, it could be possible to determine the total costs of a program and the revenue required to offset these costs.

POTENTIAL CHANGES IN THE ECONOMIC EFFICIENCY OF FISHING OPERATIONS

The committee heard testimony that the implementation of IFQ programs has provided an opportunity for improved economic efficiency in fishing operations. Measuring how the change in management regime might affect the economic efficiency of fishery operations requires reasonably accurate information on changes in the costs and revenues of fishing operations. Since many of these data are proprietary and not available, precise estimates of the potential effects of IFQ management on economic efficiency are not possible. However, the committee heard testimony from both proponents and opponents of IFQ management that fishing operations had changed with the implementation of IFQ programs. The increased revenues generated by the improved economic efficiency of fishing operations will be one of the principal benefits of an IFQ program; however, quantifying these benefits is difficult.

Many participants in IFQ fisheries indicated that under IFQ management, fishing operations could be modified to take advantage of more favorable weather conditions, and better exvessel prices or to accommodate opportunities in other fisheries, reducing the costs or increasing the revenues of operating a vessel. A

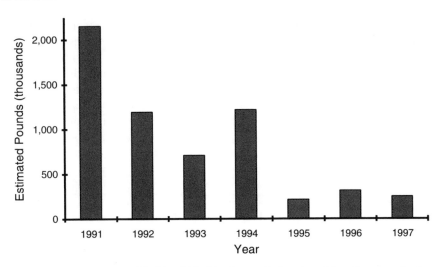

FIGURE H.9 Estimate of halibut killed by lost and abandoned longline gear in the 1991-1997 Alaskan commercial halibut fishery (IPHC regulatory Areas 2C-4). SOURCES: Gilroy et al. (1996); IPHC (1997).

number of individuals in the Alaskan halibut and sablefish IFQ fisheries noted that because of these changes, they were able to use less gear, lost less gear, and realized significant savings over pre-IFQ gear expenditures. This information seems to be corroborated by data from IPHC, where the ratios of gear (skates) lost to gear used (skates hauled) in Alaskan halibut regulatory Areas 2C and 3A are similar to the ratios observed in the Canadian halibut individual vessel quota (IVQ) fishery in regulatory Area 2B (IPHC, 1997). Figure H.9 summarizes the estimates provided by IPHC on the pounds of halibut killed in the Alaskan halibut fishery from lost or abandoned longline gear. Since the inception of the halibut IFQ program in 1995, there appears to be a significant decrease in the number and pounds of halibut killed due to ghost fishing. These values may be indicative of savings in the Alaskan sablefish fishery, which uses similar gear and techniques.

Some vessel owners in the fishery stated that costs for fuel, food, and other supplies had also been reduced due to the longer time for preparation, improved planning opportunities, and ability to travel to port to obtain supplies as needed. A number of vessel owners stated that in the derby fisheries, spares of all types of equipment had to be carried aboard so that repairs could be made at sea and fishing operations continued because the day-long open season had to be utilized fully. Some vessel owners stated that with the advent of the IFQ programs, less nonessential equipment had to be carried aboard and the costs and space required to carry this equipment had been reduced and unneeded equipment could be sold. Some of this testimony is corroborated by surveys of vessel owners in the halibut and sablefish IFQ program conducted by the CFEC, indicating that after the first year of the IFQ program (1995), costs for repairs, gear, and insurance had dropped and costs for fuel, ice, and food had increased slightly (Knapp and Hull, 1996).

The most significant increase in operating costs appears to be the purchase of quota itself. The committee was not able to gather sufficient data to compare the costs of quota purchase to the savings reported in equipment, fuel, and gear costs. Assessment of both the 3% exvessel fee and the 0.5% registration and transfer fee will also add to the overall cost of harvesting halibut and sablefish in the IFQ programs. Once these fees are assessed, their effects on the profitability of fishing operations will have to be considered.

Other changes in the efficiency of fishing operations seem to have developed in some of the fisheries. Participants in the wreckfish and Alaskan halibut and sablefish IFQ fisheries stated that because of the changes in fishing operations and the ability to provide more marketable fresh product more consistently, opportunities existed to expand operations into the wholesaling and marketing of product.

As might be anticipated, some of the changes in the patterns of product delivery negatively affected the operations of processors. The committee heard testimony that some processors in the SCOQ and the Alaskan halibut and sablefish IFQ programs felt that the shift from large pulses of product to a more even product flow had disadvantaged some of their operations. However, it appears that not all types of processing operations were equally affected by these shifts in product flow. Some processors in the Alaskan halibut and sablefish IFQ fisheries were not significantly disadvantaged by the changes in product flow. The committee did not hear specific testimony on those operations that appeared to have experienced reduced revenue with the shift to an IFQ fishery; however, it is reasonable to assume that processing operations that had relied on large pulses of product for processing for the frozen market and had failed to modify their operations to accommodate the longer IFQ season would be less profitable as season lengthened. Similarly, processing operations that shifted to accommodate longer seasons and the increasing availability of fresh product could maintain more profitable operations. From testimony received from the processing industry, it is not clear that all processing operations will be economically disadvantaged with the shift to an IFQ-managed fishery. Moreover, with changes in both exvessel and wholesale prices, it is unclear whether processor revenues have increased or decreased.

The North Pacific Fishery Management Council conducted an analysis of the potential economic effects of implementing IFQ programs for halibut and sablefish prior to their implementation (NPFMC, 1997b). These analyses indicated that the implementation of IFQ programs for these fisheries could result in significant increases in revenue due to (1) increased product quality and exvessel price; (2) decreased processing and marketing costs; (3) decreased fishing mortality due to lost gear; (4) reduction in the redundant gear costs; (5) improved efficiency of harvesting operations; (6) decreases in discard mortality of bycatch (specifically rockfish and sublegal halibut); and (7) decreased discard of incidental harvests of halibut in the sablefish fisheries.

POTENTIAL CHANGES IN EMPLOYMENT IN THE FISHERY AND ASSOCIATED INDUSTRIES

The committee received testimony that the implementation of IFQ programs can affect the number of crew positions and total employment in a fishery. A more detailed discussion of the findings concerning the changes in crew employment and number of vessels in the fishery is found in Appendix G. Several vessel owners and crew members mentioned that the number of crew members had been reduced with the implementation of the halibut and sablefish IFQ programs. Although fewer crew members may improve the economic efficiency of the vessel operation, this may have significant adverse affects on the individual crew members. The number of crew members employed is at least partially balanced by an increase in the number of days of employment for those remaining. Information from the Deep Sea Fishermen's Union provided to the committee seems to indicate that the crew members responding to an informal survey had increased incomes, but it is difficult to determine if this is due principally to the implementation of the halibut and sablefish IFQ programs. Studies have also been conducted by the CFEC (Knapp, 1997b) in an attempt to compare possible changes in personal income with the implementation of IFQs. However, these studies compare changes in personal income during the first year of the program, and subsequent studies looking at the longer-term effects of IFQ management on crew income are not available.

The committee did not hear specific testimony on how the shift to an IFQ-managed fishery might affect other industries involved in the maintenance or provisioning of fishery operations. As previously mentioned, reduced gear needs could affect those industries supplying gear, and reduced fuel and supply costs could affect other industries. Since gear expenditures and other provisions may be affected by a wide range of factors, it is not clear that IFQ management will either positively or negatively affect these industries. Although it is not unreasonable to assume that IFQ management may alter season length and fishing operations, and therefore affect industries that rely on the fishing industry, the effects of IFQ management on these other industries is not clear. Also, it is difficult to determine how these industries had responded to potential changes in demand caused by the IFQ programs. It is possible that some have modified their business operations to account for these changes.

Another issue on which there is limited information is the effect of IFQ management regimes on consumer demand. The committee received information from both the CFEC and the Canadian Department of Fisheries and Oceans that the implementation of IFQ programs in the halibut fisheries resulted in a substantial increase in the percentage of product delivered to the market fresh, probably reflecting consumer preferences. If IFQs significantly expand the fresh market by providing higher-quality product for a longer period of time, this could have substantial economic effects including the expansion of marketing and wholesale operations and potentially a reduction in more traditional markets for frozen halibut. In general, it is reasonable to assume that if IFQs provide greater opportunities for fulfilling consumer preferences, IFQ management could provide significant net benefits to the nation in terms of new marketing efforts and increased consumer satisfaction. It would be worthwhile to examine in more detail the ways in which IFQ management affects the marketing of fish product.

POTENTIAL LONG-TERM EFFECTS ON FISHERY-DEPENDENT COMMUNITIES

The committee has provided some description of the potential effects of IFQ implementation in Appendix G, which is reflected in several of the recommendations in Chapter 6. In most cases, the lack of socioeconomic data makes it impossible to characterize precisely how communities may be affected by the implementation of an IFQ program and the potential costs and benefits to the nation from these effects on fishing communities. Because a wide range of factors—including the status of other fisheries, market conditions, and alternative employment opportunities—affect fishery-dependent communities, attributing specific conditions in a community to the effects of IFQ management is difficult.

One factor that could be measured, which may provide some indication about the ability of fishery-dependent communities to respond to potential changes with the implementation of an IFQ program, is the distribution of quota in certain communities from the inception of the IFQ program to the present. Only the Alaskan halibut and sablefish IFQ fisheries have sufficient data regarding the purchase and transfer of quota in specific communities. These effects can be typified by five communities with a history of dependence on fishery resources that have typically been active in both the halibut and the sablefish fisheries: Homer, Kodiak, Petersburg, Seward, and Sitka, Alaska. The Commercial Fisheries Entry Commission and the NMFS RAM Division have tracked the total number or quota shareholders and the amount of quota held in these communities, and the total number and amount of quota held by crew members in these communities (Tables H.8-H.12).

Since the inception of the IFQ programs, four of these communities have maintained or increased their absolute amount and total percentage of quota shares. In some fishing areas, the amount and percentage of quota held may have dropped since the inception of the program, but in general, the numbers of total quota units held has increased in four of the five communities. Data indicate that residents of Kodiak and Seward hold slightly less halibut quota now than at the time of implementation; however, both communities have increased their hold-

TABLE H.8   Initial Allocation and Recent Holdings of Quota Share by Residents of Homer, Alaska in the Alaskan Halibut and Sablefish IFQ Programs

Halibut

	Quota Initially Issued			Quota Holdings as of December 31, 1997		
Area	Persons Issued	1997 IFQ Pounds	% of Area	Persons Holding	1997 IFQ Pounds	% of Area
2C	20	20,071	0.20	8	15,122	0.15
3A	316	2,610,341	10.40	238	2,161,113	8.65
3B	113	617,679	6.84	82	640,812	7.12
4A	55	289,420	9.82	41	345,135	11.70
4B	15	168,667	6.05	11	95,034	3.41
4C	2	10,881	1.88	1	5,005	0.86
4D	1	11,060	1.36	0	0	0.25
4E	1	0	0.00	1	0	0.96
Total	320	3,728,117		258	3,262,221	

Sablefish

	Quota Initially Issued			Quota Holdings as of December 31, 1997		
Area	Persons Issued	1997 IFQ Pounds	% of Area	Persons Holding	1997 IFQ Pounds	% of Area
AI	8	187,970	11.84	6	173,761	10.95
BS	8	108,222	11.16	5	79,161	8.16
CG	61	422,459	3.73	53	318,368	2.81
SE	9	76,197	0.57	5	38,138	0.47
WG	10	176,861	5.37	7	139,452	4.25
WY	30	94,005	1.86	14	71,147	1.41
Total	62	1,065,713		58	820,027	

NOTES:

1. "Persons" includes all entities (individuals, corporations, partnerships, etc.) holding QS by area and species; each such person may hold QS in more than one administrative area and for both species. Because of this, the total number of persons holding quota is not a sum of the persons holding quota in each area.

2. "Issued" quota is all QS that had been issued to residents through 1997. Initial issuance of QS is almost complete; however, very small amounts of QS may continue to be initially issued through resolution of appeals to initial determinations and other administrative actions.

3. Residency is determined by current address supplied by applicants and permit holders. Resident is defined as a resident of the particular city and does not include individuals whose residences are in neighboring communities.

4. IFQ pounds are displayed in 1997 equivalent pounds, using 1997 Quota Share Pools and TACs.

5. Halibut pounds are displayed in net (head off and gutted) pounds; sablefish pounds are displayed in round pounds.

SOURCES: CFEC, NMFS RAM Division.

TABLE H.9   Initial Allocation and Recent Holdings of Quota Share by Residents of Kodiak, Alaska in the Alaskan Halibut and Sablefish IFQ Programs

			Halibut			
	Quota Initially Issued				Quota Holdings as of December 31, 1997	
Area	Persons Issued	1997 IFQ Pounds	% of Area	Persons Holding	1997 IFQ Pounds	% of Area
2C	31	24,079	0.24	18	9,492	0.09
3A	385	5,675,162	22.69	295	5,458,522	21.86
3B	180	1,673,027	18.53	148	2,044,212	22.71
4A	62	518,270	17.59	56	609,392	20.73
4B	26	435,748	15.64	20	344,313	12.37
4C	8	68,658	11.84	7	80,448	13.87
4D	10	35,227	4.34	11	63,185	7.78
4E	2	0	0.00	2	0	0.00
Total	388	8,430,170		331	8,609,563	

			Sablefish			
	Quota Initially Issued				Quota Holdings as of December 31, 1997	
Area	Persons Issued	1997 IFQ Pounds	% of Area	Persons Holding	1997 IFQ Pounds	% of Area
SE	22	46,271	0.58	18	39,372	0.49
WY	30	303,601	6.02	30	299,897	5.95
CG	74	1,052,034	9.28	72	1,265,744	11.19
WG	24	72,617	2.21	28	222,498	6.78
AI	12	23,945	1.51	11	22,059	1.39
BS	16	84,709	8.73	18	98,791	10.18
Total	74	1,583,178		77	1,948,360	

NOTES: See notes for Table H.8.

SOURCES: CFEC, NMFS RAM Division.

TABLE H.10   Initial Allocation and Recent Holdings of Quota Share by Residents of Petersburg, Alaska in the Alaskan Halibut and Sablefish IFQ Programs

				Halibut			
Quota Initially Issued				Quota Holdings as of December 31, 1997			
Area	Persons Issued	1997 IFQ Pounds	% of Area	Persons Holding	1997 IFQ Pounds	% of Area
2C	255	1,854,685	18.41	231	2,352,833	23.35
3A	66	1,188,194	4.75	73	1,580,709	6.32
3B	9	190,532	2.09	6	186,809	2.07
4A	6	48,454	1.64	5	50,286	1.71
4B	3	76,669	2.76	2	86,236	3.10
4D	3	15,382	2.66	2	14,889	2.57
4E	2	0	1.58	2	0	1.58
Total	266	3,373,916		255	4,271,812	

				Sablefish			
Quota Initially Issued				Quota Holdings as of December 31, 1997			
Area	Persons Issued	1997 IFQ Pounds	% of Area	Persons Holding	1997 IFQ Pounds	% of Area
SE	52	1,083,684	13.38	51	1,076,846	13.39
WY	31	417,783	8.28	34	482,460	9.60
CG	37	1,121,114	9.89	29	1,351,733	11.93
WG	8	113,970	3.46	5	100,195	3.04
AI	5	29,114	1.83	4	25,734	1.74
BS	4	84,158	8.67	4	65,401	6.74
Total	57	2,849,824		68	3,102,370	

NOTES: See notes for Table H.8.

SOURCES: CFEC, NMFS RAM Division.

TABLE H.11   Initial Allocation and Recent Holdings of Quota Share by
Residents of Seward, Alaska in the Alaskan Halibut and Sablefish IFQ
Programs

	Halibut					
Quota Initially Issued				Quota Holdings as of December 31, 1997		
Area	Persons Issued	1997 IFQ Pounds	% of Area	Persons Holding	1997 IFQ Pounds	% of Area
2C	4	19,777	0.20	0	0	0.00
3A	63	428,831	1.71	43	408,811	1.64
3B	11	209,872	0.39	10	35,751	0.40
4A	2	13,962	0.47	2	3,799	0.13
4B	0	0	0.00	2	13,383	0.48
4E	0	0	0.00	1	0	5.83
Total	66	497,607		44	461,744	

	Sablefish					
Quota Initially Issued				Quota Holdings as of December 31, 1997		
Area	Persons Issued	1997 IFQ Pounds	% of Area	Persons Holding	1997 IFQ Pounds	% of Area
SE	3	52,823	0.65	1	41,249	0.51
WY	6	77,741	1.54	4	112,948	2.24
CG	25	199,860	1.76	18	261,042	2.31
WG	2	3,892	1.18	2	4,761	0.15
AI	1	375	0.02	1	1,172	0.07
BS	0	0	0.00	1	2,531	0.26
Total	25	334,691		18	423,703	

NOTES: See notes for Table H.8.

SOURCES: CFEC, NMFS RAM Division.

TABLE H.12   Initial Allocation and Recent Holdings of Quota Share by
Residents of Sitka, Alaska in the Alaskan Halibut and Sablefish IFQ Programs

Halibut

Area	Quota Initially Issued			Quota Holdings as of December 31, 1997		
	Persons Issued	1997 IFQ Pounds	% of Area	Persons Holding	1997 IFQ Pounds	% of Area
2C	328	1,681,263	16.68	275	1,737,254	17.21
3A	130	801,740	3.21	115	992,008	3.97
3B	21	254,369	2.82	16	243,353	2.70
4A	16	103,349	3.51	10	111,990	3.81
4B	8	114,685	4.12	5	97,001	3.48
4C	2	3,721	0.64	2	3,721	0.64
4D	2	6,861	0.85	1	2,981	0.37
Total	337	2,965,988		303	3,188,308	

Sablefish

Area	Quota Initially Issued			Quota Holdings as of December 31, 1997		
	Persons Issued	1997 IFQ Pounds	% of Area	Persons Holding	1997 IFQ Pounds	% of Area
SE	118	1,826,071	22.55	104	2,044,985	25.44
WY	38	316,597	6.28	36	410,737	8.15
CG	34	711,411	6.28	30	733,224	6.48
WG	13	198,916	6.04	10	189,701	5.78
AI	7	85,932	5.41	8	86,335	5.44
BS	6	47,031	4.85	6	48,541	5.00
Total	120	3,185,958		114	3,513,523	

NOTES: See notes for Table H.8.
SOURCES: CFEC, NMFS RAM Division.

ings of sablefish quota significantly (Tables H.9 and H.11). All five communities seem to have fewer individual holders of quota now than at the beginning of the program (Tables H.8-H.12). The data from Homer indicate that there has been a substantial decrease in both the number of quota shareholders and the amount of quota held. The committee did not analyze where these quota shares were transferred, nor the employment status of individuals that had sold quota and left the fishery.

These trends may not extend to other communities in the region or to a wider range of communities. Additionally, it is difficult to determine the number of individuals who had been active in these fisheries prior to implementation of the IFQ programs who may have chosen to purchase quota if they were not allocated quota during the initial allocation process. CFEC has produced a number of studies analyzing the trends in participation in the Alaskan halibut and sablefish fisheries prior to implementation of the IFQ programs. In some communities, the total amount of quota held in halibut and sablefish has increased by up to 20% above the initial quota allocations (e.g., sablefish IFQ in Kodiak, halibut IFQ in Petersburg; Tables H.9 and H.10). In the case of Homer, the total amount of quota held decreased by nearly 15% in both the halibut and the sablefish fisheries (Table H.8). These data suggest that in four of the five communities, substantial investments have been made in the IFQ programs, and following these trends will help provide further information about the ability of similar communities dependent on fishing to become actively involved in IFQ programs.

The CFEC, in response to a request from the NMFS RAM Division, has also compiled data tracking the gross revenue in many of the Alaskan communities where IFQ is held. Data from the CFEC indicate that for all five of the communities analyzed in this section, the gross revenues of the IFQ holders in these communities for both the halibut and the sablefish IFQ fisheries have increased since the inception of the program (CFEC, 1998). Further, these data indicate that the gross revenue per individual quota holder has increased in both the halibut and the sablefish fisheries in all five of these communities since the inception of the program (CFEC, 1998).

The CFEC also provides data on the amount of investment by crew members, and the changes in the allocation of quotas in these communities, in order to provide some perspective on the potential economic effects of the implementation of IFQ programs (Tables H.13-H.17). However, the results may not be representative of all communities, and a more thorough analysis may be necessary. Tables H.13-H.17 show that investment in the IFQ program is increasing in the five communities examined, and in some cases, crew members from specific communities now comprise a small, but apparently increasing, portion of the overall amount of quota held by the community in some areas (e.g., Kodiak crew members in Area 3B and 4A halibut, Petersburg crew members in Area 2C halibut; Tables H.14 and H.15). However, overall, these data indicate that, generally, crew members do not hold a large percentage of quota share in the

TABLE H.13 Quota Share Holdings for Crew Members with a Homer Designated City Address in the Alaskan Halibut and Sablefish IFQ Programs as of December 31, 1997

Area	Number of Crew	1997 IFQ Pounds	Crew IFQ Pounds as a Percent of City Pounds	Crew IFQ Pounds as a Percent of Area Pounds
		Halibut		
2C	1	9	0.0	0.00
3A	34	289,080	13.7	1.16
3B	18	151,916	24.0	1.69
4A	17	97,772	28.3	3.33
4B	3	10,190	10.7	0.37
Total	43	548,965	17.1	1.07
		Sablefish		
CG	5	18,720	5.9	0.17
WG	1	5,717	3.7	0.17
WY	2	734	1.0	0.01
Total	6	25,171	3.0	0.08
Grand Total	45		14.2	0.71

NOTES:

1. The grand total represents unique crew members; if the grand total is less than the sum of crew holding halibut or sablefish quota this is due to some crew members holding quota for both species.

2. Halibut pounds are net (head off, gutted); sablefish pounds are round weight.

3. Percent of whole area data uses CFEC "1997 Pounds Equivalent of QS for Entire Area," which differs from the actual TAC.

4. Total percent of city (or area) pounds compares crew holdings to holdings for the entire city.

SOURCES: CFEC, NMFS RAM Division

TABLE H.14   Quota Share Holdings for Crew Members with a Kodiak Designated City Address in the Alaskan Halibut and Sablefish IFQ Programs as of December 31, 1997

Area	Number of Crew	1997 IFQ Pounds	Crew IFQ Pounds as a Percent of City Pounds	Crew IFQ Pounds as a Percent of Area Pounds
Halibut				
2C	1	280	3.0	0.00
3A	37	519,310	9.5	2.08
3B	28	379,915	18.6	4.22
4A	12	119,795	19.7	4.07
4B	5	47,433	13.8	1.70
4C	1	3,178	4.0	0.55
4D	3	17,694	28.0	2.18
Total	59	1,087,605	12.6	2.13
Sablefish				
AI	1	137	0.6	0.01
CG	5	34,533	2.7	0.31
SE	1	16	0.0	0.00
WG	4	28,077	12.6	0.86
WY	4	17,054	5.7	0.34
Total	8	79,816	4.1	0.26
Grand Total	59	1,167,422	11.1	1.43

NOTES: See notes for Table H.13.

SOURCES: CFEC, NMFS RAM Division.

TABLE H.15   Quota Share Holdings for Crew Members with a Petersburg Designated City Address in the Alaskan Halibut and Sablefish IFQ Programs as of December 31, 1997

Area	Number of Crew	1997 IFQ Pounds	Crew IFQ Pounds as a Percent of City Pounds	Crew IFQ Pounds as a Percent of Area Pounds
		Halibut		
2C	50	367,216	15.49	3.64
3A	17	171,900	10.81	0.69
3B	1	2,324	1.24	0.03
4A	2	24,274	48.27	0.83
Total	60	565,715	13.18	1.87
		Sablefish		
CG	3	63,154	4.67	0.56
SE	9	69,883	6.58	0.87
WY	6	34,534	7.16	0.69
Total	14	167,570	5.43	0.33
Grand Total	66	733,285	9.9	0.90

NOTES: See notes for Table H.13.
SOURCES: CFEC, NMFS RAM Division.

TABLE H.16   Quota Share Holdings for Crew Members with a Seward Designated City Address in the Alaskan Halibut and Sablefish IFQ Programs as of December 31, 1997

Area	Number of Crew	1997 IFQ Pounds	Crew IFQ Pounds as a Percent of City Pounds	Crew IFQ Pounds as a Percent of Area Pounds
		Halibut		
3A	8	31,701	7.8	0.13
3B	2	4,722	13.2	0.05
Total	8	36,423	7.9	0.07
Grand Total	8	36,423	4.1	0.04

NOTES: See notes for Table H.13. There were no crew sablefish IFQ held in the community of Seward.
SOURCES: CFEC, NMFS RAM Division.

TABLE H.17   Quota Share Holdings for Crew Members with a Sitka Designated City Address in the Alaskan Halibut and Sablefish IFQ Programs as of December 31, 1997

Area	Number of Crew	1997 IFQ Pounds	Crew IFQ Pounds as a Percent of City Pounds	Crew IFQ Pounds as a Percent of Area Pounds
Halibut				
2C	53	202,425	11.7	2.01
3A	24	95,634	9.6	0.38
3B	5	42,622	17.5	0.47
4A	3	28,462	25.4	0.97
Total	69	369,143	11.6	0.72
Sablefish				
CG	6	25,347	3.5	0.22
SE	13	90,229	4.4	1.12
WY	6	15,210	3.7	0.30
Total	19	130,787	3.7	0.43
Grand Total	76	499,930	7.5	0.61

NOTES: See notes for Table H.13.
SOURCES: CFEC, NMFS RAM Division.

communities examined. Examining these trends in crew investment in the IFQ programs may help to improve the understanding of the incentives or effects of investment in these programs and could provide guidance for future IFQ programs or investment policies. **Trends in crew holdings of quota share with the implementation of the North Pacific Loan Program should be monitored by NMFS, CFEC, or the North Pacific Council to provide more data on crew investment and the efficacy of loan programs in increasing the share of quotas held by crew members and smaller quota holders.**

Based on the information on the distribution of quota, it appears that in all of these communities, vessel owners, and to a limited extent, crew members have made substantial investments to participate in the Alaskan halibut and sablefish IFQ fisheries. Overall, crew members held 11.2% of the halibut TAC and 4.6% of the sablefish TAC in 1997, and these percentages appear to be increasing (NMFS, 1998). Even in those communities where the total quota share may not have increased, or has decreased (e.g., halibut in Kodiak, halibut and sablefish in Homer), there appear to be shifts in the areas where quota is held, possibly representing preferences in location, regional variations in exvessel price, or the

TABLE H.18 Percentage of Landings in Weight for the Alaskan Halibut
Fishery (1991-1996)

Year	Total Landings (pounds)	% Catcher Vessels (Alaska)	% Catcher-Processors	% Catcher Vessels (other states)
1991	49,535,011	91.6	4.9	3.5
1992	51,829,522	92.6	5.1	2.2
1993	48,449,185	88.0	8.9	3.2
1994	44,449,185	87.2	9.1	3.7
1995	32,151,518	90.0	7.6	2.4
1996	35,386,715	89.2	8.3	2.6

SOURCE: CFEC.

affordability of quota share in other areas. These data indicate that fewer vessel owners, and fewer participants overall, hold quota than at the implementation of the program. Based on testimony from other sources, it is likely that the reduction in quota shareholders represents a reduction in the number of individual fishing operations in these fisheries. Note, however, that even with these reductions, the current number of quota shareholders is greater that the maximum number of fishermen who participated in any given year of the derby fishery. Additionally, based on testimony received from these communities and other regions, it appears that the overall crew employment in the remaining fishing operations has probably decreased. However, the gross revenue per quota shareholder appears to have increased since the inception of the program (CFEC, 1998).

What is much more difficult to determine is the overall effects of these shifts on the local economies. Investment in the IFQ programs has increased, the amount of product landed in Alaska has not changed significantly, and the general ranking of these communities in terms of landed product from these fisheries does not appear to have significantly changed since the implementation of the program (Tables H.18-H.21). However, there are certainly shifts within the landings of product, although it is not clear if these changes are due to factors other than the implementation of IFQ management and drawing conclusions on the long-term distributions of landings is difficult. Overall exvessel price for the product has increased, although prices have fallen recently due to substantial increases in the TAC and possibly other factors. Data for 1997 were not available at the time of the committee's analysis; however, additional years of harvest landings data will provide a more accurate indication of whether there are any changes in the relative rankings of ports in terms of landings.

Some communities have also been affected by changes in other fisheries, such as low exvessel prices for salmon and herring, and it is not easy to separate the effects of changes in other fisheries from the potential effects of the IFQ

TABLE H.19 Percentage and Relative Ranking of Landing (in Weight) for Alaskan Ports in the Alaskan Halibut Fishery (1991-1996)

Port Region	Percentage (Relative Ranking) of Harvest Landed in Alaskan Ports					
	1991	1992	1993	1994	1995	1996
Kenai Peninsula-Anchorage	20.2%(2)	21.9%(2)	20.4%(2)	22.9%(1)	19.5%(2)	21.4%(1)
Kodiak Island Borough	25.1%(1)	26.0%(1)	21.2%(1)	21.2%(2)	23.0%(1)	20.3%(2)
Aleutians-Alaska Peninsula Bering Sea	16.8%(3)	15.1%(3)	14.4%(3)	13.6%(3)	12.6%(3)	12.2%(3)
Wrangell-Petersburg area	6.1%(5)	7.7%(4)	8.5%(4)	7.0%(5)	10.1%(4)	11.0%(4)
Sitka Borough	6.0%(6)	6.1%(6)	6.2%(6)	6.3%(6)	8.8%(5)	8.0%(5)
Skagway-Yakutat-Angoon area	7.9%(4)	6.4%(5)	6.6%(5)	7.5%(4)	8.5%(6)	7.5%(6)
Valdez-Cordova area	4.5%(6)	4.9%(7)	4.5%(8)	4.6%(7)	3.5%(7)	3.4%(7)
Ketchican-Prince of Wales	3.9%(7)	3.3%(8)	4.8%(7)	3.3%(8)	2.6%(8)	2.7%(8)
Juneau Borough	1.1%(8)	1.0%(9)	1.2%(9)	0.6%(9)	1.3%(9)	2.6%(9)
Haines Borough	0.1%(10)	0.2%(10)	0.2%(10)	0.1%(10)	0.1%(10)	0.1%(10)

NOTE: Some areas are aggregated in order to protect the confidentiality of individual processors. Typically, the areas aggregated represent similar types of processing facilities.

SOURCE: CFEC.

TABLE H.20   Percentage of Harvest Landings in Weight for the Alaskan Sablefish Fishery (1991-1996)

Year	Total Landings (pounds)	% Catcher Vessels (Alaska)	% Catcher- Processors	% Catcher Vessels (other states)
1991	51,209,634	87.7	12.1	0.2
1992	48,400,987	85.3	13.6	1.1
1993	49,313,981	78.0	21.3	0.7
1994	44,827,268	81.1	16.7	2.2
1995	40,628,028	84.3	13.6	2.1
1996	33,143,809	82.4	15.8	1.8

SOURCE: CFEC.

fisheries. Additionally, the change in processing operations with the shift to a larger percentage of the market sold as fresh product has probably reduced employment at processing plants that produced frozen product. However, other employment opportunities could have been created to market fresh product.

The fact that these communities have increased their overall quota share seems to indicate a desire by some residents to make the investments necessary to continue in the halibut and sablefish fisheries. These data indicate that there are opportunities for members of these communities to invest in quota; however, the relative ease of access to capital and the possible constraints that this may have on the overall quota holdings in these communities could not be determined. Undoubtedly, some members of the communities have benefited from the initial allocation of quota shares whereas others, including crew, have lost employment in the fishery as a result of fewer fishing operations or the consolidation of fishing operations. The committee received testimony from these communities that some vessel owners and crew members who had not received quota had "bought into" the fishery and felt that fishing under IFQ management was preferable to the previous derby fisheries. Others, typically those who did not receive an initial allocation or a significant initial allocation, stated that price of the quota prohibited them from participating in the fisheries and expressed great dissatisfaction with the IFQ programs.

Based on the testimony received and trends in the allocation of quota in these communities, and from testimony from other communities, it is difficult to determine precisely the economic effects of IFQ management on the overall economic conditions in these communities, and the net benefits and costs of IFQs to the nation. The fact that some members of these communities have purchased quota and that crew members appear to be increasing their share of quota seems to indicate that residents of these communities can continue to be involved in these fisheries. If quota share holdings in all or most of these communities decreased significantly, this would be a strong indication that these communities were

TABLE H.21 Percentage and Relative Ranking of Harvest Landed (in Weight) for Alaskan Ports in the Alaskan Sablefish Fishery (1991-1996)

Port Region	Percentage (Relative Ranking) of Harvest Landed in Alaskan Ports					
	1991	1992	1993	1994	1995	1996
Kenai Peninsula-Anchorage	26.0%(1)	21.4%(1)	20.6%(1)	18.4%(1)	24.8%(1)	26.8%(1)
Sitka-Juneau-Haines	7.6%(5)	7.9%(5)	7.3%(4)	12.6%(3)	14.8%(2)	16.2%(2)
Skagway-Yakutat-Angoon area	11.9%(3)	12.4%(3)	15.7%(2)	17.5%(2)	13.7%(3)	10.4%(3)
Kodiak Island Borough	14.8%(2)	13.3%(2)	15.5%(3)	12.3%(4)	10.3%(4)	8.0%(4)
Aleutians-Alaska Peninsula	11.3%(4)	10.2%(4)	4.0%(7)	2.6%(8)	8.9%(5)	7.9%(5)
Wrangell-Petersburg area	4.3%(7)	6.0%(7)	6.6%(5)	9.0%(5)	5.0%(6)	5.3%(6)
Valdez-Cordova area	6.4%(6)	5.3%(8)	4.5%(6)	4.4%(6)	4.2%(7)	3.9%(7)
Ketchican-Prince of Wales	2.9%(8)	2.2%(9)	2.5%(8)	4.0%(7)	1.5%(8)	2.0%(8)
Floating processor	2.5%(9)	6.7%(6)	1.2%(9)	0.4%(9)	1.1%(9)	1.8%(9)

NOTE: Some areas are aggregated in order to protect the confidentiality of individual processors. Typically, the areas aggregated represent similar types of processing facilities.

SOURCE: CFEC.

unable to participate in the fisheries. The fact that gross revenues appear to have increased since the inception of the programs suggests that the implementation of IFQs has improved the profitability of fishing operations in these communities, based on the assumption that costs have not increased considerably.

Implementation of the IFQ programs has altered the ways in which communities participate in these fisheries; however, it is not clear that this change has been detrimental to these communities as a whole from an economic perspective. Analysis of trends in quota share and crew participation could provide a clearer indication of the long-term effects of the halibut and sablefish IFQ programs on communities. However, the increasing quota share in four of the five communities reviewed indicates that the means for purchasing quota and participating in these fisheries exists and substantial investments are being made by some members of these communities. These data suggest that significant changes in the Alaskan halibut and sablefish IFQ programs could have profound economic effects on those members of the community that hold quota and participate in these fisheries, as well as on the overall economic stability of these communities.

OVERALL CONSIDERATION OF THE POTENTIAL ECONOMIC COSTS AND BENEFITS TO THE NATION

It appears that substantial savings have been realized by vessel operators due to improved harvesting conditions, but the amount of these savings cannot be precisely quantified. Overall, it appears that the economic efficiency of fishing operations has improved by allowing greater freedom for harvesters to time their harvests and select fishing techniques to minimize costs and increase the profitability of their operations. In the case of the Alaskan halibut IFQ program, it appears that fresh markets are expanding and exvessel prices are increasing. Expenditures for search and rescue operations have been significantly reduced. Nevertheless, expenditures for enforcement operations for the U.S. Coast Guard and NMFS appear to have increased.

There appear to have been consolidation of quota and reduction in employment opportunities for some crew members. Some crew members have purchased quota, and in some regions the share of quota held by crew members may become significant if current trends continue. Gross revenue appears to have increased for quota shareholders. Processing operations have also been affected, although the economic benefits or costs of these changes are not clear. Improved monitoring and data collection regarding the effects of income distribution in the processing sector and among crew members could further clarify the relative impacts of IFQ management on these sectors. Assessment of the 3% exvessel fee and the 0.5% registration and transfer fee provided for in the Magnuson-Stevens Act will further increase the revenue transferred to the Treasury and available to fund IFQ management. As stated earlier, it is likely that if IFQ programs were implemented for larger fisheries with higher exvessel value, the assessment of capital gains taxes, the 3% exvessel fee, and the 0.5% registration and transfer fee could generate significant net revenues for the nation, or for individuals and communities disadvantaged by IFQs.

In analyzing these potential benefits and costs, it is important to consider that there are continual changes in the structure of fishery management. In some cases, the long-term economic benefits of shifting management to IFQs may be significant. In other fisheries, these long-term benefits may not be as significant. **It is important that no matter what the management regime used, the economic costs and benefits of this management system be considered in comparison to other management options and their potential costs and benefits. If IFQ programs are proposed for other fisheries, the committee recommends that improved socioeconomic data collection and analysis be conducted to provide a clearer indication as to both the national and the local costs and benefits of this and alternative management tools.**

Care must be taken in distinguishing between real benefits and costs on the one hand and fiscal effects, or transfers, on the other. The real net benefits of an IFQ program consist of increases in revenues over costs, including management and enforcement costs. Increased tax revenues are important, however, due to the fact that management and enforcement costs are currently being paid almost exclusively from public revenue and increased tax revenues from capital gains taxes. Other sources also transfer funds to pay for the administration of these programs. The economic benefits of an IFQ program will be reflected in the value of quota, which appears to have increased since the inception of the program. Economic benefits will also be reflected in better quality of product and higher demand for the product.

In summary, this appendix provides some indication of the range of issues that must be considered when comparing IFQ management to other forms of management. A worthwhile exercise in the development of future IFQ programs is to identify the range of economic costs and benefits associated with IFQ management. The collection of data on the price of transfer, the value of the quota, and other economic factors will greatly improve the ability to assess the overall costs and benefits of IFQs on the regional and national levels. The only U.S. IFQ programs that could be analyzed in any depth are those for the Alaskan halibut and sablefish fisheries. The SCOQ and wreckfish IFQ programs collect very few of the data necessary for a thorough analysis of net economic benefits and costs. **The committee recommends mandatory reporting of transfer price information for all existing and new IFQ programs.**

Index